MODERN DRUG RESEARCH

MEDICINAL RESEARCH
A Series of Monographs

1. Drugs Affecting the Peripheral Nervous System
   edited by Alfred Burger *(out of print)*
2. Drugs Affecting the Central Nervous System
   edited by Alfred Burger *(out of print)*
3. Selected Pharmacological Testing Methods
   edited by Alfred Burger *(out of print)*
4. Analytical Metabolic Chemistry of Drugs
   edited by Jean L. Hirtz *(out of print)*
5. Biogenic Amines and Physiological Membranes in Drug Therapy
   edited by John H. Biel and Leo G. Abood
6. Search for New Drugs
   edited by Alan A. Rubin
7. Clofibrate and Related Analogs
   by Donald T. Witiak, Howard A. I. Newman, and Dennis R. Feller
8. Quantitative Drug Design: A Critical Introduction
   by Yvonne Connolly Martin
9. The Chemistry of the Tetracycline Antibiotics
   by Lester A. Mitscher

10. Physical Chemical Properties of Drugs
edited by Samuel H. Yalkowsky, Anthony A. Sinkula, and Shri C. Valvani

11. SAR: Side Effects and Drug Design
by Eric J. Lien

12. Modern Drug Research: Paths to Better and Safer Drugs
edited by Yvonne Connolly Martin, Eberhard Kutter, and Volkhard Austel

Additional Volumes in Preparation

MODERN DRUG RESEARCH
Paths to Better and Safer Drugs

edited by

YVONNE CONNOLLY MARTIN
Computer-Assisted Molecular Design Project
Pharmaceutical Products Division
Abbott Laboratories
Abbott Park, Illinois

EBERHARD KUTTER
Department of Research and Development
Boehringer Ingelheim KG
Ingelheim, Federal Republic of Germany

VOLKHARD AUSTEL
Department of Chemistry
Dr. Karl Thomae GmbH
Biberbach, Federal Republic of Germany

MARCEL DEKKER, INC. New York and Basel

ISBN 0-8247-7902-9

This book is printed on acid-free paper.

MARCEL DEKKER, INC.
270 Madison Avenue, New York, New York 10016

Current printing (last digit):
10 9 8 7 6 5 4 3 2 1

PRINTED IN THE UNITED STATES OF AMERICA

We dedicate this book to Professor Corwin Hansch in gratitude for all that he has contributed to the world of drug research. Many readers of this book will recognize that a quarter of a century ago his new approaches heralded the age of rational drug design: they were developed and applied to ultimately influence each of the disciplines covered in this book. However, his contributions are not limited to scientific ones. Two of us had the opportunity to work directly with Corwin and learn by example how to make science exciting and rewarding. His enthusiasm for research, his energetic pursuit of creative and useful answers to scientific questions even in the face of seemingly impossible odds, and his open-mindedness to new ideas are characteristics that all scientists would do well to emulate.

About the Series

This series of monographs was conceived by the late Dr. Fred Schueler of Tulane University. His untimely death came prior to the completion of the first volumes of the series, and his friend and very able colleague, Dr. Alfred Burger of the University of Virginia, took on the task of editing the first three volumes of the series. Dr. Burger's heavy responsibilities forced him to withdraw from further active participation with the series, and the next three volumes were selected from books already under contract to Marcel Dekker, Inc., with Dr. Gary Grunewald of the University of Kansas as Consulting Editor. At this point an Editorial Advisory Board was selected and Grunewald became the Series Editor. The traditions established under the leadership of Schueler and Burger will be continued.

It is hoped that books for the series will serve a useful role in the areas of medicinal chemistry, chemical pharmacology, and biochemistry. It is the intent of the Editor and the Editorial Advisory Board that books selected for the series should be timely and fill a definite need in the general areas mentioned. We welcome suggestions for future monographs in the series.

Gary L. Grunewald

Preface

A new era is dawning in drug research. New ways of developing better and safer drugs are becoming apparent. Major changes are taking place both in how and which chemical compounds are tested for their therapeutic usefulness and in the possibilities for interpreting biological effects at the molecular level.

This new era has arisen partly from the exciting advances in the scientific knowledge of the molecular basis of bodily functions and of how these functions change in illness. Technological innovations, particularly in the fields of molecular and structural biology and computer technology, are making significant contributions to modern drug research. Furthermore, the substantial advances in the capabilities of theoretical and synthetic chemistry, as well as macromolecular structure determination by X-ray diffraction and nuclear magnetic resonance spectroscopy, have led to an increasing understanding of the interactions of active substances with their biological target molecules. Because of the progress in all of these fields, it is now sometimes possible to study in atomic detail the binding sites for ligands on DNA, enzymes, and other proteins such as antibodies. In such cases, the variation of the affinity within a series of test compounds can be interpreted in the light of how the individual molecules fit into the macromolecular binding site.

The most dramatic advance of modern biology is undoubtedly that of genetic engineering. With these techniques scientists can now produce large quantities of specific DNA, RNA, or protein molecules; molecules that form the structural and functional basis of living things. Furthermore, specific changes can be made to these molecules to probe the effect of such changes in structure on changes in function. As a consequence of these and related developments in biochemistry, the strategies and experimental methods used in the search for new drugs are changing fundamentally. Where once testing of potential new drugs started with studies in organ or whole-animal preparations, such tests are now often secondary to biochemical tests.

Technological advances in polymers, on the other hand, provide a vital basis for innovation in pharmaceutical dosage forms that are safer and more convenient for the patient or that expand the types of molecules suitable for therapeutic use.

However, it is not only scientific and technical developments that are changing the face of drug research: the research and development environment at pharmaceutical companies has also experienced change. Special disciplines no longer stand alone as independent units; research and development processes now center around collaboration within interdisciplinary project groups. Better and safer drugs will be discovered and developed by a close collaboration of pharmacologists, biochemists, molecular biologists, medicinal and theoretical chemists, toxicologists, pharmacists, and physicians.

What are the therapeutic challenges for which new drugs are needed? Bacterial infections, once the plague of mankind, have now been largely abated. We can now treat the symptoms of many other common diseases such as cardiovascular, metabolic, respiratory, and mental illnesses. However, the chief goal of future drug research must be the search for a cure or prevention of these diseases. Moreover, drugs must be developed for diseases for which no drug treatment has yet been successful: diseases of the immunological system; viral infections, especially AIDS; malignant tumors; and diseases of the central nervous system, especially Alzheimer's disease. Successful cure of these diseases will not only prolong life but improve the quality of life as well.

Because of the revolutionary changes taking place, we decided to organize a book that outlines the scientific, technical, organizational, and social context of current and future drug research. It is designed to give readers an insight into the possibilities of this field and help those currently involved in it gain an insight into the complexity of the various disciplines. We expect that at least parts of this book will prove useful for courses in medicinal chemistry or those that touch on matters associated with the research and development of drugs.

To accomplish these goals we chose authors who have worked many years in the field of their contribution, applying their research to create newly marketed drugs. Their knowledge and experience give the chapters a flavor of the excitement and hard work involved in drug research. Additionally, the authors have thought carefully about the practical application of the various new technologies and sometimes sound a cautionary note not obvious to someone more distant from the problem.

For didactic reasons we divided the book into three parts. Chapters 1-5 describe the modern scientific basis, Chapters 6-9 describe the strategic aspects, and Chapter 10 describes the social environment of modern drug research. We included Chapter 10 because we are convinced that without the basic approval of society, which appears to be decreasing in some countries, the huge financial commitment for the development of revolutionary new drugs would not be possible. Furthermore, Chapter 10 discusses the appropriate corporate culture to foster creative drug research and development.

We appreciate the patience of the authors who carefully worked over their chapters and listened to our comments to make them a part of a coherent picture. We also are indebted to Mrs. Gudrun Schwaar for valuable organizational assistance and to Miss Diana Schlichthaerle for drawing the chemical formulae.

Yvonne Connolly Martin
Volkhard Austel
Eberhard Kutter

Contents

About the Series - Gary L. Grunewald v
Preface vii
Contributors xv

Chapter 1
DRUG ACTION AND RECEPTOR THEORY 1

I. Functional and Structural Aspects of Organisms 1
II. Interaction of Drugs with Receptors and Enzymes 8
III. Signal Transduction 19
IV. Conclusions 31
References 33

Chapter 2
PROGRESS IN MOLECULAR PHARMACOLOGY AND BIOCHEMISTRY AS A LINK BETWEEN DISEASES AND NEW DRUGS FOR THEIR TREATMENT 35

I. Introduction 35
II. Cardiovascular Diseases: Atherosclerosis 37
III. Diseases of the Central Nervous System 39
IV. Viral Infections 47
V. Immunological Diseases 53
VI. Cancer 59
VII. Outlook 71
References 72

Chapter 3
PRINCIPLES OF PHARMACOKINETICS AND DRUG METABOLISM 77

I. The Availability of the Drug at Its Sites of Action 77
II. Physiological Considerations in Pharmacokinetics 93

III. Biotransformation of Xenobiotics 118
IV. The Role of Pharmacokinetics and Biotransformation in the Development of Better and Safer Drugs 146
References 158

Chapter 4
THEORETICAL BASIS OF MEDICINAL CHEMISTRY: STRUCTURE-ACTIVITY RELATIONSHIPS AND THREE-DIMENSIONAL STRUCTURES OF SMALL AND MACROMOLECULES 161

I. Forces With Which a Biological Macromolecule Binds a Small Molecule 162
II. Structure-Activity Relationships (SAR's) in Terms of Sub-Structural Features 171
III. QSAR: Correlation between Physical Properties and Biological Potency 176
IV. The Analysis of Three-Dimensional Property-Activity Relationships to Map Binding Sites on Biomacromolecules 190
V. Structure of the Biological Macromolecule-Drug Complex 201
VI. Summary and Outlook 212
References 213

Chapter 5
THE IMPORTANCE OF BIOTECHNOLOGY FOR THE DISCOVERY OF BETTER AND SAFER DRUGS 217

I. Introduction 217
II. Production of Human Proteins, the First Generation of New Drugs from Recombinant DNA 218
III. Natural Proteins as Leads for Structurally Modified Proteins: The Second Generation of Products from Biotechnology 233
IV. Examination of the Structure and Function of Proteins as Tools for the Evaluation of Pathological Processes: The Third Generation of Products from Biotechnology 236
V. Summary 239
References 240

Chapter 6
THE MEDICINAL CHEMIST'S APPROACH 243

I. General Aspects 243
II. Procedures That Lead to Better and Safer Drugs 250

III. Summary and Outlook 303
References 306

Chapter 7
EVALUATION OF THE THERAPEUTIC POTENTIAL OF NEW COMPOUNDS 309

I. Introduction 309
II. Pharmacology and Microbiology as Links Between Biochemistry and Therapy 310
III. Ethical Considerations on Animal Experiments 313
IV. Stages of Biological Testing 314
V. Basic Considerations for Test Procedures 316
VI. Explorative Pharmacological or Microbiological Investigations 318
VII. Profiling of Selected Compounds 330
VIII. Extrapolation of Results From Animal Experiments into Human Therapy 348
References 350

Chapter 8
SAFETY EVALUATION OF NEW DRUGS 355

I. Introductory Remarks 355
II. Animal Welfare 356
III. Prerequisites to Toxicological Studies 358
IV. Veterinary Care of Animals in Toxicology 361
V. Toxicological Studies in Laboratory Animals 362
VI. Test Parameters in Toxicological Studies 381
VII. Evaluation of Toxicity Studies 384
VIII. Toxicology of Biotechnological Products 389
IX. Expenditure in Toxicology 392
X. Summary 394
References 395

Chapter 9
THE CONTRIBUTION OF PHARMACEUTICS TO IMPROVEMENTS IN DRUG THERAPY 401

I. General Aspects of Pharmaceutics in Drug Development 401
II. Principal Working Areas in Pharmaceutical Research 406
III. Specific Examples of New Drug Delivery Systems 421
IV. Outlook for the Contribution of Pharmaceutics to Drug Development 434
References 439

Chapter 10
SEARCH FOR BETTER AND SAFER DRUGS: IMPACT OF EXTERNAL FACTORS - THE SOCIAL, FINANCIAL, AND WORKING ENVIRONMENT 441

I. Drug Therapy and Public Opinion 442
II. Drug Research and Financial Constraints 448
III. Working Climate for Innovative Drug Research 458
References 468

Abbreviations 469

Index 471

Contributors

V. AUSTEL Department of Chemistry, Dr. Karl Thomae GmbH, Biberach, Federal Republic of Germany

G. BOZLER Department of Biochemistry, Dr. Karl Thomae GmbH, Biberach, Federal Republic of Germany

PETER R. FARINA Department of Biochemistry, Boehringer Ingelheim Pharmaceuticals, Inc., Ridgefield, Connecticut

D. HARTMANN Pharmaceutical Research and Development, F. Hoffmann-La Roche AG, Basel, Switzerland

D. H. HINZEN Pharmaceutical Research and Development, F. Hoffmann-La Roche AG, Basel, Switzerland

H. M. JENNEWEIN Department of Pharmacology, Boehringer Ingelheim KG, Ingelheim, Federal Republic of Germany

JAMES J. KEIRNS Department of Biochemistry, Boehringer Ingelheim Pharmaceuticals, Inc., Ridgefield, Connecticut

A. KIESER University of Mannheim, Institute of Organizational Behavior, Mannhein, Federal Republic of Germany

W. KOBINGER Department of Pharmacology, Ernst-Boehringer-Institut für Arzneimittelforschung, Bender & Co GesmbH, Dr. Boehringer-Gasse, Vienna, Austria

E. KUTTER Department of Research and Development, Boehringer Ingelheim KG, Ingelheim, Federal Republic of Germany

C. LILLIE Department of Pharmacology, Ernst-Boehringer-Institut für Arzneimittelforschung, Bender & Co GesmbH, Dr. Boehringer-Gasse, Vienna, Austria

YVONNE CONNOLLY MARTIN Computer-Assisted Molecular Design Project, Pharmaceutical Products Division, Abbott Laboratories, Abbott Park, Illinois

D. S. MEYER Department of Experimental Pathology and Toxicology, Boehringer Ingelheim KG, Ingelheim, Federal Republic of Germany

J. SCHMID Department of Biochemistry, Dr. Karl Thomae GmbH, Biberach, Federal Republic of Germany

P. SWETLY Ernst-Boehringer-Institut für Arzneimittelforschung, Dr. Boehringer-Gasse, Vienna, Austria

B. ZIERENBERG Department of Pharmaceutics, Boehringer Ingelheim KG, Ingelheim, Federal Republic of Germany

1

Drug Action and Receptor Theory

James J. Keirns and Peter R. Farina

Department of Biochemistry
Boehringer Ingelheim Pharmaceuticals, Inc.
Ridgefield, Connecticut

I. FUNCTIONAL AND STRUCTURAL ASPECTS OF ORGANISMS

A. INTRODUCTION

Therapeutics began with earliest man as a totally empirical science. Over the ages it has helped many people, harmed others and contributed to the advance of scientific knowledge. As our knowledge of biological processes has progressed, we have produced more effective and safer therapeutic agents. The future search for better and safer drugs will require an even better understanding of biological processes and even more clever technology.

Modern pharmacological research, either in academia or industry, requires the study of the effects of the compound on both functional (physiological) and mechanistic (biochemical) responses of the organism for an adequate understanding of the biological effects of the compound and for an effective assessment of the potential of the compound to be a useful drug.

Until the past two decades, the effects of chemical compounds were measured in whole animals, isolated organs, tissues or bacteria. The endpoint measured was a functional response of the animal (e.g., a change in blood pressure) or tissue (e.g., contraction), when the compound to be tested was administered to the animal or added to the bathing fluid for the tissue. In the past 20 years, there has been an explosive growth in biochemical or mechanistic models. These molecular level studies are a great help to the medicinal chemist because they permit the more detailed understanding of the factors that govern the interaction of the synthesized compound with the target macromolecule. They aid the biologist because they suggest new approaches to the cure of a disease and they highlight differences in potential mechanisms of producing the same biological endpoint.

B. STRUCTURE OF BIOLOGICAL MEMBRANES AND PROTEINS

Proteins provide the cellular machinery for the transport of molecules into and out of cells, for the regulated transformation of one molecule into another, and for the unified response of an organism to an internal or external stimulus. One of the quests of modern biochemistry is to understand how proteins accomplish these tasks.

Although all proteins are polymers of amino acids, each protein has a specific sequence of these amino acids and each folds into one or a few specific three-dimensional structures. This topic will be discussed in more detail in Chapter 4. It is these three-dimensional structures that are responsible for the functions of the protein.

Cells are separated from their external environment by biological membranes; membranes also divide the insides of cells into compartments. These membranes are able to maintain concentration gradients exceeding 1000-fold between the phases they separate. In addition, most receptor proteins and many enzymes that are targets for drugs are components of biological membranes. A detailed description of biological membranes is given in Chapter 3.

Proteins can be associated with the membrane in one of several different ways. They can loosely associate with the polar groups, often with the aid of divalent metal ions. Proteins can bind firmly to the hydrocarbon interior of the membrane by means of hydrophobic interactions of either an alpha helix of hydrophobic amino acids or a fatty acid attached to an amino group of the protein. Proteins can be partially immersed in the lipid phase of the membrane by means of hydrophobic interactions of a large area of hydrophobic residues. Finally, proteins can span the membrane by means of a hydrophobic waist and hydrophilic head and feet.

Since usually a large portion of a membrane-associated protein is still exposed to water, it can have any of the enzymatic or binding activities exhibited by water soluble proteins. In addition, proteins that span the membrane can transduce signals across the membrane. It is thought that this occurs as follows: binding a regulator on the extracellular side of the receptor embedded in the membrane produces an allosteric modification of the enzymatic activity or binding properties of a site on the other side of the membrane. Proteins that span the membrane can also form pores or transport systems to allow ions or hydrophilic molecules to pass through the hydrophobic membrane.

The same thermodynamic factors that drive the assembly of biological membranes also drive proteins into their tertiary and quarternary structures. Overall, the most important factor is the gathering of hydrophobic residues into the core of the protein and exposure of charged and hydrophilic residues to water on the outside.

Most drugs interact with specific proteins (receptors or enzymes) to exert their action. Several ways of evaluating these specific interactions will be discussed. A few drugs exert their effects by interacting with biological structures other than proteins. General anesthetics interact with the lipoproteins of plasma membranes; soaps and detergents also interact with membranes; ionophores dissolve in membranes to form pores or carriers of ions and polar molecules; and a number of antibacterials and anti-proliferative agents interact directly with DNA. Nonetheless, the concept of drugs interacting with specific proteins has been the basis of the recent explosive growth of biochemical pharmacology.

C. REGULATION OF BIOLOGICAL PROCESSES

A living system is characterized by the delicate balance maintained between the levels of various compounds and also between the relative rates of various reactions. When this balance goes awry, a person has a disease; the purpose of drug therapy is to restore the balance to normal.

Even the simplest unicellular organisms possess many mechanisms for regulating their metabolism to efficiently use external sources of energy in the form of light or chemicals, to propagate, and to sense and respond to changes in the external milieu. One classic regulatory mechanism is an allosteric enzyme: this is an enzyme whose three-dimensional structure, and thus catalytic activity, is regulated by the binding of another molecule at a position remote from the active site. Often, the regulator molecule is a product or precursor to the metabolic pathway of which the enzyme is a part. Chemoreceptors are proteins that sense the concentration of a particular chemical in the external environment and transmit this information to the response mechanisms inside of the cell. Pumps and channels form specific pathways for the movement of molecules and ions into and out of the cell. Repressors are proteins that interact with DNA to increase or decrease the rate of transcription and translation of a DNA message into functioning protein.

More complex unicellular organisms use all of these regulatory mechanisms plus compartmentation within their cells. There are, thus, additional mechanisms to coordinate activities between different compartments.

Multicellular organisms use all the above regulatory mechanisms plus those that are needed to coordinate the activity of the many different specialized cells. Many of these regulatory mechanisms are aimed at homeostasis, i.e., the maintenance of a constant internal equilibrium. Others are needed to respond to changes in food supply and other factors in the external environment.

In a healthy organism all of these regulatory processes operate at the correct rate to maintain the constant internal equilibrium as well as the ability to respond effectively to changes from the outside. When one of these regulatory mechanisms is absent or is operating either too slowly or too rapidly, disease occurs. Ideal diagnosis will identify the specific deranged mechanism; ideal therapy will restore this mechanism to normal. To reach such an ideal we need to totally understand human physiology and the cell biology and biochemistry of the constituents of the body.

D. RECEPTOR THEORY

The concept of receptors for drugs, the paradigm for research in pharmacology and medicinal chemistry, arose long before the development of the biochemical techniques to demonstrate their existence (Parascandola, 1980). In

1874, Langley was experimenting with pilocarpine, an alkaloid that had recently been isolated from the leaves of a South American shrub. He found that the effects of pilocarpine on salivary glands could be antagonized by atropine, a poison of unknown chemical structure that had been isolated from belladonna 50 years earlier. In the first volume of the Journal of Physiology, Langley concluded: "We may assume that there is a substance or substances in the nerve endings or gland cells with which both atropine and pilocarpine are capable of forming compounds. On this assumption then, the atropine or pilocarpine compounds are formed according to some law of which their relative mass and chemical affinity for the substance are factors." Twenty-five years later, he found that curare antagonized the effects of nicotine on skeletal muscle. The site of action was directly on muscle, since denervation had no effect. Since the muscle could still respond to direct electrical stimulation, nicotine and curare were not binding to the "contractile substance," but rather to a "receptive substance."

Meanwhile, Ehrlich had developed the "side chain" or "receptor" theory of the interaction of antibodies with toxins. Langley's data persuaded Ehrlich to extend the theory to drug interactions with cells. In the following decade Dixon and Dale showed that the effects of parasympathetic nerve stimulation could be mimicked by acetylcholine. In the 1920's, Loewi and Bain showed that stimulation of the vagus nerve caused release of acetylcholine. Some effects of acetylcholine could be mimicked by nicotine and others by muscarine. These data could be integrated by proposing two different types of receptor for acetylcholine:

Receptor	Agonist	Antagonist
Nicotinic	Nicotine	Curare
Muscarinic	Muscarine	Atropine

By definition an agonist acts on the receptor to produce a functional response. An antagonist causes no response on its own, but prevents the response to an agonist. Both agonists and antagonists are ligands, substances that bind to a receptor. There is a conceptual parallel between an agonist and the substrate for an enzyme and between an antagonist and an inhibitor of an enzyme. In receptor systems a biochemical or physiological response is measured; in enzyme systems the conversion of substrate to product is measured.

The next receptor system to be physiologically characterized, especially through the work of von Euler and of Ahlquist (1979), was the adrenergic system. It mediates the effects of epinephrine and norepinephrine. The adrenergic system was divided into two types (alpha and beta), each of which was further subdivided into two subtypes, based on the differential actions of agonists and antagonists. These compounds also proved to be useful drugs for a variety of therapeutic indications, especially in the cardiovascular and respiratory fields.

Today it is realized that there are dozens of receptors with which drugs can interact. The amino acid sequence of several receptors has been determined. It appears now that the muscarinic and adrenergic receptors are very similar in three-dimensional shape. They are proposed to have seven membrane-spanning helices with the binding site for the agonist located at the intersection of these helices. The receptors for peptide hormones, on the other hand, seem to have one membrane-spanning region and the ligand binding site located in an extra-cellular domain of the protein. The normal physiological function of receptors is to mediate the cellular response to external stimuli, such as neurotransmitters and hormones.

II. INTERACTION OF DRUGS WITH RECEPTORS AND ENZYMES

A. DOSE RESPONSE CURVES (SCHILD ANALYSIS)

The concept of a drug-receptor or enzyme-inhibitor binding suggests that to characterize the properties of any pharmacological agent, it is necessary to apply the agent at several doses (or concentrations) and measure the biochemical or physiological response to each.

The pharmacological response to an agonist can be characterized by efficacy, the maximum magnitude of the response, and potency, the minimal dose of the agonist that produces the response (Ross and Gilman, 1985). Efficacy is usually defined by the ratio of the maximum effect of the test compound to that of the natural agonist. For an agonist, potency is usually measured as the dose or concentration which produces half of the maximal response (abbreviated as ED_{50} or EC_{50}). To accurately determine this value it is necessary to define the maximal response by two or more high doses that define a plateau, as well as doses that bracket the half maximal response.

In vivo, the organism sometimes provides the agonist, but sometimes *in vivo* and usually *in vitro*, the investigator provides both the agonist and antagonist and can control the doses of both. Certain compounds, called partial agonists, have properties intermediate between an agonist and an antagonist. By itself, a partial agonist will produce a response, but at all concentrations this response is smaller than the maximal response to a "full" agonist (Figure 1).

Determination of potency of an antagonist is complicated since a response is a function of the doses of both the agonist and the antagonist. The usual approach is to fix the concentration of agonist and vary the concen-

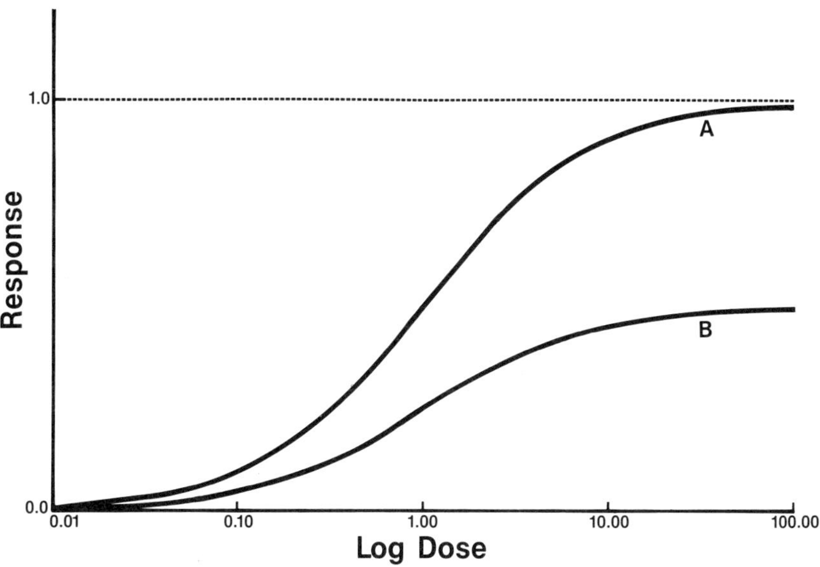

Figure 1 Log dose response curve for full and partial agonists. The maximum response for the full agonist is 1 and for this particular partial agonist 0.5. This partial agonist also has the same ED_{50} as the full agonist. In both cases at dose = ED_{50}, the response is 50% of maximum.

tration of antagonist to determine the dose of antagonist that reduces the response to 50% (ID_{50} or IC_{50}). If the experiment is repeated at a different concentration of agonist, a different ID_{50} will usually be obtained. Schild assumed that the agonist and antagonist both bind to the same receptor following the law of mass action (Tallarida et al., 1979). Since pharmacological dose response curves are conveniently plotted as response versus negative logarithm of the dose of agonist, analogous to pH titration curves, Schild defined a variable called pA (negative logarithm of agonist dose). In the presence of a given concentration of antagonist (B), the agonist dose response curve will be shifted to the right by a factor x (Figure 2). If the agonist and antagonist compete for the receptor following law of mass

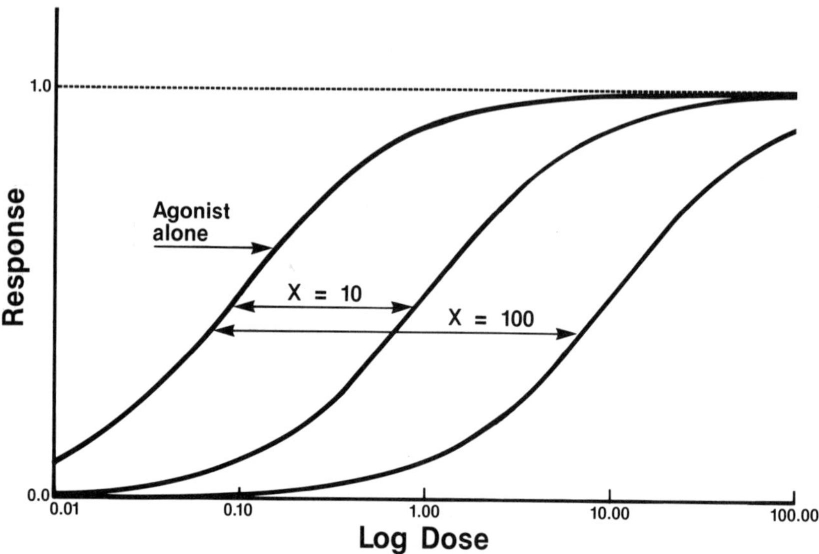

Figure 2 Competitive antagonist. The left curve is the dose response of the agonist alone with an ED_{50} = 0.1. The center and right curve are in the presence of increasing concentrations of the antagonist which produce 10-fold and 100-fold shifts of the agonist curve.

action, a plot of log(x-1) versus -log B will give a straight line of slope -1, which intersects the abscissa at a point called pA_2 (Figure 3). pA_2 is equal to the logarithm of the association constant of the antagonist for the receptor.

B. ANALYSIS OF ENZYME INHIBITION

Enzyme reactions are measured by monitoring the appearance of the product of the reaction or the disappearance of substrate. It is desirable to design the assay to obtain an initial rate. The initial rate is more useful than is an average rate over a period of time since it is not complicated by exhaustion of substrate, back reaction of product to substrate, or feedback regulation by product.

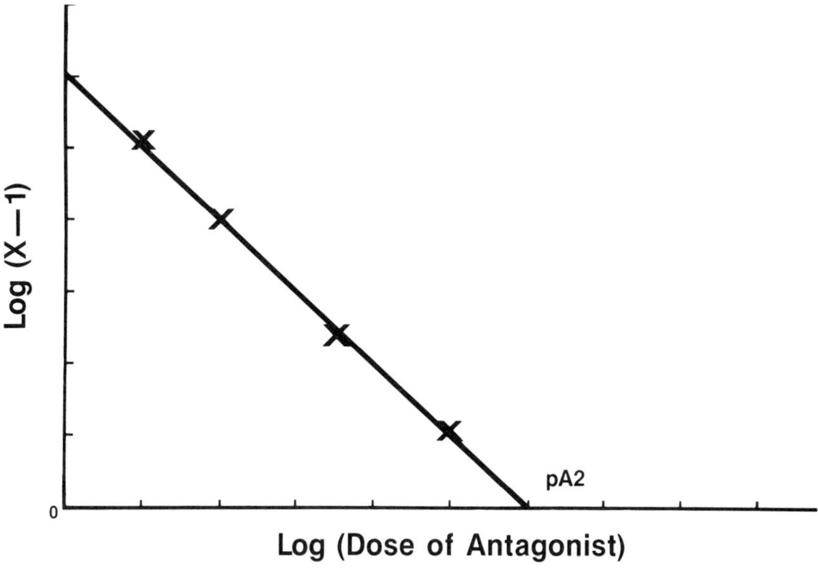

Figure 3 Schild plot. Data from an experiment similar to Figure 2 but with four doses of antagonist are plotted.

In many cases it is not possible to continuously monitor product or substrate, so the reaction must be stopped at a particular time and product separated from substrate or chemically modified before quantitation. Then, it is important to study the time dependence of the reaction so that the rates measured approach the initial rate as closely as practical. Frequently, there is a trade-off between stopping the reaction soon enough to approximate initial rate conditions, and letting the reaction proceed long enough to have sufficient product to accurately quantitate.

Enzyme reactions should be studied at various concentrations of substrate. Usually at low substrate concentrations the reaction rate increases proportionally to substrate concentration (i.e., is first order), while at high substrate concentrations the rate is constant

(V_{max}) and independent of substrate concentration (i.e., zero order), Figure 4. The parameter K_m defines the substrate concentration around which this transition from first order to zero order occurs; more precisely K_m is the substrate concentration at which the rate is equal to one-half of V_{max} (Figure 4). This behavior can be rationalized by the Michaelis-Menten model in which the substrate rapidly forms a reversible complex with the enzyme, and is then slowly converted to product. The hyperbola in Figure 4 can be linearized by plotting rate/(substrate) versus rate (Figure 5).

If we consider the substrate as analogous to the agonist in a receptor system, an inhibitor of the enzyme is analogous to the antagonist. If the inhibitor competes

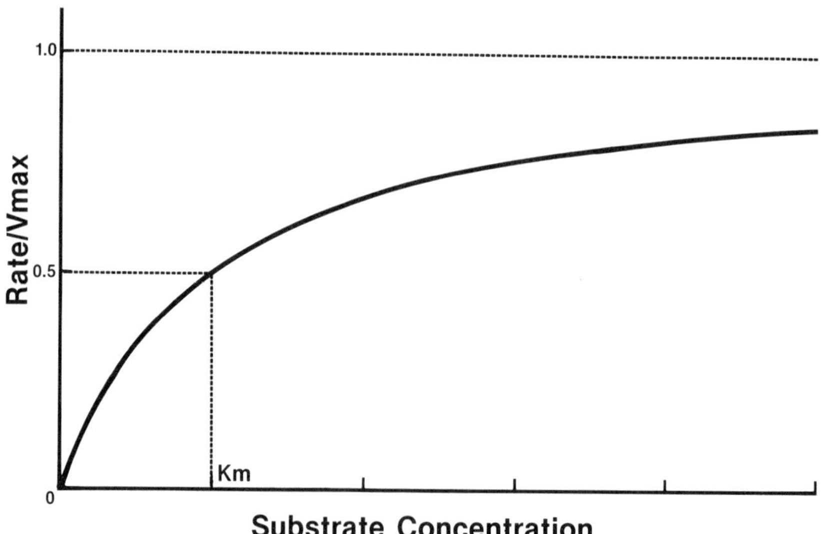

Figure 4 Michaelis-Menten kinetics. The maximum rate (V_{max}) has been normalized to 1. At a substrate concentration = K_m, the rate = 0.5. The reaction is nearly first order at substrate concentration less than 0.2 x K_m and nearly zero order at substrate concentrations much greater than 5 K_m.

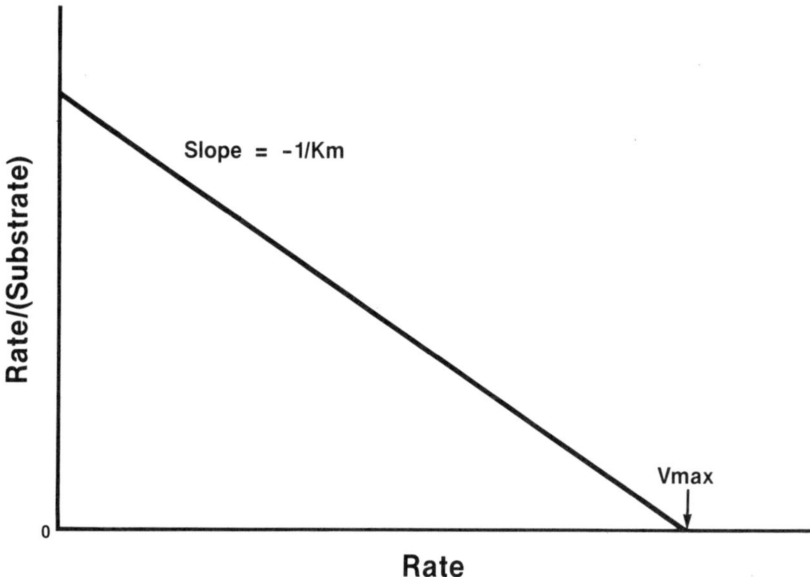

Figure 5 Eadie-Hofstee Plot. The intercept on the abscissa is V_{max}. The slope is $-1/K_m$.

with substrate to bind reversibly to the same binding site, or at least to an overlapping binding site, then a given concentration of inhibitor will shift the rate versus substrate curve to the right. An analysis similar to Schild analysis results in a parameter K_i. It can be considered to be a dissociation constant for the inhibitor-enzyme complex. An inhibitor which behaves this way is called competitive (Figure 6).

Alternatively, an inhibitor can bind irreversibly, or it can bind at another site on the enzyme remote from the substrate binding site. In such cases, the inhibition cannot be overcome by high concentration of substrate, and the inhibitor is called non-competitive.

In trying to relate studies of the inhibition of isolated enzymes to the situation in the intact cell,

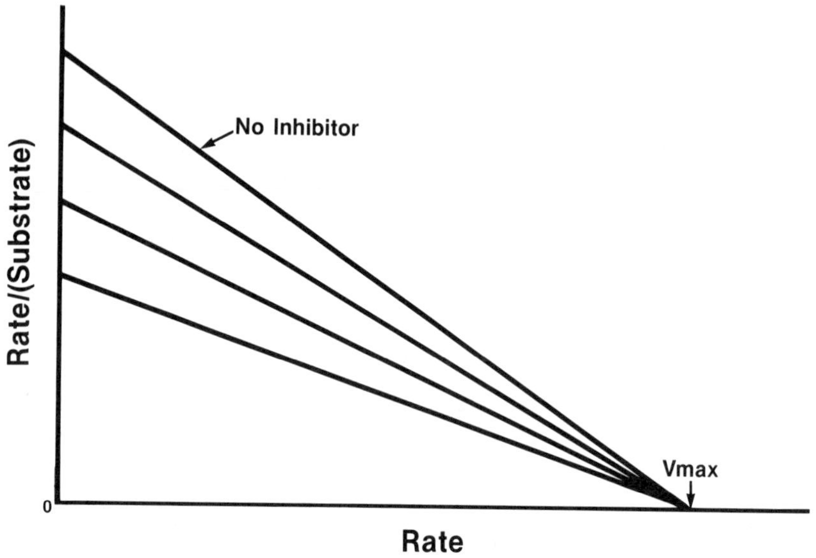

Figure 6 Competitive inhibitor. The top curve is without inhibitor and the lower curves with increasing concentrations of inhibitor. V_{max} does not change. K_m increases with increasing concentration of inhibitor.

consideration should be given to the concentration of the substrate and the possible presence of other regulatory factors.

C. RECEPTOR BINDING METHODS (SCATCHARD ANALYSIS)

It is possible to measure directly the affinity of a drug for its receptor. A molecule bound to a receptor will be isolated along with it. Since receptors are generally located on cell membranes, they may be separated from the soluble cellular fraction by physical techniques. If the receptor is then resuspended in a solution with radioactively labeled drug, then one may reisolate the receptor to separate bound from free radioactive drug. This provides a direct assay of drug-receptor interaction.

Furthermore, a second, unlabeled, drug will compete with the first in a manner predicted by the law of mass action.

In designing a binding assay for compounds to a receptor, one must decide what ligand to label. Usually, the ligand chosen is a potent, non-lipophilic antagonist of the postulated receptor. The ligand must be labeled to a very high specific activity. Tritium or iodine-125 are usually used since, in most cases, carbon-14 does not give sufficiently high specific activity. The ligand must also be chemically and metabolically stable.

To separate receptor from solution and hence bound ligand from free, one may use ultracentrifugation, dialysis, or ultrafiltration; however, most commonly one uses special filters to which proteins including receptors adhere, but which allow free ligand to wash through.

The fraction of drug bound as a function of the concentration of ligand has a steep slope at low ligand concentrations and a more shallow slope at high concentrations. According to mass action, a plateau should be reached at high concentrations, but in practice binding usually asymptotically approaches a line with a small positive slope (Figure 7). If the experiment is repeated in the presence of a large excess of unlabeled ligand, one sees only the shallow slope parallel to the upper portion of the binding curve of labeled ligand alone. This binding, seen in the presence of a large excess of unlabeled ligand, is usually called non-specific binding, although it may represent a second binding site with lower affinity for the ligand. Usually the non-specific binding curve is subtracted from total binding before further analyzing specific binding.

Scatchard analysis is based on the following assumptions: ligand binds reversibly to the receptor; this process is at equilibrium; and both ligand and receptor

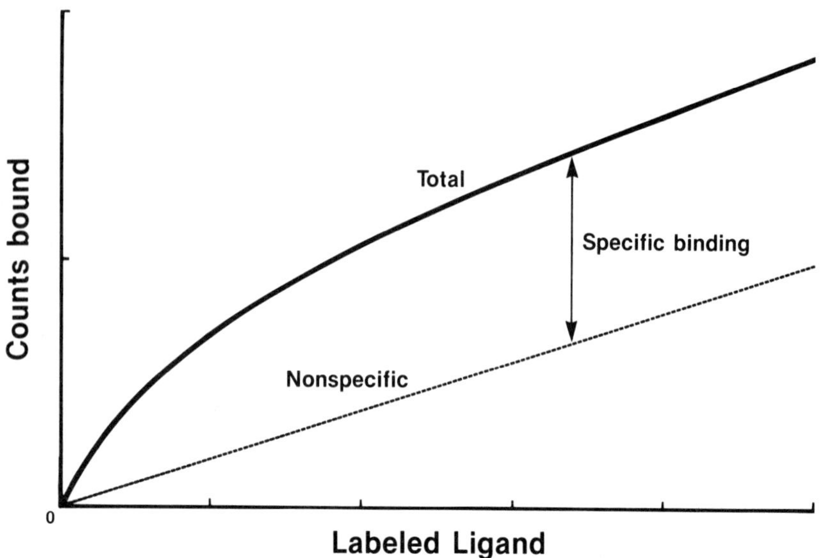

Figure 7 Plot of bound versus total ligand. The upper curve with labeled ligand alone is total binding. The lower curve (dotted) with labeled ligand plus a large excess of unlabeled ligand is nonspecific binding. The gap between the two curves is specific binding.

are stable (Boeynaems and Dumont, 1980). In the simplest case, it is further assumed that all receptors are equivalent and do not interact. The mass action equation is

$$RF/B = K$$

where R and F are the concentrations of free receptor and ligand respectively, and B is the concentration of complex. K is the dissociation constant of the complex. The equation for conservation of receptor is

$$R_t = R + B$$

where R_t is the total concentration of receptor. Using the second equation to substitute for R in the first and rearranging:

$$\frac{B}{F} = \frac{R_t}{K} - \frac{1}{K} B$$

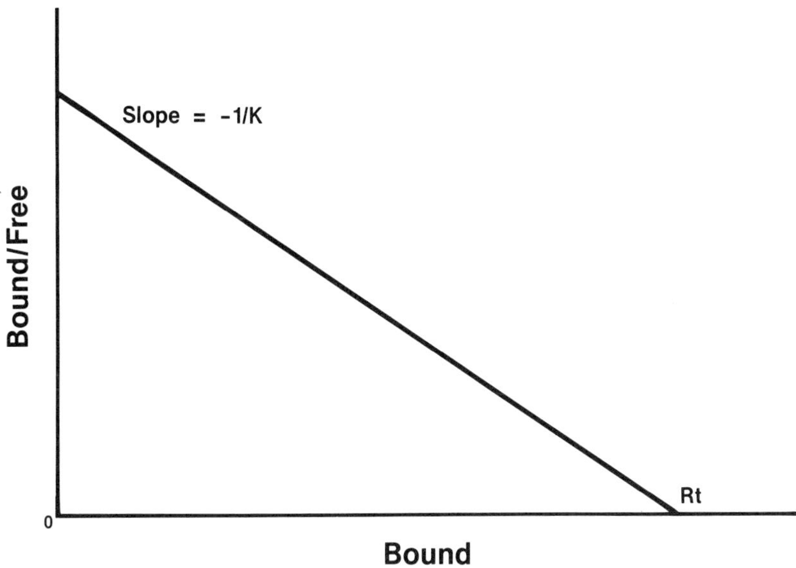

Figure 8 Scatchard plot of specific binding for a single site with mass action behavior. The intercept on the abscissa is total concentration of receptor. The slope is -1/K.

In this simple case, the Scatchard plot gives a straight line (Figure 8). The receptor-ligand dissociation constant can be obtained from the slope, and the total number of receptors from the intercept.

In more complex cases, such as multiple receptors with different K values or interaction of the receptors, the Scatchard plot will be curved. The complex cases are often analyzed using the Hill plot; $\log[B/(R_t - B)]$ versus $\log F$ which has a slope (Hill coefficient) equal to 1.0 for the simple case and a slope different from 1.0 in the more complex cases. With multiple receptors or "negative cooperativity" (i.e., binding of the first ligand makes binding of the second more difficult), the Hill coefficient is less than 1.0. With "positive cooperativity" (i.e., binding of the first ligand facilitates binding of the second), the Hill coefficient is greater than 1.0.

D. TIME DEPENDENCE OF EFFECTS

To be a useful therapeutic agent a drug must have a reasonable duration of effect. If the effect in man lasts only a few seconds or minutes, the compound will probably not be useful. In addition, a drug which can be administered once or twice a day to give continuous around-the-clock effect usually will be more attractive to the patient than an equivalent drug which must be administered more frequently. Usually, the clinical duration of effect is governed by pharmacokinetic factors which are discussed in Chapter 3. A notable exception is a drug such as aspirin, which irreversibly alters the target biomacromolecule. In such a case the kinetics of the regeneration of the target rather than the kinetics of the clearance of the drug from the body are relevant.

A second type of time-dependence of drug effects is tachyphylaxis. Frequently, the response to an agonist fades away even though the agonist continues to be present. Tachyphylaxis can be due to cellular feedback mechanisms, such as: receptor internalization (endocytosis of the membrane which carries the receptor); modification of the receptor by phosphorylation or other regulatory mechanisms; depletion of the substrate for an enzyme or second messenger system. The simplest assay systems (ligand binding or enzyme inhibition) generally do not show tachyphylaxis even when it occurs *in vivo* with the same receptor. Isolated cells or tissues sometimes exhibit tachyphylaxis which is relevant to the *in vivo* situation, but also sometimes give artifacts which are due to depletion of substances *in vitro* which are maintained *in vivo* or are simply due to the limited viability of the system *in vitro*.

The time course of the effect of anti-infective compounds is of course governed by both the time course of the drug in the body and the time course of the effect of the drug on the growth of the bacteria or virus.

E. HOW DO UNDESIRABLE SIDE EFFECTS COME ABOUT?

A side effect of a compound could result from binding to a different receptor or interacting with a different biological structure. This is often amenable to molecular modification. Simply seeking compounds with greater potency is likely to improve the ratio of desired-to-undesired activity. If the receptor for the side effect is known, new compounds can be screened for both the desired and undesired activities and optimized on the basis of the ratio of the two.

A side effect of a compound can also result from binding to the same receptor sub-type, but in different tissues. This problem can be approached pharmacokinetically, by choosing a route of administration which delivers higher concentrations to the organ where the desired effect occurs, or by looking for compounds which simply distribute differentially in various organs (e.g., compounds that do not readily cross the blood-brain barrier). Careful biochemical pharmacology can also reveal evidence for previously unsuspected receptor sub-types. This may lead to identification of new compounds, selective for one sub-type, that are efficacious but lack the side effect.

III. SIGNAL TRANSDUCTION

A. INTRODUCTION

Cells are continually engulfed in a milieu of electrical and chemical stimuli. Receptors enable cells to differentiate among the stimuli and relay a specific message. This generally leads to a defined cellular response. The process that translates the stimulus-receptor interaction into an intra-cellular chemical message is termed signal transduction. The molecules that are formed in response to the stimulus-receptor interaction are called second messengers. Although there are many identified receptors, there are only a few signal transduction pathways.

B. STIMULATION OF ADENOSINE-3',5'-CYCLIC-MONO-PHOSPHATE (cAMP)

This was the first transduction system to be well characterized. In the 1950's, Sutherland (Robison et al., 1971) was studying the mechanism by which the hormones glucagon and epinephrine stimulate the breakdown of glycogen to glucose in the liver. He investigated phosphorylase, the enzyme that catalyzes the conversion of glycogen plus inorganic phosphate to glucose-1-phosphate. Phosphorylase is turned off by dephosphorylation and turned on by phosphorylation. The enzymes involved in phosphorylation are all soluble; however, hormone regulation of glycogen breakdown occurs only in the presence of membranes. Incubation of membranes with hormone and adenosine triphosphate (ATP) produced a heat stable factor that stimulated the phosphorylation of phosphorylase and consequent breakdown of glycogen (Figure 9). The heat stable factor was found to be cAMP; it is the second messenger of the glucagon and epinephrine receptor activation. The enzyme that catalyzes the formation of cAMP from ATP is called adenylate cyclase (Figure 10). Krebs showed that cAMP stimulates a protein kinase (now called protein kinase A), which phosphorylates, among other substrates, phosphorylase kinase which in turn phosphorylates and activates phosphorylase. This sequence of enzymatic reactions connects the binding of

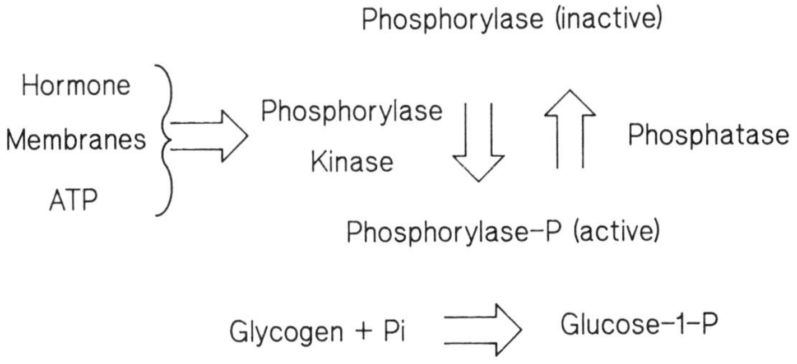

Phosphorylase (inactive)

Hormone
Membranes ⇒ Phosphorylase Kinase ⇓ ⇑ Phosphatase
ATP

Phosphorylase-P (active)

Glycogen + Pi ⇒ Glucose-1-P

Figure 9 Activation of glycogen breakdown in liver by hormones. The first arrow on the left encompasses the steps in Figure 10.

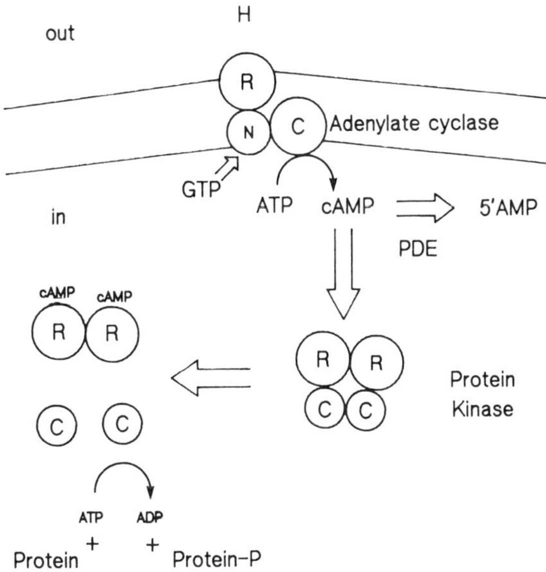

Figure 10 Cyclic AMP cascade. Activation of adenylate cyclase and subsequently of protein kinase A by hormone.

hormone to its receptor on the plasma membrane to the rate controlling step of glycogen breakdown.

How is this process turned off when the hormone dissociates from its receptor? cAMP is hydrolyzed by an enzyme called phosphodiesterase, each of the phosphorylations is reversed by a phosphatase, and adenylate cyclase is inactivated by a mechanism that involves GTP. GTP (guanosine triphosphate) is an essential cofactor for activation of adenylate cyclase by hormone. GTP binds to a protein called G_s-protein or N_s-protein. G_s-protein binds to the receptor. The binding of GTP to the G-protein is enhanced by the binding of hormone to receptor. The GTP-N_s complex stimulates the catalytic unit of adenylate cyclase. The N_s-protein also can hydrolyse GTP to terminate the activation of adenylate cyclase. More persistent activation can be achieved with the synthetic analogs of GTP that cannot be hydrolyzed (Lefkowitz et al., 1983) and by cholera enterotoxin. The latter catalyzes the covalent attachment of an ADP-ribose group from nicotinamide adenine dinucleotide (NAD) to the binding site for GTP on the N-protein.

Certain hormones lead to the inhibition of adenylate cyclase by binding to a receptor that binds to a second GTP-binding protein, N_i. The toxin from *Bordetella pertussis* catalyzes a covalent modification of N_i analogous to the effect of cholera toxin on N_s (Ueda and Hayaishi, 1985).

Table 1 lists some of the hormones and transmitters to which the physiological responses to hormone-receptor interaction are mediated by stimulation or inhibition of adenylate cyclase.

The unraveling of the mysteries of the signal transduction process mediated by the G proteins is opening new doors in our understanding of cancer-causing oncogenes. These oncogenes are altered or mutated forms of genes that

Table 1 Stimulation of cAMP accumulation.

Stimulus	Tissue	Response
Catecholamines	Liver	Phosphorylase activation
Catecholamines	Muscle	Phosphorylase activation
Catecholamines	Adipose	Phosphorylase activation
Catecholamines	Cardiac	Positive inotropism
Histamine	Cardiac	Positive inotropism
Catecholamines	Smooth muscle	Relaxation
Catecholamines	Parotid gland	Amylase release
Catecholamines	Pancreas	Insulin release
Glucagon	Pancreas	Insulin release
Corticotropin	Pancreas	Insulin release
Corticotropin	Adrenal	Steroidogenesis
Luteinizing hormone	Ovary	Steroidogenesis
Angiotensin	Adrenal	Steroidogenesis
Vasopressin	Toad bladder	Increased permeability
Vasopressin	Ventral frog skin	Increased permeability
Melanocyte stimulatory hormone	Dorsal frog skin	Melanocyte dispersion
Thyrotropin stimulatory hormone	Thyroid	Thyroid hormone release
Histamine	Gastric mucosa	HCl production
Acetylcholine	Exocrine pancreas	Amylase release
Thyrotropin releasing hormone	Anterior pituitary	Thyrotropin release
Catecholamines	Nervous	Facilitation of neuromuscular transmission
Catecholamines	Cerebellar	Reduced discharge frequency
Catecholamines	Liver	Enzyme synthesis
Prostaglandins	Platelets	Inhibition of aggregation
Parathyroid hormone	Bone	Calcium mobilization

functionally regulate important cellular processes. A common oncogene, *ras*, is found in a high percentage of colon cancers removed through surgery. The *ras* family of genes regulates cell growth and development. Evidence is mounting that the *ras* gene protein product is structurally related to G proteins and performs in an analogous manner in signal transduction. The normal *ras* protein binds GTP in an activation step and cleaves the nucleotide to GDP and inorganic phosphate which temporarily inactivates the protein. This on-off switching mechanism is now suspected of being left in the on-position by a *ras* oncogene protein which cannot cleave GTP. The result appears to be an unregulated signal which can lead to uncontrolled cell growth and division. Recently, the three-dimensional structure of a 21,000 molecular weight *ras* oncogene protein has been determined (de Vos et al., 1988). It appears that the oncogene protein and the normal protein are structurally quite similar, except for changes in the catalytic site.

C. STIMULATION OF THE PHOSPHATIDYLINOSITOL PATHWAY

More recently a second signal transduction pathway has been defined. It appears to play an important role in many receptor events that involve mediator release, chemoattractants, hormones, growth factors, secretagogues and mitogens (Table 2). The key elements of this system are inositol phospholipids, a phosphodiesterase (phospholipase C, PLC) and protein kinase C (PKC).

The current view of this receptor transduction process is shown in Figure 11 (see Berridge, 1984). The cellular response to receptor activation is mediated by two second messengers, inositol triphosphate and diacylglycerol. The former apparently exerts its effect by mobilizing calcium, the latter by activating phosphorylation of proteins via protein kinase C.

Table 2 Summary of the tissues which respond to external stimuli by a change in the hydrolysis of phosphatidylinositol-4,5-bisphosphate.*

Stimulus	Tissue
Acetylcholine	Rabbit iris smooth muscle
Acetylcholine	Brain
Acetylcholine, substance P	Parotid
Acetylcholine, cerulein	Pancreas
Acetylcholine	Avian salt gland
Adrenaline, ATP, vasopressin, angiotensin, platelet activating factor	Liver
Serotonin	Blowfly salivary gland
ADP, thrombin, platelet activating factor	Blood platelets
Vasopressin	Sympathetic ganglion
Thyrotropin-releasing hormone	GH_3, pituitary tumor cells
Angiotensin	Adrenal cortex
Glucose	Pancreatic islets
f-Methionyl-leucyl-phenylalanine	Neutrophils
f-Methionyl-leucyl-phenylalanine	HL-60 leukemic cells
Pseudomonal leukocidin	Leukocytes
Phytohemagglutinin	T-lymphoblastoid cell
Antigen	Leukemic basophils
Platelet-derived growth factor, bombesin, vasopressin	Swiss 3T3 cells
Bradykinin	Neuroblastoma-glioma hybrid NG108-15
Photons	Photoreceptors
Spermatozoa	Sea urchin eggs

*Reprinted (with modifications) by permission from Nature, 312:315. Copyright (C) MacMillan Journals Limited (1984).

An important aspect of cellular regulation is communication between different regulatory systems. An example of this is seen in links between the phosphatidylinositol (PI) pathway and that for metabolism of arachidonic acid via the cyclooxygenase and lipoxygenase pathways. There is also some evidence that guanine nucleotide regulatory proteins modulate PLC activity.

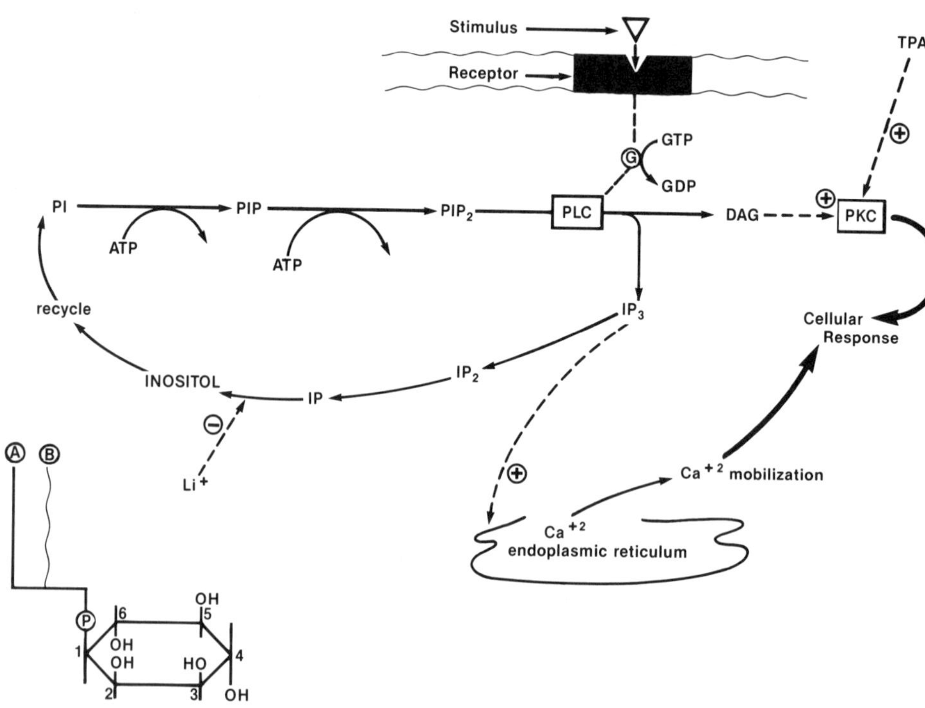

Phosphatidylinositol (PI)

A generally a saturated fatty acid, i.e., stearic acid
B generally an unsaturated fatty acid, i.e., arachidonic acid

Figure 11 Signal transduction via the phosphatidylinositol pathway. Abbreviations are: **TPA**, 12-O-tetradecanoylphorbol-13-acetate (PKC activator); **PI**, phosphatidylinositol; **PIP**, phosphatidylinositol-4-phosphate; **PIP_2**, phosphatidylinositol-4,5-bisphosphate; **DAG**, 1,2-diacylglycerol; **IP_3**, inositol-1,4,5-trisphosphate; **IP_2**, inositol-1,4-bisphosphate; **IP**, inositol-1-phosphate.

No specific inhibitors of PKC are as yet known. However, a number of agents, notably local anesthetics and psychotropic agents, appear to inhibit the enzyme by interacting with surrounding phospholipids (Nishizuka, 1984).

Retrospectively, we may already have one drug which modulates this signal transduction process - lithium. Lithium ion has been well accepted as an important drug in the control of manic-depressive illness. It is known that lithium ion lowers the level of inositol in rat brain, probably through the inhibition of inositol-1-phosphatase (Figure 11). Perhaps the therapeutic action of lithium ion is through depletion of the phosphatidylinositol pool, thereby interrupting hormone-receptor signal transduction in the brain (Hallacher and Sherman, 1980).

D. GROWTH FACTOR RECEPTORS AS TYROSINE PROTEIN KINASES

A variety of proteins stimulate cellular growth or proliferation. Insulin, insulin-like growth factor I (somatomedin C), epidermal growth factor (EGF), platelet derived growth factor, and nerve growth factor (NGF) have been the most extensively studied. Several oncogenes code for fragments of these factors or their receptors. In 1980 Ushiro and Cohen showed that the receptor for epidermal growth factor is a tyrosine protein kinase, the activity of which is stimulated by EGF. Subsequently, the other receptors, except for NGF, were also found to have tyrosine protein kinase activity (Hunter and Cooper, 1985). Thus far, the subsequent steps linking phosphorylation to the ultimate response of the cell have not been elucidated.

E. ION CHANNELS

The composition and concentration of ions within a cell differ considerably from its external environment. Nature has used this to advantage by employing cations such as sodium, potassium, calcium and magnesium and the chloride anion as inorganic second messengers to regulate and activate cells. The flow of these ions into the intracellular environment is generally through gates or channels that traverse the membrane bilayer. Most of our knowledge about their existence has come about indirectly, particularly by measurements of ion conductances or fluxes which cannot be explained by carrier mechanisms or non-permeable membranes. We have also learned from Nature's defensive mechanisms that blocking ion transport can have devastating effects on the organism. This can lead to paralysis or in the extreme to death. Those who fancy eating the Japanese puffer fish should be aware that the liver and ovaries contain the potent neurotoxin named tetrodotoxin. It blocks the inward flux of sodium by blocking an ion channel. Careful removal of the offending organs from the puffer fish insures that the diner will once again have the opportunity to enjoy this gastronomical delight. Fortunately we have begun to harness the fundamental biochemical knowledge about ion transport to develop better cardiac drugs, anesthetics, anxiolytics, and antibiotics.

Drugs can interact with ion channels by exerting indirect allosteric effects or by direct physical hindering of the entry of ions. The allosteric effects can arise from a hydrophobic drug interacting with the adjacent lipid bilayer or with an integral membrane protein to alter the three-dimensional shape of the channel. General anesthetics are believed to act in this manner. Most local anesthetics, however, appear to act by directly plugging the ion channel. These drugs generally hinder sodium or potassium transport into the channel by binding to the outside if they are charged species or to the interior if they are not ionized.

Ion channels are often under receptor-mediated control. The nicotinic acetylcholine receptor, which plays an important role in neurotransmission, regulates monovalent cation permeability. Acetylcholine, the agonist, stimulates the opening of an ion channel to allow the influx of sodium ions for several milliseconds. The toxic effects seen with some snake venoms are apparently exerted by a reaction between the venom and sulfhydryl groups on the receptor. This stabilizes disulfide bonds which keeps the ion channel closed. Local anesthetics and barbiturates can also interact with this receptor ion channel as described above.

GABA (γ-aminobutyric acid) is an inhibitory neurotransmitter that binds to a specific receptor to regulate the flow of chloride ions into and out of the cell. The anxiolytic drugs of the benzodiazepine type, of which Valium® is the best known, are now believed to exert their activity by way of a specific receptor adjacent to the GABA binding site. In effect, these drugs act as co-agonists making the chloride channel more efficient.

Some microorganisms produce substances, some of which have potent antibiotic activity, that work by facilitating the transport of ions across lipid membranes. These ion carriers are called ionophores. Some are pore formers; they provide the ion with a path through the membrane and behave in a manner similar to cellular ion channels. Other ionophores incorporate the ion into a cage-like structure and transport it across the lipid membrane. Gramicidin A is an example of a pore former and valinomycin is a cage carrier.

A23187 is an ionophore that is selective for calcium. This compound has been used extensively in biological research to by-pass receptor activation processes that require calcium; for example, the A23187 stimulated release of histamine and leukotriene biosynthesis in mast cells. In these assays, the ionophore circumvents the classic receptor mediated IgE-antigen secretory response.

Alternatively, agents that block calcium channels have also contributed to our understanding of the role of this inorganic ion in cellular regulation. Two well known anti-anginal/anti-arrhythmic drugs, nifedipine and verapamil, are examples. These compounds inhibit calcium influx into the cell and are generally referred to as calcium channel blockers. Although they are particularly effective on cardiac cells, these drugs have also served as valuable tools in studying calcium transport in a variety of other cellular systems.

F. PROTEINS AS SECOND MESSENGERS OF RECEPTOR ACTIVATION

Although the ultimate biological response to various steroids is quite diverse, the cellular mode of action of these hormones is quite similar. In contrast to the receptors described above, steroid receptors are intracellular proteins. They have high selectivity and affinity for their ligands. They are generally found in the cytosol of cells, although there is evidence that unoccupied receptors may be associated with the nucleus. Steroids bind to the receptor by hydrophobic interactions and by hydrogen bonding to the carbonyl groups on the steroid at the C-3 and C-20 positions. Analogs with rapid rates of dissociation from the receptor do not reside long enough on the receptor to cause activation and may act as antagonists at low concentrations. Certain steroid receptors are substrates for specific protein kinases. The resulting phosphorylation may regulate steroid binding affinity (Dougherty et al., 1985). In mouse fibroblasts, two receptor proteins have been identified: a major protein of 92,000 molecular weight and a minor one of 100,000 molecular weight. A rat liver glucocorticoid receptor has also been found to have two components consisting of 90,000 and 24,000 molecular weights. The steroid-receptor complex is translocated into the nucleus of the cell, where it enhances or depresses the transcription of specific genes. Generally, the binding leads to newly synthesized m-RNA and subsequent protein synthesis. These new proteins are thus the second messengers of steroid receptor occupation.

A class of steroids called glucocorticoids inhibits pro-inflammatory events such as prostaglandin and leukotriene production, lysosomal enzyme release, chemotaxis, cell adherence and vasopermeability. Some of these actions are now thought to be mediated through *de novo* synthesis and release of protein second messengers. Three potential anti-inflammatory proteins that have been identified are lipocortin (Hirata et al., 1980); inhibitor of plasminogen activator; and vasocortin (DiRosa et al., 1985). Isolation, purification and characterization of these proteins are of considerable importance in examining in more detail the myriad number of biochemical events which glucocorticoids modulate. In addition, some of these proteins may become new selective therapeutic agents in their own right.

IV. CONCLUSIONS

Traditionally screening for new drugs was dominated by the use of *in vivo* pharmacological models. One notable exception to this was the early search for antibiotics in which microbiological assays were and still are heavily employed. Assays which permit the measurement of readily quantifiable parameters such as blood pressure, heart rate, pulmonary function, reflexes, body temperature, weight and edema have guided biologists and medicinal chemists in the discovery and development of many of the therapeutic agents introduced in the sixties and

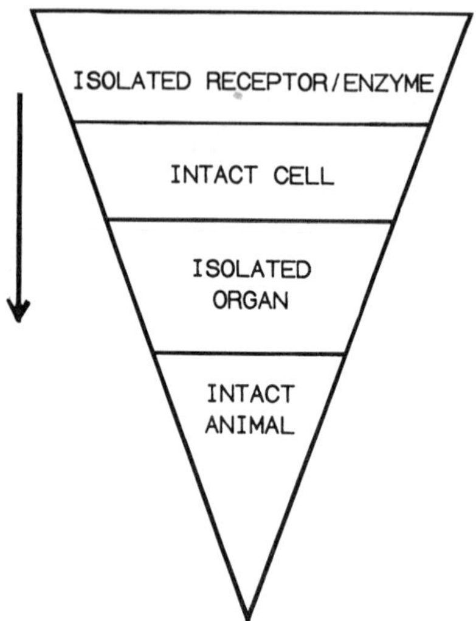

FIG. 12 Test flowchart for the discovery and evaluation of new therapeutic agents.

seventies. This approach, which was successful in defining major classes of therapeutic agents, has limitations; animal models often are not related to human disease, and underlying mechanisms of action frequently remained obscure.

In vitro mechanistic models have the following advantages over *in vivo* models in drug screening: high throughput, more varied experimental approaches, and generally less expense. However, *in vitro* models also have shortcomings and investigators must be cognizant of them. First, biological systems are not inherently simple. Often, autoregulation or redundant pathways are operative *in vivo*. Second, the relationship of the hypothesis to the disease is, at best, a guess and, at worst, a wild goose chase. Last, compounds which may be highly active *in vitro* may show little or no activity *in vivo* as a consequence of metabolism, pharmacokinetics or bioavailability.

Approaches to the design and discovery of new drugs with fewer side effects and novel modes of action are clearly shifting in the direction of a more balanced use of in vivo and in vitro assays (Figure 12). For screening purposes, in vitro assays are more important; however, for profiling compounds in the development phase in vivo methods still remain essential.

REFERENCES

Ahlquist, R. P., Adrenoceptors, Trends in Pharmacological Sciences, pp. 16-17 (1979).

Berridge, M. J., Inositol Trisphosphate and Diacylglycerol as Second Messengers, Biochem. J., 220, 345-360 (1984).

Boeynaems, J. M. and J. E. Dumont, "Outlines of Receptor Theory," Elsevier, Amsterdam (1980).

de Vos, A. M., L. Tong, M. V. Milburn, P. M. Matias, Jarmila Jancarik, Chigeru Noguchi, Susumu Nishimura, Kazunobu Miura, Eiko Ohtsuka and Sung-Hou Kim, Three-Dimensional Structure of an Oncogene Protein: Catalytic Domain of Human c-H-*ras* P21, Science, 239, pp. 888-893 (1988).

DiRosa, M., A. Calignano, R. Carnuccio, A. Ialenti and L. Sautebin, Multiple Control of Inflammation by Glucocorticoids, Agents and Actions, 17, 284-289 (1985).

Dougherty, J. J., R. K. Puri and D. O. Toft, Phosphorylation of Steroid Receptors, Trends in Pharmaceutical Sciences, Feb. 83-85 (1985).

Hallcher, L. M., and W. R. Sherman, The Effects of Lithium Ion and Other Agents on the Activity of myo-Inositol-1-Phosphatase from Bovine Brain, J. Biol. Chem., 255, 10896-10901 (1980).

Hirata, F., E. Schiffmann, K. Venkatasubramanian, D. Solomon and J. Axelrod, A Phospholipase A_2 Inhibitory Protein in Rabbit Neutrophils Induced by Glucocorticoids, Proc. Natl. Acad. Sci. USA, 77, 2533-2536 (1980).

Hunter, T. and J. A. Cooper, Protein Tyrosine Kinases, Ann. Rev. Bioch., 54, 897-930 (1985).

Lefkowitz, R. J., J. M. Stadel, and M. G. Caron, Adenylate Cyclase-Coupled Beta-Adrenergic Receptors, Ann. Rev. Bioch., 52, 159-186 (1983).

Nishizuka, Y., Turnover of Inositol Phospholipids and Signal Transduction, Science, 225, 1365-1370 (1984).

Parascandola, J., Origins of the Receptor Theory, Trends in Pharmacological Sciences, pp. 189-192 (1980).

Robison, G. A., R. W. Butcher, and E. W. Sutherland, "Cyclic AMP," Academic Press, New York (1971).

Ross, E. M., and A. G. Gilman, "Pharmacodynamics," in Goodman and Gilman's The Pharmacological Basis of Therapeutics (A. A. Gilman, L. S. Goodman, T. W. Rall and F. Murad, eds.), 7th ed., Macmillan, New York, pp. 35-48 (1985) .

Tallarida, R. J., A. Cowan, and M. W. Adler, pA_2 and Receptor Differentiation: A Statistical Analysis of Competitive Antagonism, Life Sciences, 25, 637-654 (1979).

Ueda, K. and O. Hayaishi, ADP-Ribosylation, Ann. Rev. Biochem., 54, 73-100 (1985).

2

Progress in Molecular Pharmacology and Biochemistry as a Link Between Diseases and New Drugs for Their Treatment

H. M. Jennewein

Department of Pharmacology
Boehringer Ingelheim KG
Ingelheim, Federal Republic of Germany

D. H. Hinzen and D. Hartmann

Pharmaceutical Research and Development
F. Hoffmann-La Roche AG,
Basel, Switzerland

I. INTRODUCTION

Milestones of medicine in the past century include vaccines and antibiotics for the prevention and treatment of infections as well as countless drugs for the symptomatic and sometimes even causal treatment of degenerative diseases such as hypertension, congestive heart failure, gastrointestinal ulcers, diabetes, asthma, chronic bronchitis and disorders of the central nervous system.

The rapid surge in innovative drugs after World War II was followed in the 1970s and, particularly, the 1980s by a slow down in drug discovery. The more stringent regulations for the registration of new drugs was not the only reason for this. The main reason for the lack of dramatically new drugs is that we do not yet have the necessary knowledge to overcome the unsolved problems in drug therapy. Only if there is relevant information about the pathophysiology of the disease will scientists be able to test proposed cures for it.

Pharmacology of the last few decades has been characterized primarily by widespread screening of numerous chemical compounds in various empirical biological models. The early successes from the molecular modification of endogenous compounds, natural products, or synthetic compounds were followed by the testing of these in high-capacity pharmacological test systems. There has now been a fundamental change in the type of problems presented in pharmaceutical research and in therapy.

Physicians are concluding that the drugs currently available drugs for the symptomatic treatment of many chronic diseases are still inadequate. They want drugs that will treat the causes of disorders such as rheumatism, atherosclerosis, depression and schizophrenia. However, traditional pharmacological models are not suitable for the development of this innovative type of drug, since they are not based on an understanding of the condition. Furthermore, there are no models for many diseases that have recently gained in significance due to changes in the age structure of the population, for instance, Alzheimer's disease.

Thus, classical approaches to finding new drugs are unsuitable for meeting the challenges faced by research today. If the ambitious goals are to be reached, new scientific findings and new strategies for drug discovery are essential. Fortunately, there has been a recent burst of knowledge in biomedical-pharmaceutical science, that offers a firm basis for the development of more progressive drugs. We have also seen the advent of new technologies. Together these provide an excellent breeding ground for a second pharmacological revolution.

The opportunities available for developing better drugs on the basis of new biochemical and pharmacological findings are summarized below in the light of pressing therapeutic problems.

II. CARDIOVASCULAR DISEASES: ATHEROSCLEROSIS

Atherosclerosis, a disease of the large arteries, is one cause of cardiovascular diseases, such as myocardial infarction, angina pectoris and stroke. These diseases are the main leading cause of death in highly developed countries.

Atherosclerosis is a degenerative, chronic disease of unknown etiology. A variety of pathological changes to the arteries are categorized as atherosclerosis: it is not one disease. The knowledge of the many risk factors for inducing atherosclerosis, especially increased total cholesterol, LDL cholesterol and low HDL cholesterol content in the blood; hypertension; diabetes mellitus; and cigarette smoking has not clarified its etiology.

Earlier, there were two complementary theories of the pathology of atherosclerosis: on the one hand, the permeation and depositing of blood plasma components (especially lipids and fibrin) through the intima of the arterial cell wall. On the other hand, the embedding of a thrombus in the artery wall and subsequent development into an atherosclerotic plaque with lipid accumulation leading to the continued growth of plaque.

The objective of traditional drug therapy of atherosclerosis was to normalize blood lipid content.

This is still an important approach. The finding that hydroxymethylglutaryl-coenzyme-A-reductase (HMG-CoA) is a key enzyme in the biosynthesis of cholesterol paves the way for the development of inhibitors of this enzyme and thus for more specific pharmacotherapy.

Just as important, however, is the pursuit of other innovative therapeutic approaches. The "response-to-injury" hypothesis (Glomiset and Ross, 1976) suggests that following damage to the arterial wall, blood platelets adhere to the wall and release platelet-derived growth factors. This provokes local intimal accumulation and proliferation of smooth muscle cells. Platelet-derived growth factor (PDGF) is considered to be one of the main factors in inducing the migration and proliferation of smooth muscle cells. Finding an inhibitor of PDGF might thus be considered a primary goal of pharmacological studies. The deposits of fibrin, connective tissue, and above all lipids then induce the development of plaque.

While smooth muscle cells are involved in the pathogenesis of atherosclerosis, it is questionable if they play as pivotal a part as previously thought. On the other hand, recent studies support the hypothesis that macrophages are vital to atherosclerosis (Mitchinson and Ball, 1987). Macrophages are particularly capable for creating atherosclerotic damage because they -like endothelial cells- have a special mechanism allowing them to absorb modified LDL, the so-called "Scavenger receptor pathway" (Brown and Goldstein, 1985; Schaefer, 1981).

In addition, the secretion of growth factors, cytotoxins, lymphokines, enzymes and oxygen radicals is seen as the cause of atherosclerosis in various animal models. New approaches to better drugs can follow from correcting the disturbed interaction between various blood cells, blood cell mediators such as growth factors, and the vascular endothelium. The transforming growth factor ß, in particular, delays DNA-synthesis and the migration and replication of endothelial cells. This impairs endothelial regeneration and stimulates the development of the atherosclerotic plaque. Future drug research into atherosclerosis could therefore aim at finding antagonists to such mediators.

Finally, atherosclerosis should also be regarded as a pathological immune or autoimmune reaction. There is a suggestion that the T-lymphocytes are involved in the development of atherosclerotic lesions. The progress already made or expected in the near future in the field of atherosclerosis gives reason to hope that new approaches can be found to eliminate this scourge of mankind.

III. DISEASES OF THE CENTRAL NERVOUS SYSTEM

Diseases of the central nervous system can have a particularly vicious effect on a person as they can cause irreversible damage to body functions, break down the intellect and destroy the personality. Only two of these disorders are reviewed below.

A. SCHIZOPHRENIA

Although schizophrenia is divided into several sub-categories, all are characterized by a loss of the sense of reality, sometimes combined with anxiety, paranoia, hallucinations, excitation, etc. It comes in bouts which can be followed by a personality breakdown, general deterioration and intellectual collapse. Structural changes in the brain have been found, but, little is known about the pathophysiology of schizophrenia.

Consequently it is extremely difficult to discover new antipsychotic compounds.

Theories on the pathophysiology of this disease include genetic predisposition, viral etiology, disturbance of various neurotransmitters, social environmental factors, chemical poisoning from the environment, or poisoning resulting from metabolic disturbances. Most of these hypotheses cannot be confirmed. There is one hypothesis that may be strengthened by certain arguments: that is that hyperactivity of the dopaminergic system plays a role in the development of certain symptoms. Evidence of this is provided by the fact that dopamine antagonists have an antipsychotic effect, while amphetamine, which releases dopamine, can lead to a paranoid-like psychosis after long-term administration. Positron emission tomography studies have also revealed that there are far more dopamine receptors present in untreated schizophrenic patients than from normal persons. The mesolimbic dopaminergic pathway of the brain is said to be the one most involved in certain symptoms of schizophrenia. However, clinical success has not been achieved with antagonists more specific to the mesolimbic vs the nigrostriatal regions of the brain nor with those specific to D_1 or D_2 receptors.

One new approach to the treatment of schizophrenia is to inhibit not the postsynaptic receptors, but the activity of the dopaminergic neurons themselves via dopamine auto-receptor agonists, such as 3-PPP, (1) and BHT 920, (2) (Jennewein et al., 1986). (Figure 1 explains these terms.) This imitates the natural feedback mechanism that terminates the activity of the dopaminergic neurons. Presynaptic stimulation results in a reduction in the neuron firing rate, inhibition of dopamine synthesis and reduced dopamine release, all character-

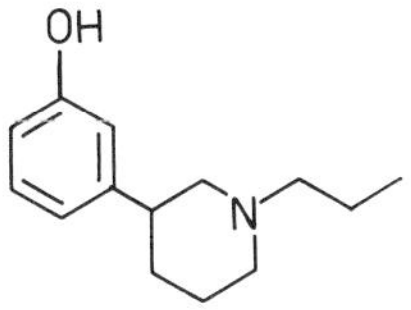

1

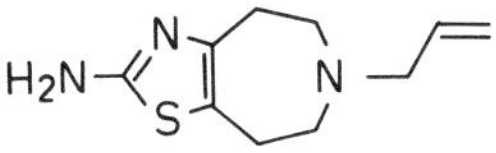

2

istic of a reduction in dopaminergic neuron activity. Since this inhibition is observed primarily in the mesolimbic system, auto-receptor agonists could represent a new form of treatment for schizophrenia. The range of side effects is much narrower than that of dopamine antagonists, and more importantly, there is no risk of adverse effects to the motor system seen with traditional agents. There is yet another unexpected effect of dopamine auto-receptor agonists in a model of Parkinson's disease produced with the neurotoxin MPTP. MPTP was discovered by chance as a contaminant of an illegally synthesized opioid compound. Addicts who injected this compound developed the symptoms of Parkinson's disease. These cases led to the discovery of the chemical structure of this neurotoxin and that it destroys the nigrostriatal neurons. As MPTP also triggers Parkinson's disease in monkeys, we now have a test for potential anti-Parkinson substances. A number of auto-receptor agonists proved to be extremely effective (Schingnitz and Hinzen, 1986). This was surprising insofar as, under normal circumstances, these compounds have no postsynaptic

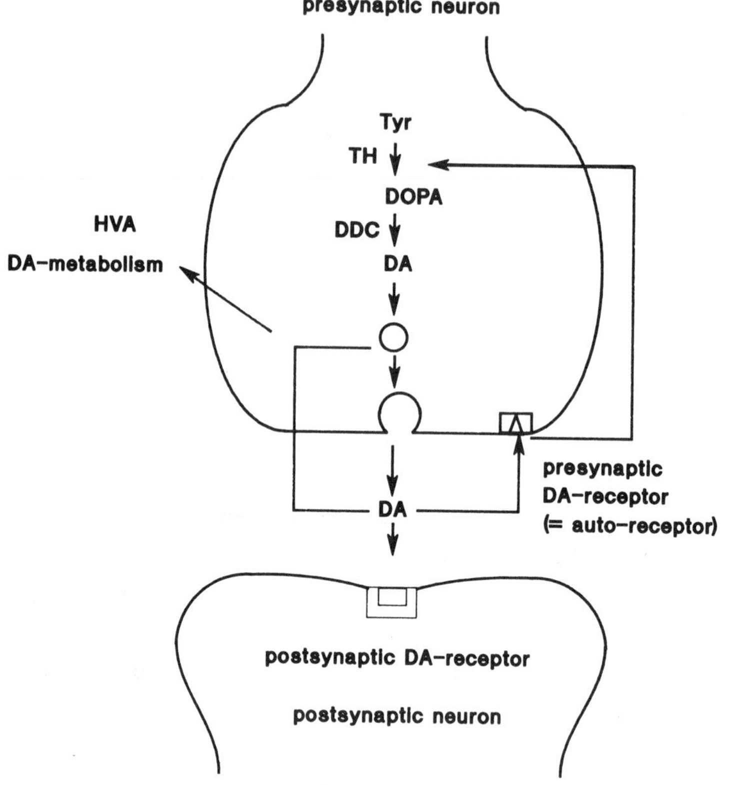

Figure 1 Diagram of a dopaminergic synapse with pre- and post-synaptic receptors.

effects on the nigrostriatal system. Following nerve degeneration the receptors seem to undergo a change which increases sensitivity to the dopamine auto-receptor agonists. This example of dopamine auto-receptor agonists shows how modern drug research combines traditional pharmacology methods, animal models of the diseases, with the methods of molecular pharmacology and pathophysiological theories to uncover new approaches to better drugs.

Besides the dopamine hypothesis, other as yet unvalidated approaches are being used in the search for new drugs effective against schizophrenia. For example, some are working on influencing specific serotonin receptors, inhibiting opioid sigma receptors, or influencing the activity of cholecystokinin, a co-transmitter of dopamine.

The above examples show the benefit of molecular pharmacology and biochemical approaches to drug discovery; however, in the very case of schizophrenia, much basic research has still to be done if new pathophysiological paths are to be revealed. Drug research can contribute significantly by providing innovative substances with original or selective biochemical modes of action.

B. ALZHEIMER'S DISEASE

Alzheimer's disease or senile dementia of the Alzheimer type (SDAT) is the most common cause of dementia in the elderly. The clinical syndrome of brain failure resulting from SDAT is characterized by a global and progressive deterioration of intellect and personality.

The average life expectancy of man - especially in the highly developed industrialized countries - has risen dramatically over the last 100 years. This is due, not least of all to the success of medical and pharmaceutical research. At the beginning of this century only approximately 25 % of the population, in the USA for example, reached the age of 65. Today, about 70 % of the American population reaches the age of 65 (Brody, 1984). The over-80 age group is currently the fastest growing section of the population.

This increase in life expectancy parallels an increase in age-dependent diseases. This applies particularly to the central nervous system. About 1/4 of people over 65 years of age suffer from cerebral diseases, especially dementia and/or depression. Whereas there are drugs showing some benefits in the treatment of depression, there is no effective treatment for the most common central nervous disorder in the elderly, namely senile dementia.

Effective treatment for senile dementia, however, is an urgent medical, social and economic problem as its incidence increases. As many as 20 % of people over 80 are already suffering from it and some expect that in the USA alone there will be a jump in the number of people over 65 years of age from the present 2 million to over 16 million by the year 2050 (Brody, 1985).

Senile dementia is a pathological progressive loss of the intellect, memory and emotional sensitivity as well as motor functions of conscious patients. It can be triggered by a number of different diseases. However, SDAT alone accounts for more than half of the dementia cases.

What efforts are being applied to SDAT by modern drug research? Since the etiology of SDAT is unknown, we must take as our premise the pathophysiological data (for excellent review, see Reisberg, 1983).

The histopathology of SDAT shows degeneration of the nerve cells, particularly in the association centers of the cerebral cortex, and accumulation of intraneuronal neurofibrillary tangles and neuritic (senile) plaques. Not specific to but typical of these histological changes is the appearance of abnormal proteins in the tangles as well as in the central amyloid core of the plaques, and as amyloid deposits around cerebral blood vessels. Further pathophysiological indicators are the lack of central neurotransmitters, especially acetylcholine and a restricted glucose utilization and reduced cerebral blood flow. The latter indicators could be secondary to restricted neuronal function.

What new and promising approaches to discovering effective therapy seem feasible?

As already mentioned, cholinergic deficiency might be at the root of the central nervous system biochemical disturbance in SDAT. However, the number of postsynaptic cholinergic receptors is unchanged even once Alzheimer's disease has progressed. This is similar to Parkinson's disease, which is characterized by a lack of the transmitter dopamine but intact postsynaptic dopamine receptors. It was upon this basis that the currently most effective therapy of Parkinson's disease with L-dopa, a metabolic precursor of dopamine, was devised. A similar approach was also attempted in SDAT, unfortunately without success. Therapy using synthetic cholinergic agonists, cannot be used as the cholinergic agonists available have serious side effects that arise partly from stimulation of cholinergic receptors outside the central nervous system. One way to achieve selective stimulation of the central cholinergic transmitter system, and hence fewer adverse effects, would be to design compounds that stimulate postulated specific subtypes of the cholinergic receptor. Research into centrally-selective, cholinergic modulators holds much promise for the treatment of Alzheimer's disease.

A certain amount of success in the treatment of Alzheimer's disease has already been achieved with inhibitors of the enzyme acetylcholinesterase. This enzyme metabolizes any natural acetylcholine present; This inhibition of acetylcholinesterase leads to an accumulation of acetylcholine. However, an active inhibitor could be used therapeutically only if it acted exclusively within the central nervous system since otherwise the side effect would be comparable to those of synthetic cholinergic agonists.

The abnormal proteins in the brains of Alzheimer patients are, perhaps, of pathogenic significance. The protein amyloid is encoded by a gene on chromosome 21. This is particularly interesting as the central nervous disorder Down's syndrome occurs in persons who have inherited a duplicate copy of chromosome 21. When Down's patients reach the age of 30 to 40, the histopathological changes of Alzheimer's disease are observed in their brains. Clinically too, patients with Down's syndrome suffer from mental retardation or dementia. The similarities with Alzheimer's disease are therefore obvious. Furthermore, there is an autosomal, predominantly hereditary form of Alzheimer's disease. A few years ago the gene encoding for the ß-amyloid was thought to be located immediately next to or on the same site as the gene for the hereditary form of Alzheimer's disease. Thus it was hoped that this would explain the etiology of Alzheimer's disease. However, several research groups have since demonstrated that the ß-amyloid gene and the gene for familiar Alzheimer's disease are distinct and do not occupy the same position on chromosome 21. Neither duplication of the ß-amyloid gene nor overexpression of the amyloid protein has been demonstrated in any case of Alzheimer's disease (Tanzi et al., 1987; St. George-Hyslop et al., 1987). Nevertheless, the origin and the significance of the amyloid protein in Alzheimer's disease merit special attention. Once the abundant presence and chemistry of the amyloid protein has been explained and the origin of this protein examined, new approaches to the testing of potential drugs for the therapy of Alzheimer's disease may be possible. Especially finding a domain containing a protease-inhibitor sequence, within the precursor of the amyloid ß-protein is exciting (Carrell, 1988). Speculations arise that plaque formation in SDAT is a consequence of proteolysis of the amyloid precursor protein releasing a self-aggregating amyloid protein (A_4 peptide). Alzheimer's disease should therefore derive particular benefit from the progress made in protein chemistry and in molecular biology.

Advances in developmental biology, such as in the study of synaptogenesis, growth factors and the functions of the cytoskeleton will also provide potential starting points for the discovery of effective drugs. It is only now that Alzheimer's disease is more widely known that studies of phenomena such as memory and cognition have assumed a new importance. Basic neurobiological research will profit from this, and again may lead to new approaches to therapy for other central nervous system diseases.

IV. VIRAL INFECTIONS

Viral infections can provoke a wide range of diseases, from mild ailments such as colds to serious or fatal diseases such as AIDS, herpes encephalitis and smallpox.

AIDS (acquired immune deficiency syndrome), serves as an example of the medico-social problems raised by viral infections. It has now spread into the average population and is no longer confined to the original high-risk groups. It can be assumed that infection of the population will eventually reach the same degree as in this case of hepatitis B: 5 - 15 % of the population would be infected which would mean around 500 million people. The WHO expects 5 to 10 million people to contract the disease in 1988. This figure will double every year so that in five years we can expect about 200 million cases of infection. Of those people infected, between 10 and 100 % will develop the full-blown disease. AIDS is fatal and there is currently no cure. The number of people infected by AIDS, the increasing costs incurred to society, and the human suffering brought about by the disease are tantamount to world catastrophe. Hygienic measures to combat this epidemic could slow it down, but of course not cure those already infected.

The AIDS infection develops from a retrovirus that contains lymphotropic RNA. This virus is known as HIV-1 (human immune deficiency virus), LAV-1 (lymphadenopathy associated virus), or HTLV-3 (human T-cell lymphotropic virus).

Three other types of the AIDS virus have since been discovered. This creates problems with respect to future types of treatment.

AIDS reveals the significance and medical-ethical challenges presented by viral infections. Unlike in the combat against bacteria, however, it took a long time before any effective antiviral chemotherapeutic agents were developed, probably because we were searching with an inadequate knowledge of the pathophysiology of viral infections. Although there are still gaps in our knowledge of the pathological mechanisms of viral infections, there has been such significant progress in this field that the search for effective antiviral drugs now holds much promise of success.

The infection of human cells by viruses can be broken down into four stages, each of which offers starting points and opportunities for testing potential antiviral substances.

The initial stage is the invasion of the host cell by the virus. Three processes occur at this stage: adsorption and penetration into the cell and uncoating, or removal of the protein coat that covers the viral nucleic acid. Specific receptors are required on both the host cell and the surface of the virus for adsorption. In HIV-1 infection of T-cells in AIDS, special glycoproteins of the virus coat, ENV, react with a glycoprotein of the T-cells, the CD_4 receptor. An octapeptide that contains four threonine aminoacids, peptide T, can block the CD_4 receptor, and thus inhibit adsorption of the HIV virus. CD_4 receptor protein produced by genetic engineering could, perhaps, have the same effect on the HIV virus. Thus, there is preliminary evidence that the adsorption process can be delayed and that this might be a mechanism for antiviral therapy. We have still found no way of influencing penetration. Inhibition of uncoating is another possible mechanism for antiviral compounds. There is evidence that substances that combat influenza or the common cold such as amantadine, isoxazol-oxazoline derivatives, and rimantadine do so by inhibiting uncoating. By protein crystallography (discussed in Chapter 4) it has been shown that the isoxazol-oxazolines react specifically with a hydrophobic pocket of the viral protein VP_1, to stabilize the coat and inhibit virus uncoating.

In the case of many viruses, the second stage in infection of human cells involves making DNA transcripts of viral RNA using viral reverse transcriptase but host cell energy- and substrate- generating processes. This stage includes the actual reverse transcription including activation by a viral protease, circularization of viral DNA and integration in the host DNA. In the case of retroviruses, the reverse transcriptase complex plays a pivotal role in viral proliferation. It ensures that the viral RNA is converted into viral DNA, thus allowing it to be incorporated into the host cellular DNA. Accordingly there has been great interest in reverse transcriptase inhibitors. Suramine was the first: thereafter, substances with specific effects were found,

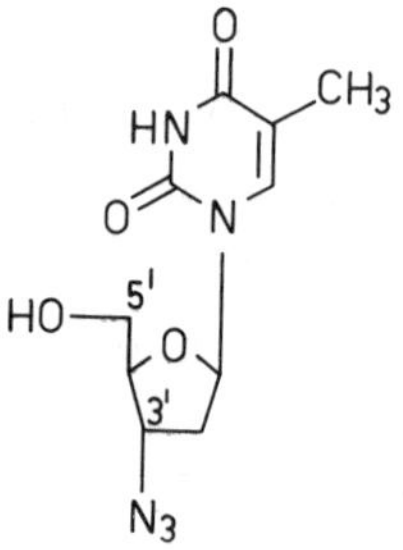

3

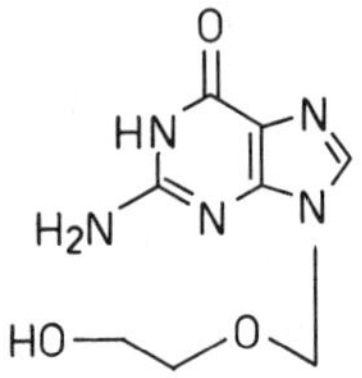

4

one of the most effective of these being azidothymidine (AZT, zidovudine, Retrovir, 3).

Since it is a nucleoside derivative, AZT is phosphorylated by host kinases to the 5'triphosphate, the reverse transcriptase inhibitor. Furthermore, the triphosphate can be incorporated into the viral DNA. This causes chain termination, since the AZT does not contain the 3'hydroxy group to which the next nucleoside would attach.

AZT is effective in AIDS patients although its adverse effects on blood-forming organs means that it is far from ideal. Owing to its mode of action, AZT has no virocidal effect. Patients are therefore still carriers of the virus. The disease can, however, be attenuated and long-term therapy may be one way of preventing the disease from appearing in infected individuals. The necessary clinical experience is, however, still lacking.

Acyclovir (Zovirax, 4) is another nucleoside analogue with a particularly specific and potent effect. To a certain extent, it could serve as a model for antiviral chemotherapeutics. The mode of action is shown in Figure 2.

Like guanosine, acyclovir is monophosphorylated by a viral thymidinekinase at the hydroxy group whilst the cellular, non-viral kinases cannot catalyse this first

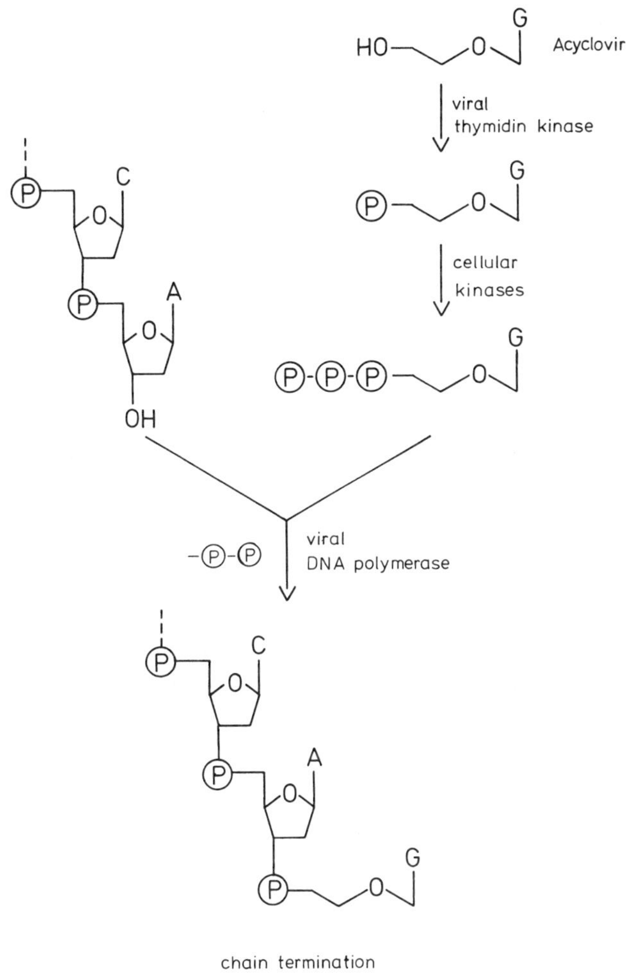

Figure 2 The metabolic transformations of acyclovir involved in its anti-viral activity.

step in the mechanism of action of acyclovir to the same extent. However, the cellular enzymes convert acyclovir mono-phosphate to the di- and finally triphosphate. Viral DNA polymerase has 200 times greater affinity for acyclovir triphosphate than has the cell-specific DNA polymerase. It integrates acyclovir triphosphate into the virus DNA. As the acyclovir does not contain a 3′

hydroxyl group, chain termination results. Acyclovir monophosphate also forms a strong complex with the viral DNA polymerase to inhibit it. These special mechanisms using viral enzymes mean that acyclovir has an effect in infected cells only and is thus selective.
Acyclovir acts on DNA viruses such as varicella zoster viruses, but particularly the type I herpes simplex viruses. Use in complications such as herpes encephalitis has reduced mortality from 70 to 25% of the cases.

The third stage in viral reproduction includes transcription, splicing, processing and translation: processes that lead to production of functional viral proteins. Some of these processes are controlled by transactivating factors that are potential targets of anti viral substances. It is possible that Avarol interferes in these processes.

Another substance, ribavirin, acts at the end of the translation process. Ribavirin, which is a guanosine analogue, specifically prevents a modified guanosine from being attached to the viral mRNA. This prevents binding to the ribosome and thus inhibits biosynthesis of the viral proteins. In addition, phosphorylated ribavirin, inhibits the enzyme that synthesizes guanosine triphosphate. Ribavirin inhibits the respiratory syncitial virus (RSN) and Lassa fever virus.

The fourth and final stage of viral reproduction is elimination of the virus particles from the host cell. This stage is characterized by the assembly, budding and release processes. It is possible that interferons intervene in this stage, although interferons also have other effects such as stimulating the formation of proteins that halt cell protein synthesis and activating an enzyme that destroys viral DNA. Only relatively limited clinical utility has been demonstrated for interferons.

Finally, brief mention must be made of vaccination as a means of combating viruses. As we know, vaccinations have proven extremely successful against certain viral diseases, the most important examples being smallpox and polio. Equally high expectations are currently being set on the development of a vaccine against AIDS. There is indeed much activity in this area. For example, the development of vaccines based on cleavage products of viral proteins is in progress. Certain HIV surface proteins have been produced by genetic engineering and injected with an adjuvant designed to stimulate the formation of antibodies. One possible adjuvant is $Al(OH)_3$, but more recently a spherical immunostimulating protein molecule has been tested as an adjuvant and the HIV surface proteins are covalently linked to it. The efficacy of any vaccine against AIDS will be extremely difficult to test because were it to fail, it would be fatal.

To sum up, it can be said in view of the most recent biochemical findings of viral diseases that we are on the verge of a new era. It therefore appears certain that by applying modern drug research a number of invaluable antiviral chemotherapeutic agents can be developed.

V. IMMUNOLOGICAL DISEASES

The progress in research into the immunological system is so remarkable and so significant that there is a wealth of new starting-points and opportunities to discover new substances in drug research today. It must be said, however, that there are considerable gaps in the range of drugs available to treat diseases of the immunological system.

Diseases of the immunological system can result when the immunological system overreacts and when it does not function adequately. The first case includes such diseases as rheumatism, allergies, asthma, while the second includes conditions that arise following radiation, cytostatic treatment or infections. Possibilities for the treatment of the latter conditions are seen in the colony stimulating factors that stimulate production of granulocytes, macrophages or several types of blood cells. We can now manufacture one of these by genetic engineering to make sufficient quantities available for clinical studies following favorable results in animals.

However, it is often necessary to slow down the immunological system as in allergies and hypersensitivity. In these very common diseases, the release of biogenic amines, especially histamine, from the mast cells is the direct result of the triggering antigen-antibody reaction. Antihistamines and cromoglycate are substances currently used to stop the effects of histamine. However, the later phases of allergic reaction are more promising for opening up approaches to novel drugs. Phospholipid mediators and various cells may play a role here. The important phospholipid mediators are prostaglandins, leukotrienes and the platelet activating factor (PAF). Figure 3 shows very roughly how these mediators are produced. Here it can also be seen where there are possible new mechanisms for drug action. The formation of a mediator can be prevented by inhibiting its biosynthesis. Alternatively one can block the effect of a mediator at its site of action by a receptor antagonist.

Initial research into inhibition of mediator biosynthesis was stimulated by the knowledge that cyclooxygenase inhibitors, such as non-steroidal antiinflammatory drugs, inhibit the formation of prostaglandins. These mediators, especially the E_2 prostaglandins, play an

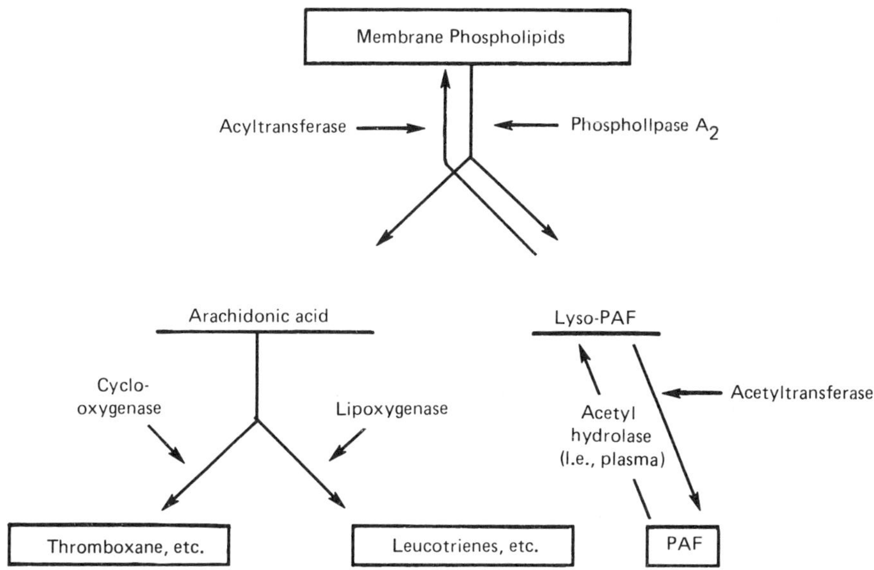

Figure 3 Diagram of the relationship and brosynthesis of prostaglandins, leukotriens and PAF.

important part in rheumatic diseases of the joints whilst bearing little significance for asthma. It still cannot be said to what extent thromboxane and leukotrienes are involved in allergies, asthma or rheumatism. However, the development of specific inhibitors could certainly provide tools to answer this question. The usefulness of thromboxane- and leukotriene-antagonists as antiallergic agents is being investigated intensively. Of the leukotrienes, it is chiefly peptidoleukotrienes, LT_{C4}, LT_{D4} and LT_{E4}, upon which antagonists should act. LT_{D4} appears to have the most marked bronchospastic effect and most antagonists are described in this connection. LT_{B4}, on the other hand, has less bronchospastic effect, showing rather inflammatory effects, especially of the airways. There are certain similarities between LT_{B4} and the platelet activating factor (PAF) which is currently the object of considerable research interest.

The reasons for this similarity and, by way of example of a modern approach to drug discovery, PAF research will be outlined here in more detail. PAF is a potent phospholipid mediator that is believed to play a role in allergies, inflammation, shock, graft rejection, platelet aggregation and ischaemic conditions. A PAF-antagonist is therefore of high scientific, medical and commercial potential (for review see Braquet et al., 1987).

PAF is located in, and can be synthesized by, many cells: platelets, leucocytes, endothelial cells and some kidney and lung cells. PAF is formed from membrane phospholipids. The first step is deacylation of the phospholipid by phospholipase A_2. This forms arachidonic acid and lyso-PAF which lacks the biological activity. Lyso-PAF is converted to the biologically active compound by acetylation by a CoA-dependent acetyl-transferase. A second pathway to PAF has, as its final step, not an acetylation but the introduction of a phosphocholine group. The cells that can synthesize PAF are either involved in inflammatory processes or they belong to shock target organs such as the lungs, kidneys or vasculature.

Most cell types are affected by PAF. It induces marked systemic effects when given intravenously: arterial hypotension, pulmonary hypertension, bronchoconstriction, gastrointestinal erosions and increased vascular permeability. Usually these effects are parallelled by a reduction in circulating leucocytes and platelets, which aggregate in the lung. Higher doses of PAF cause animals to die from shock.

Although the physiological and pathophysiological role of PAF is not clear, evidence suggests involvement in diseases such as asthma, shock, renal diseases and inflammation. PAF may also play a role in urticaria, psoriasis and skin graft rejection. To clarify the pathophysiology of PAF, careful clinical investigations will have to be performed with a specific, highly active and non-toxic PAF-antagonist.

Numerous PAF-antagonists have been described but only two groups seem to be of any value so far. These are the gingko biloba extracts, the so-called gingkolides, and certain triazolobenzodiazepines. The latter were found by using a high-capacity screening procedure in which various chemically different agents were tested.

Systematic structure-activity analysis showed that the PAF-antagonism of the triazolobenzodiazepines is independent of benzodiazepine receptor affinity. WEB 2086-BS (5) was the most potent and specific PAF-antagonist with no sedative effects. As WEB 2086-BS is a strong and specific PAF-antagonist, this compound will help to clarify the pathophysiology of PAF. There is experimental evidence to suggest efficacy in bronchial asthma and other diseases. WEB 2086-BS may be clinically useful in a number of diseases in which present treatment is unsatisfactory.

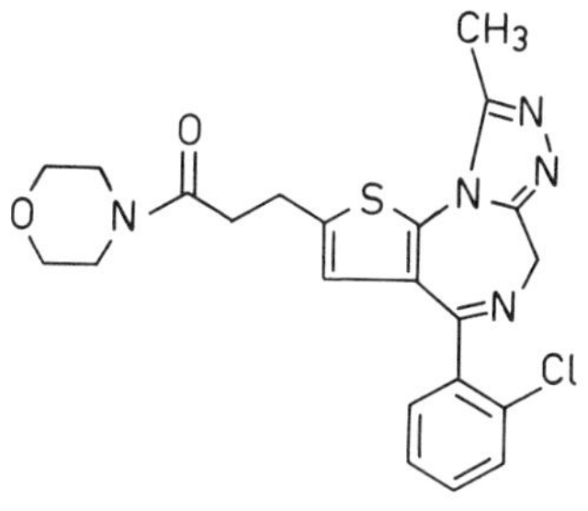

CH_3 O O N S N N N N Cl

5

Since cyclooxygenase inhibitors are the most widely used drugs for treatment of arthritis, inhibition of the enzymes involved in lipid mediator synthesis seems to be a good way to discover new drugs. For leukotrienes, inhibition of 5-lipoxygenase is especially attractive. Inhibition of phospholipase A_2 also seems interesting as this would attack the cascade at a very early stage. Phospholipase A_2 is activated in many inflammatory diseases and inhibited in the organism by endogenous peptides that are induced by glucocorticoids. Besides extracellular humoral factors, intact cells accumulating at the inflamed target organ also play a vital role in most allergic and inflammatory processes. While the eosinophils act late in the allergic process, macrophages in particular dominate the initial inflammatory processes. During conflict with bacteria or antigens macrophages express intercellular adhesion molecules on their surface. They use these molecules to adhere to T-lymphocytes via other adhesion molecules, such as leucocyte function-associated antigen. The recruitment, accumulation and activation of cells important in inflammatory diseases such as rheumatism or asthma are regulated by cytokines or interleukins, and the adhesion process is necessary for this. Six interleukins are now known which, in chemical terms, are usually glycoproteins.

Thus, there are several targets for new drug design: one could block the adhesion molecule with the result that one of the first steps in the immunological conflict would be inhibited. It should also be possible to block the signal protein receptors. Without such antagonists it is of course difficult to decide how important the individual cytokines are in the disease. However, we do know that cyclosporin A inhibits the transcription of the interleukin 2 receptor. Since cyclosporin A has marked immunosuppressive effects, it seems that interleukin 2 plays a pivotal role in immune response. As a further example, interleukin 4 appears to enhance specifically IgE synthesis which is an imunoglobulin directed against parasites and pollen. By inhibiting interleukin 4, i.e. via a receptor antagonist, IgE synthesis may be specifically suppressed. Consequently other antibodies important for defence against infections are not influenced by such an intervention. Therefore such an antagonist could be of value in treating atopic allergies.

These speculations are designed to show that many different openings have resulted from recent immunological research. This knowledge has provided tools for discovery of specific agents, and these agents in turn will serve as tools to probe the cause of the blue disease. Accordingly this demonstrates the synergistic results of basic research and drug research. This can be used to patients' benefit.

VI. CANCER

The first decisive steps in the treatment of cancer were taken in the 1890s. In 1894 William Halsted suggested enbloc resection in cancer surgery, e.g. as radical mastectomy in the case of carcinoma of the breast, Wilhelm Röntgen discovered X-rays which were then used to treat tumors and Paul Ehrlich discovered the first alkylating agent.

It was 50 years before it was discovered that alkylating agents have a cytotoxic effect in man: in World War II there was an accident in Naples harbor during which alkylating compounds in war gas were released. The seamen exposed to the substance developed serious bone marrow and lymphatic tissue damage. This observation led, at Yale University in 1943, to the first successful therapeutic studies in patients with Hodgkin's disease, thus marking a breakthrough in the treatment of advanced metastasizing tumors (De Vita jr., 1982). It was soon found that the occasionally spectacular recovery was usually very shortlived. This led to combination chemotherapy which had already been applied successfully in the treatment of acute leukemia. De Vita set a milestone by treating Hodgkin's disease with combination of mustargen, oncovin, procarbazine and prednisone (De Vita Jr. and Serpick, 1967). 80% of the patients showed complete remission which in most cases went hand in hand with cure. In the early 1970's a study by Einhorn et al. produced further important progress, this time with a solid carcinoma, cancer of the testicles. A combination of cisplatin, velbe and bleomycin led to complete remission in 70% of cases; again most of the patients were completely cured (Einhorn and Donohue, 1977). Anyone practising oncology in the 1960s and 1970s would have found it difficult not to share the widespread optimism in medical oncology and duly expect combination chemotherapy to lead to soon breakthroughs in the treatment of more common tumors such as breast, lung and intestinal cancer. Although a number of other tumors such as choriocarcinoma, Burkitt's lymphoma, Wilm's tumor, histiocytic lymphoma and acute lymphatic leukaemia are now potentially curable malignant diseases, which in itself obviously represents major progress in cancer treatment, the optimistic forecasts for the major tumors have not proven accurate. On the contrary, for about 10 years now there has been stagnation with occasional indications of resignation, as reflected in a recent article from the daily press which raised the following question: a crisis in cancer therapy? (Brunner, 1988). This change of mood is due partly to the fact that the molecular modification of tried and tested cytostatics such as cisplatin, anthracyclin derivatives, alkylating agents, antimetabolites, etc. did not produce more active or at least less toxic compounds. Nor did any new, revolutionary substances of different chemical structure find their way into clinical practice. The recent publication by Harrap and Connors (1987) takes a glance at current activities in this field. Below is an overview of a few new preclinical research approaches which will perhaps have medium- to long-term consequences for the treatment and diagnosis of malignant diseases.

A. ONCOGENES

Oncogenes are sequences of genes found in tumour viruses that can lead in suitable hosts to the development of malignant tumors. There are remarkable similarities between these sequences and certain gene sections of human DNA which explains why these human DNA sequences are sometimes referred to as protooncogenes. For about 20 protooncogenes there are corresponding sequences in retroviruses. However, for about 20 other protooncogenes this has not been established. Usually protooncogenes are strictly regulated and their activation during DNA synthesis or cell proliferation is shortlived and does not lead to malignant transformation. However, changes in the genome, such as point mutation, a shift of genetic material from one chromosome to another, or gene amplification, can lead to the malignant transformation of the cell.

The exact mechanism by which oncogenes and their corresponding proteins promote the development of tumors has still not been fully explained; it appears that two genetic changes are required before a cell undergoes malignant transformation. For instance, the HL-60 cell, derived from a promyelocyte leukaemia in man, contains both an amplified myc gene and an activated N-ras gene. It is also known that oncogenes whose genetic products are observed in the nucleus of the cell, e.g. myc or myb, are supplemented during the malignant transformation of cells by oncogenes whose genetic products are found in the cytoplasm, e.g. ras, src or abl.

The proteins coded for by oncogenes have various functions. For example c-myc, N-myc, L-myc, myb, fos, ski and p 53, code for nuclear proteins that may regulate DNA synthesis. The products of erb-b, fms, src, yes, fes, ras, fgr and raf are cytoplasmic enzymes that phosphorylate proteins and hence play a part in the regulation of enzyme activity. The products of ras oncogenes are cytoplasmic GTP-binding proteins, while the genetic product of v-sis has a very high sequence homology with the B chain of the growth factor PDGF.

From the discussions on signal transductions in Chapter 1, Section III, it can be appreciated that any of these gene products could potentially disturb regulated cell growth.

So far 3 oncogenes or groups of oncogenes have been established as responsible for the development of malignant diseases in man: (1) ras in the case of lung, colonic and bladder cancer, leukaemia and sarcomas; (2) myc in the case of Burkitt's lymphoma, neuroblastoma, retinoblastoma and small cell cancer of the lung; and (3) abl in the case of chronic myeloid leukaemia. The ras family is the most common, accounting for about 15% of all tumors (Aaronson, 1985; Barbacid, 1986; Minden, 1987).

No new therapeutic approaches based on the above information are apparent, so it will be a long time before revolutionary drugs can be expected. However, currently much research is being done on the ras oncogenes. One possible starting point would be to restore the altered GTP-ase activity. Blocking tyrosine phosphorylation would be one approach to antagonists of oncogenes that code for proteins with this type of kinase function. Finally, monoclonal antibodies for oncogenic products could be important tools for the diagnosis of tumors and for increasing our knowledge of the extent to which various oncogenes are involved and cooperate in the development of various tumors.

B. GROWTH FACTORS

As discussed in Chapter 1, growth factors are essential for the proliferation and differentiation of normal cells. Growth factors can be secreted by either the endocrine or the paracrine route (cells in the immediate vicinity) or, by the cells themselves, i.e. autocrine route. Under physiological conditions, release, concentration and receptor binding are under stringent control. A loss of efficacy of this feedback system sometimes leads to malignant transformation. Under experimental conditions, fibroblasts can be transformed by stimulation with growth factors such as TGF-alpha (transforming growth factor-), TGF-beta, PDGF (platelet derived growth factor), EGF (epidermal growth factor) or IGF II (insulin growth factor II). On the other hand cells infected with oncogenic retroviruses produce growth factors (Sporn and Roberts, 1985). There is less evidence that these proteins are responsible for malignant transformation in man. For instance, PDGF is produced by the cancer cells even though they have no receptors for PDGF. The role of EGF is not clear although there are receptors for EGF in many human tumors, such as glioblastomas, small cell carcinoma of the lung and ovarian, cervical, renal and bladder cancer. Bombesine-like peptides (BLP) may be an important form of autocrine stimulation of small cell cancer of the lung. Antibodies to BLP can prevent or delay growth of this tumor type in the athymic mouse (Cuttitta et al., 1985). Similarly monoclonal antibodies for the EGF receptor have been reported to delay growth of a human squamous epithelium cell line (Masui et al., 1984).

Either monoclonal antibodies or other blockers of specific growth factor receptors may be promising avenues for development of new cancer therapies. This approach could be particularly promising for tumors in which autocrine stimulation by one or a few growth factors is the established cause of malignancy.

C. SIGNAL TRANSDUCTION

Recent work has enhanced our knowledge of the biochemical regulation and processing of signals which - received on the surface of the cell, transmitted through the cytoplasm and finally reaching the nucleus-lead to cell proliferation (Berridge, 1983; Berridge; 1984). One important biochemical system is the phosphatidylinositol (PI) pathway discussed in Chapter 1, Section III. Stimulating the sodium-proton pump leads to an increase in intracellular pH (Nishizuka, 1986), a signal for cell proliferation. Additionally IP 3, leads to the release of calcium ions from intracellular vesicles which, in turn, triggers a proliferation stimulus. Under physiological conditions, this transmission of signals is strictly controlled. In transformed cells, however, there seems to be increased production of proliferation stimuli. Another member of the PI signal transduction protein kinase (PKC), is currently enjoying a central role in experimental oncology since it appears to be a receptor for the tumor-promoter t-PA.

We are still in the very first stages of research into the relationship between cancer and signal transduction. A strategy for drug discovery could be to find tumor-specific low-molecular blockers. However, we are not sure if the enzymes from tumor cells differ from those of normal cells, so this approach may not yield specific therapeutic agents. Monoclonal antibodies against the enzymes involved could gain importance in the diagnosis and characterization of tumors thus helping to select those tumors suitable for treatment with enzyme-blockers. Results from research in this field can only be expected in the medium to long term, if at all.

D. BIOLOGICAL RESPONSE MODIFIERS (BRM)

These are members of a large family of polypeptides that affect components of the immune system or other defence mechanisms of the body. These include various substances such as interferons, thymosine, interleukins, lymphotoxins and the tumor necrosis factor (TNF) (Tannock, 1987). Some of these compounds have been produced by genetic engineering and some have even been tested clinically. This section will only look at interferons, interleukin 2 (IL 2) and TNF.

Long before it was tested clinically, interferon was considered to be a wonder drug for the treatment of cancer. Today, we know that those expectations have not been fulfilled: it is effective in only a very few indications, such as hairy cell leukaemia, chronic myeloid leukaemia, perhaps Kaposi's sarcoma in the case of AIDS, cancer of the kidney and malignant melanoma (Quesada et al., 1986; Talpaz et al, 1987).

Much hope was also placed in interleukin 2 (IL 2) and TNF. IL 2 is produced by and stimulates T-lymphocytes. It inhibits growth of a number of tumors when given in high doses in combination with activated lymphocyte (LAK) cells (Rosenberg et al., 1985). This treatment unfortunately involves great risk and costs. Similarly, results with TNF have been disappointing since it is quite toxic (Blick et al., 1986).

Despite the very modest progress in the use of BRMs in the treatment of cancers, the potential of this type of compounds has not been fully exhausted. New clinical approaches might include carefully combined combination studies of biomodifiers with cytostatics, combination of different biomodifiers, and use of these substances in treatment after removal of the tumor or following successful induction therapy with a marked decrease in tumor volume. If these criteria are observed, the side effects of these polypeptides could perhaps also be reduced.

E. MONOCLONAL ANTIBODIES (MOAB)

An MOAB is an immunoglobulin that has a defined specificity for binding to a certain antigen, such as a specific cellular protein. Monoclonal antibodies are produced by special techniques discussed in more detail in Chapter 5. Monoclonal antibodies that recognize specific protein antigens on tumor cells could be used for either diagnostic protein or therapeutic purposes. The therapeutic effect could be based on the lysis of cells by the combination of an MOAB with other lytic proteins, or an activation of cytolytic monocytes or the natural killer cells. MOAB can be covalently linked to cytostatics, toxins, or radioisotopes for therapeutic uses or to fluorescent agents or radiosiotopes for diagnostic purposes (Tannock, 1987).

Despite these very promising prospects for the use of MOAB, there are still many problems to be solved, some of which deserve mention here. Tumors consist of heterogeneous cell populations and thus have varied antigenic properties. Additionally, many tumor cells have only a mild antigenic effect and tumor antigens are often found on other cells. Antigens on the surface of a tumor cell can be shed or absorbed into the inside of the cell with the result that they are no longer accessible for binding to MOAB. Monoclonal antibodies are proteins and hence autoimmunogenic; this can lead to neutralization of the MOAB. As MOAB are macro-molecules, it is not easy for them to leave the bloodstream to reach the site of action.

The manufacture of innovative MOAB is an attempt to tackle some of these problems, using techniques described in Chapter 5. For example, to reduce immunogenicity, chimeric antibodies have been produced by genetic engineering; the constant section of the immunoglobulin molecule is of human origin and the variable section stems from the mouse (Yokoyama et al., 1987). There have also been attempts to produce a purely human monoclonal antibody by fusing human lymphocytes with human myeloma cells, or by infecting human lymphocytes from immunized patients with the Epstein Barr virus. Another revolutionary approach is to stimulate the organism into producing antigen-specific antibodies on its own by vaccination with so-called antiidiotypical antibodies (Herlyn et al., 1986) described in Chapter 5.

The clinical results achieved so far with monoclonal antibodies have not yet led to a breakthrough in tumor therapy (Goodman et al., 1985; Spitler et al., 1987).

The diagnostic use of MOAB also raises many problems to be solved. Besides those mentioned above, there are also technical problems, e.g. the unsatisfactory properties of the radioisotopes used at present, such as iodine 131. However, we do hope that more useful MOAB methods for tumor diagnosis can be developed in the medium term.

F. METASTATIC SPREAD

Today roughly half of all tumor patients already have metastases when the diagnosis is made. Patients do not usually die from the primary tumor, but from the metastases. Individual tumors spread, according to a recent theory, as follows: (1) tumor cells adhere to the basement membrane, the interface that separates the epithelium from the connective tissue. (2) Enzymes such as collagenase, plasminogen activators, cathepsin B, hyaloronidase, and heparanase are released. (3) The tumor cells then move freely within the connective tissue until they find their way to a lymphatic or blood vessel. At this point the cells can spread until they resettle in a suitable healthy organ. Metastatic spread is therefore not a passive, but an active, highly specific process. For example laminine receptors and fibronectin have been implicated in the adherence of cells to cell membrane or connective tissue (Scanlon, 1985; Liotta, 1985; Wewer et al., 1987).

Such receptors might offer a starting-point for a therapeutic approach. Should it be confirmed that the metastatic spread depends on receptors, prevention of this process might become possible by use of blockers of these receptors. The result would not be destruction of the tumor, but inhibition of its metastasizing potential. Even where patients already had a local metastasis, some could be cured completely by surgery. Alternatively, monoclonal antibodies against these receptors would permit a histological diagnosis of the metastatic potency of tumors. This field is still very much at the experimental stage, however; results are only expected in the long range.

G. ANGIOGENESIS

The formation of new blood vessels is a basic biological process in the regeneration of tissue. This requires stimulation by factors referred to generally as angiogenins. A large number of such substances have been identified, for instance ECGF (endothelial cell growth factor), bFGF (basic fibroblast growth factor). aFGF (acidic fibroblast growth factor), TGF-alpha, TGF-beta, hyaluronic acid fragments, coeruloplasmin, prostaglandins, and nicotinamide, to mention only a few (Folkman, 1985; Crum et al., 1985; West et al., 1985). Furthermore, a series of angiostatic factors are also known, e.g. protamine and hexasaccharide from heparin in combination with glucocorticoids (Folkman J., 1985). The vascular endothelial cells are the targets for these angiogenic factors; these cells have ECGF and bFGF receptors and can be stimulated by both into proliferation.

Tumor cells also produce angiogenic factors and so they themselves can partially determine how they spread locally (Kull et al., 1987; Fett et al., 1985; Strydom et al., 1985). We do not know how many different stimuli or receptors are involved in the formation of new blood vessels. Nor is it known if there is a difference between stimuli or receptors in the regulation of the formation of new vessels in normal compared to tumor cells, given that the target is invariably the normal vascular endothelial cell. One could imagine therapy based on the blockade of the endothelial cell receptors, although it is still open as to whether or not it is possible to block only tumor-specific angiogenins. One could also look for physiological counter-regulators. On the whole, this is a very experimental field and results can be expected only in the medium to long range.

H. MULTIPLE RESISTANCE TO CYTOSTATICS

Multiple resistance (MR) is a problem of drug therapy in oncology. There are two types of resistance, primary (or intrinsic) and secondary (or acquired). In primary MR there is an inadequate clinical response to chemotherapy; this is typical of colonic or renal cancer. Secondary MR is common in tumors which do at first respond, for instance in leukaemia, lymphoma, testicular teratoma or small cell cancer of the lung. These patients become resistant to a number of drugs such as anthracyclins, vinca alkaloids, epipodophyllotoxin derivatives and actinomycin D, even though they have previously been treated with only one compound representative of these various groups. Recent work on human tumor tissue has shown that the degree of MR runs parallel to the expression of a protein (P-170). P-170 is a membrane pump that eliminates cytostatics and other toxins from the cell. The human gene that codes for this protein has now been cloned and there are already monoclonal antibodies against P-170 (Ozols and Cowan, 1986; Pastan and Gottesman, 1987; Thiebaut et al., 1987). A number of substances such as anaesthetics, detergents, amphotericin B, calmodulin antagonists and calcium blockers can overcome MR under experimental conditions. Verapamil has been most extensively investigated; it prevents certain cytostatics from binding to P-170, thus increasing their intra-cellular concentrations (Cornwell et al., 1986;

Cornwell et al., 1987). This effect of verapamil has nothing to do with its effect on calcium metabolism.

The search for new drugs is currently focusing on finding P-170 inhibitors directed against the ATP or substrate binding sites. Experiments are also being carried out with MOAB coupled to toxins (Fitzgerald et al., 1987). An unsolved problem is the specificity of such therapy since P-170 is found in normal colon, kidney, adrenal, liver and pancreas tissue. Inhibitors should in themselves be non-toxic and should not increase the toxicity of cytostatics. A predictive screening system is vital for the selection of substances. Results from this area of research can be expected in the short to medium term.

VII. OUTLOOK

Therapeutic progress with innovative drugs will occur once the basic pathophysiological and biochemical aspects of the disease can be considered at a molecular level during the development of drugs. Molecular pharmacology, biochemistry and, in particular, genetic engineering, have created the basis for a more rational drug design. Wherever possible, a mechanistic approach, tested several times over in vitro and at the cellular level, should of course be supplemented by whole animal studies. A mechanistic approach to drug research is often a long-term process and demands rethinking in many disciplines connected with drug research as well as considerable investment in research and development. A continual flow of financial resources, doing away with routine scientific work and creating atmospheres conducive to innovation, as described in chapter 10, will represent the most vital elements for successful drug research in the future.

REFERENCES

Aaronson, St. A., The role of oncogenes in human neoplasia, in "Important Advances in Oncology", (V.T. De Vita jr., S. Hellman and St.A. Rosenberg, eds.) J.B. Lippincott Company, Philadelphia, 1985, pp. 3-15.

Barbacid, M., Human oncogenes in "Important Advance in Oncology" (V.T. De Vita jr., S. Hellman and St.A. Rosenberg, eds.) J.B. Lippincott Company, Philadelphia, 1986, pp. 3-22.

Berridge, M. J. A., A novel cellular signaling system based on the integration of phospholipid and calcium metabolism, in "Calcium and Cell Function", (W.Y. Cheung, ed.), Vol. 3, Academic Press, New York, 1983, pp. 1-36.

Berridge, M. J. A., Inositol trisphosphate and diacylglycerol as second messengers, Biochem. J., 220, 345-360 (1984)

Blick, M. B., S. A. Sherwin, M. G. Rosenblum and J. U. Guttermann, A phase I trial of recombinant tumor necrosis factor (r TNF) in cancer patients, Proc. Am. Soc. Clin. Oncol., 5, 14 (1986).

Braquet, P., L. Touqui, T. Y. Shen and B. B. Vargaftig, Perspectives in Platelet-activating Factor Research, Pharmacological Reviews, 39, 97-145 (1987).

Brody, J. A., Publ. Helth Rep., 99, 468-475 (1984).

Brody, J.A., Prospects for an Ageing Population, Nature, 315, 463-466 (1985).

Brown M. A. and J. L. Goldstein, Ann. Rev. Biochem. 52: 223-261 (1985).

Brunner, K., Krebstherapie in der Krise? Neue Züricher Zeitung, 1988, p. 25.

Carrell, R. W., Alzheimer's disease, Enter a protease inhibitor, Nature, 331, 478-479 (1988).

Cornwell M. M., I. Pastan and M. M. Gottesman, Certain calcium channel blockers bind specifically to multidrug resistant human KB carcinoma membrane vesicles and inhibit drug binding to P-glycoprotein, J. Biol. Chem., 262, 2166-2170 (1987).

Cornwell M. M., A. R. Safa, R. L. Felsted, M. M. Gottesman and I. Pastan, Membrane vesicles from multidrug resistant cancer cells contain a specific 150-170 K Da protein detected by photoaffinity labelling, Proc. Nat. Acad. Sci., 3847-3850 (1986).

Crum R., S. Szabo and J. Folkman, A new class of steroids inhibits angiogenesis in the presence of heparin or a heparin fragment, Science, 230, 1375-1378 (1985)

Cuttitta F., D. Carney, J. Mulshine, T. Moody, J. Fedorko, A. Fischler and J. Minna, Bombesin like peptides can function as autocrine growth factors in human small cell lung cancer, Nature, 316, 823-826 (1985)

De Vita V.T. Jr., in "Cancer, Principles and Practice of Oncology" (V.T. De Vita jr., S. Hellman and St. A. Rosenberg, eds.), J.P. Lippincott Company, Philadelphia, 1982, pp. 132-155.

De Vita V.T. jr. and A. Serpick, Combination chemotherapy in the treatment of advanced Hodgin's disease, Proc. Amer. Ass. Cancer Res., 8, 13 (1967).

Einhorn L. and J. P. Donohue, Cis-diammine dichloroplatinum, vinblastine and bleomycin combination chemotherapy in disseminated testicular cancer, Ann. Intern. Med., 87, 293-298 (1977).

Fett J. W., D. J. Strydom, R. R. Lobb, E. M. Alderman, J. L. Riordan , B. L. Vallee, Isolation and characterization of angiogenin, an angiogenic protein from human carcinoma cells, Biochemistry, 24, 5480-5486 (1985).

Fitzergerald D. J., M. C. Willingham, C. O. Cardarelli, H. Hamada, T. Tsuruo, M. M. Gottesman and I. Pastan, A monoclonal antibody-pseudomonas toxin conjugate that specifically kills multidrug resistant cells, Proc. Nat. Acad. Sci., 4288-4292 (1987).

Folkman J., Angiogenesis and its inhibitors, in "Important Advances in Oncology", (De Vita, V.T., S. Hellman and St.A. Rosenberg, eds.), J.B. Lippincott Company , Philadelphia, 1985, pp. 42-62.

Glomiset J. A. and R. Ross, The Pathogenesis of Atherosclerosis, 2 parts, New Engl. J. Med., 295, 369-377 and 420-425 (1976).

Goodman G. E., P. Beaumier, I. Hellström, B. Fernyhough, K.-E. Hellström, Pilot trial of murine monoclonal antibodies in patients with advanced melanoma, J. Clin. Oncol., 3, 340-352 (1985).

Harrap K. R. and Th. A. Connors, New avenues in developmental cancer chemotherapy, The Institute of Cancer Research, Medical Research Council, Academic Press, London 1987.

Herlyn D., A. H. Ross and H. Koprowski, Antiidiotypic antibodies bear the internal image of a human tumor antigen, Science, 232, 100-102 (1986).

Jennewein H. M., E. A. Bruckwick, J. Hanbauer, J. Mierau and W. Lovenberg, Evidence for a specific effect of BHT 920, an azepine derivative, on tyrosine hydroxylase in the dopaminergic system of the rat, European J. Pharmacol, 123, 363-369 (1986).

Kull F. C., D. A. Brent, I. Parikh and P. Cuatrecasas, Chemical identification of a tumor-derived angiogenic factor, Science, 236, 843-848 (1987).

Liotta L. A., Mechanism of cancer invasion and metastasis, in "Important Advances in Oncology",(V. T. De Vita jr., S.Hellman and St.A. Rosenberg, eds.) J.B. Lippincott Company, Philadelphia, 1985, pp. 28.41.

Masui H., T. Kawamoto, J. Sato, B. Wolf, G. Sato and J. Mendelsohn, Growth inhibition of human tumor cells in athymic mice by antiepidermal growth factor receptor monoclonal antibodies, Cancer Res., 44, 1002-1007 (1984).

Minden M. D., Oncogenes, in "The Basic Science of Oncology" (I.F. Tannock and R.P. Hill, eds.) Pergamon Press, New York, 1987, pp. 72-88.

Mitchinson, M. J. and R. Y. Ball, Macrophages and Atherogenisis, Lancet, II, 146-149 (1987)

Nishizuka J., Studies and perspectives of proteinkinase-C, Science, 233, 305-312 (1986).

Ozols R. F. and K. Cowan, New aspects of clinical drug resistance: the role of gen anmplification and the reversal of the resistance in drug refractory cancer, in "Important Advances in Oncology" (V.T. De Vita, S. Hellman, St.A. and Rosenberg, eds.) J.B. Lippincott Company, Philadelphia, 1986, pp. 129-157.

Pastan J. and M. Gottesman, Multiple drug resistance in human cancer, New Engl. J. Med., 316, 1388-1393 (1987).

Quesada J.R., E. M. Hersh, J. Manning, J. Reuben, M. Keating, E. Schnipper, L. Itri and J. U. Gutterman, Treatment of hairy cell leukemia with recombinant a-interferon, Blood 86 (2), 493-497 (1986).

Reisberg B., Alzheimer's disease, The Free Press, New York, 1983.

Rillema J. A., Phospholipase-C activity in normal rat mammary tissues and in DMBA-induced rat mammary tumors, Proc. Soc. Exp. Biol. Med., 181, 450-453 (1986).

Rosenberg St. A., M. T. Lotze, L. M. Muul, S. Leitman, A. E. Chang, St. E. Ettinghausen, Y. L. Matory, J. M. Skibber, E. Shiloni, J. T. Vetto, C. A. Seipp, C. Simpson and Ch. M. Reichert, Observations on the systemic administration of autologous lymphokine activated killer cells and recombinant interleukin-2 to patients with metastatic cancer, New Engl. J. Med., 313 (23), 1485-1493 (1985).

Scanlon E. F., The process of metastasis Cancer, 55, 1163-1166 (1985).

Schaefer H.-E., in "Disorders of the monocyte-makrophage system", (D. Huhn, H.-E. and F. Schmalzl, eds.), Springer-Verlag, Berlin, 1981, pp. 137-142.

Schingnitz G. and D. Hinzen, MPTP-Modell of Parkinsonism, Effect of Dopaminergic Agonist B-HT 920, Naunyn-Schmiedeberg's Arch. Pharmacol. 332, R91 (1986).

Spitler L. E., M. del Rio, A. Khentigan, N. I. Wedel, N. A. Brophy, L. L. Miller, W. S. Harkonen, L. L. Rosendorf, H. M. Lee, R. P. Mischak, R. T. Kawahata, J. B. Stoudemire, L. B. Fradkin, E. E. Bautista and P. J. Scannon, Therapy of patients with malignant melanoma using a monoclonal antimelanoma antibody-ricin A chain immunotoxin, Cancer Res., 47, 1717-1723 (1987).

Sporn M. B. and A. B. Roberts, Autocrine growth factors and cancer Nature, 313, 747-751 (1985).

St. George-Hyslop P. H., Absence of Duplication of Chromosome 21 Genes in Familial and Sporadic Alzheimer's Disease, Science, 238, 664-666 (1987).

Strydom D. J., J. W. Fett, R. R. Lobb, E. M. Alderman, J. L. Bethune, J. F. Riordan and B. L. Vallee, Amino acid sequence of human tumor derived angiogenin, Biochemistry, 24 (20), 5486-5494 (1985).

Talpaz M., H. M. Kantarjian, K. B. Mc Credie, M. J. Keating, J. Trujillo and J. Gutterman, Clinical investigation of human a-interferon in chronic myelogenous leukemia, Blood, 69 (5), 1280-1288 (1987).

Tannock I. F., Immunotherapy and the potential applications of monoclonal antibodies, in "The Basic Science of Oncology", (I.F. Tannock and R.P. Hill, eds.) Pergamon Press, New York, 1987, pp. 326-336 .

Tanzi R. E., The genetic defect in familial Alzheimer's disease is not tightly linked to the amyloid ß-protein gene, Nature, 329, 156 (1987).

Thiebaut F., T. Tsuruo, H. Hamada, M. M. Gottesman, I. Pastan and M. C. Willingham, Cellular localisation of the multidrug-resistance gene product P-glycoprotein in normal human tissues, Proc. Nat. Acad. Sci., 7735-7738 (1987).

West D. C., I. N. Hampson, F. Arnold and S. Kumar, Angiogenesis induced by degradation products of hyaluronic acid, Science, 228, 1324-1328 (1985).

Wewer U. M., G. Taraboletti, M. E. Sobel, R. Albrechtsen and L. A. Liotta, Role of laminin receptor in tumor cell migration Cancer Res., 47, 5691-5698 (1987).

Yokoyama M., Y. Nishimura and T. Watanabe, Suppression of tumor growth by in vivo administration of a recombinant human mouse chimeric monoclonal antibody, Jap. J. Cancer Res., 78, 1251-1257 (1987).

3

Principles of Pharmacokinetics and Drug Metabolism

G. Bozler and J. Schmid

Department of Biochemistry
Dr. Karl Thomae GmbH
Biberach, Federal Republic of Germany

I. THE AVAILABILITY OF THE DRUG AT ITS SITES OF ACTION

The efficacy of a drug is not determined by its pharmacodynamic characteristics alone. It also depends to a large degree on the pharmacokinetic properties of the drug since pharmacokinetic processes control the rate and the extent to which the drug can reach its site of action. Pharmacokinetics mainly describes the time course of the concentration of the drug in the various parts of the body; often it includes the time course of metabolites derived from the parent drug.

In view of the high degree of structural variability of drugs and the multiplicity of kinetic and metabolic reactions known, the task of establishing clear correlations between structural characteristics of substances and their pharmacokinetic properties appears rather daunting.

However, the following paragraphs will indicate that the pharmacokinetic fate of drug molecules is a consequence of their physico-chemical properties and may therefore to some extent be predicted from their chemical structure.

A. PHARMACOKINETIC CONCEPTS

By definition, all processes by which an organism handles the drug or molecules derived from it constitute the overall pharmacokinetic properties of the drug. Viewing an organism as a black box, there is drug input and drug output. Drug input that reaches the systemic circulation other than by a direct intravascular dose requires transport through biological membranes and is termed absorption. Drug output, termed elimination, comprises excretion of unchanged drug from the body and metabolic transformation to different molecular species.

Viewed from the concentration of the drug in the blood, as has been done classically since Dost (1953) coined the term pharmacokinetics in his book "Der Blutspiegel" (blood level), one distinguishes irreversible losses by elimination and reversible losses via distribution processes. Elimination and distribution are collectively termed disposition, Figure 1.

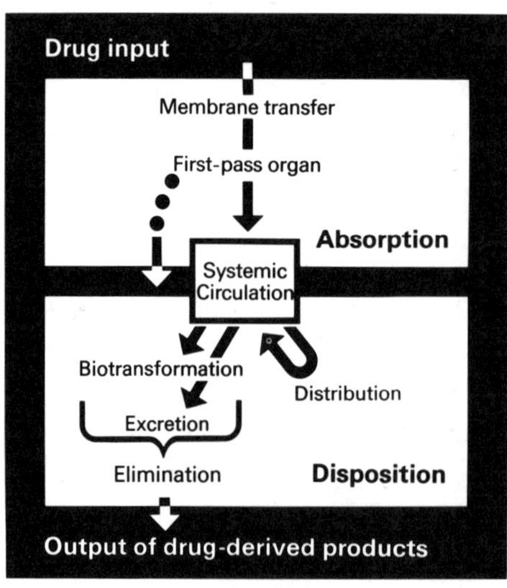

Figure 1 Pharmacokinetic concepts and definition of terms.

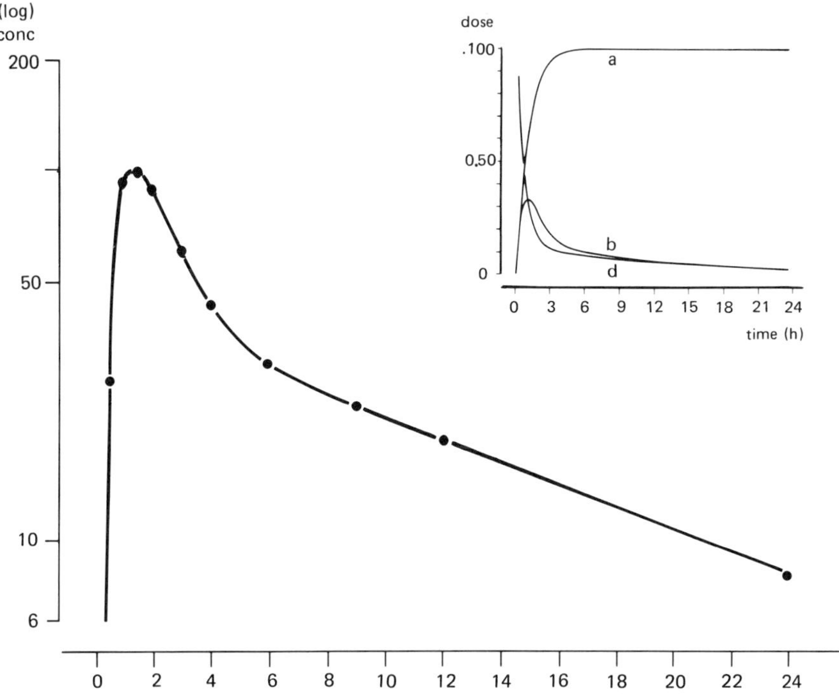

Figure 2 Typical concentration-time course in blood after oral administration of a drug showing the experimental data (•) together with the line of best fit in a semilogarithmic plot. This blood level profile (b) results from the action of disposition (d) on the amount absorbed (a) into the system (in the example complete absorption of the dose is assumed).

A typical result of the combined action of absorption and disposition is the time course of the blood concentration after oral administration of a drug, Figure 2. Conceptually, this curve (b) can be constructed (in mathematical terms: convoluted) from a hypothetical curve (a) representing the amount absorbed without concomitant disposition and a curve representing disposition (d) alone. Curve (d) may be experimentally derived from a separate dose administered intravascularly. Practically, however, the pharmacokineticist often has to confine himself to the inverse procedure (deconvolution) of extracting the kinetic information from the more complex time courses observed, when all processes act on the drug simultaneously. Computerized data fitting is used for such tasks.

In order to evaluate the factors that govern the time course of the availability of an active principle at its site of action, its overall distribution within the "black box" has to be considered. This time course is determined by all processes that compete for the drug including distribution to other tissues and elimination reactions.

All kinetic processes such as absorption, distribution, metabolism and excretion involve transport and binding steps.

B. DRUG TRANSPORT

Blood or plasma receives the drug from the site of absorption and carries it rapidly to all tissues, including both the elimination organs and the target tissue. The availability of a drug from the blood to a tissue can be limited either by perfusion or by diffusion (see Rowland and Tozer, 1980).

1. Hemoperfusion. When tissue membranes present almost no barrier to diffusion, perfusion of the tissue by blood will become the rate-limiting process in delivery of drug molecules. Perfusion varies from tissue to tissue. Well perfused tissues receive a high percentage of the cardiac output (about 5000 ml/min in man): lung, 100 %; kidneys, 22 %; liver, 27 %; and brain, 14 %. Their rate of perfusion is 10 to 150 times higher than the hypothetical average perfusion of the total body. Poorly perfused tissues, such as skin, muscle, body fat and bones, receive under basal conditions only one third of that average rate.

An eliminating organ can at the most remove the amount of drug that is presented by perfusion. The conser-

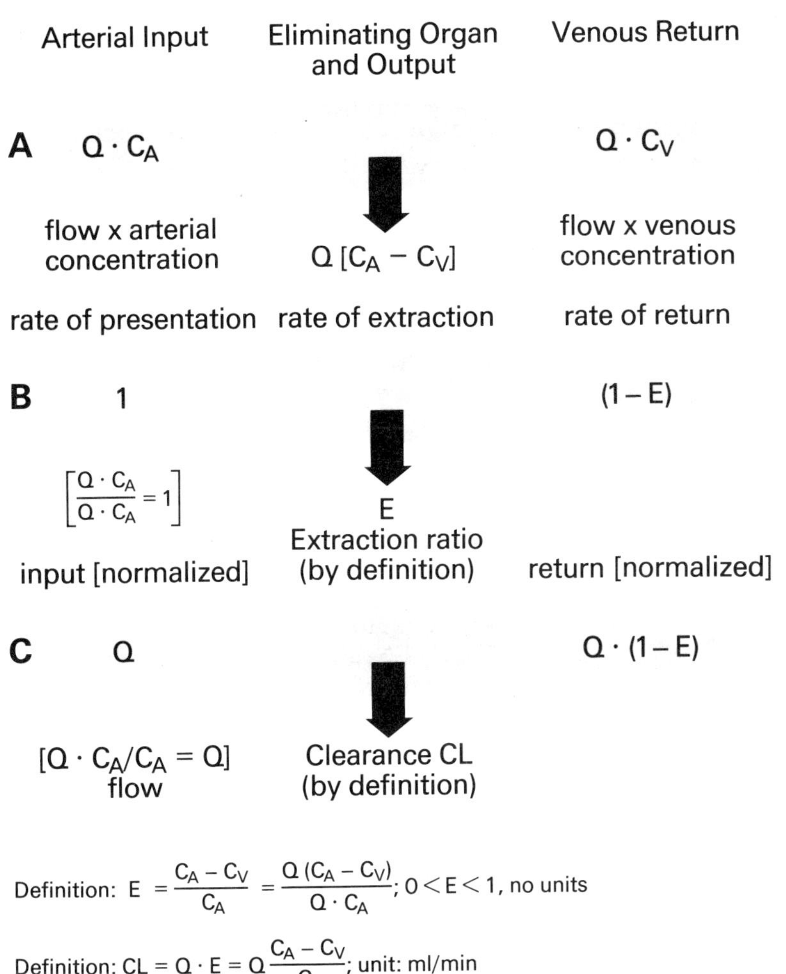

Definition: $E = \frac{C_A - C_V}{C_A} = \frac{Q\,(C_A - C_V)}{Q \cdot C_A}$; $0 < E < 1$, no units

Definition: $CL = Q \cdot E = Q\frac{C_A - C_V}{C_A}$; unit: ml/min

Figure 3 Extraction of a drug from the circulation by an eliminating organ: quantitative description in terms of mass balance (A); normalizing the rate of presentation to unity leads to the definition of the extraction ratio (B); correcting the input for concentration yields a definition of clearance.

vation of mass during this process can be expressed in physiological terms, Figure 3: clearance is defined as the product of the blood flow and the extraction ratio (a number between 0 and 1 that characterizes the extraction efficiency of the organ). In other words, clearance indicates that volume of blood from which all drug would appear to be removed per unit time (e.g., in ml/min).

Drugs may be classified according to their clearance: when the extraction ratio approaches unity, clearance is practically equal to the blood flow and, perfusion rate becomes the limiting factor. If on the other hand, the extraction ratio is low, clearance will be insensitive to changes in hemoperfusion.

2. <u>The biological membrane.</u> On its way to the site of action, a drug has to traverse a series of biological barriers made up of membranes of some kind. Irrespective of their often very different functions, membranes are composed quite uniformly of lipids and proteins. While most of the functional properties of membranes are attributed to the embedded proteins, the lipids are primarily responsible for the physico-chemical properties of membranes.

Membrane lipids include cholesterol and a variety of phospholipids. A typical phospholipid consists of apolar hydrophobic tails, i.e. fatty acid residues, and a polar hydrophilic head. The phospholipids are arranged in the biological membrane in a bilayer spanning 5 to 10 nm. The hydrophobic tails are directed towards the interior while the polar heads are in contact with the aqueous phases on either side of the membrane.

Biological membranes are dynamic rather than static. According to the fluid mosaic model, (Figure 4) the double layer of phospholipids forms a viscous liquid in which globular protein molecules are embedded and can float more or less freely. The size and structure of a protein deter-

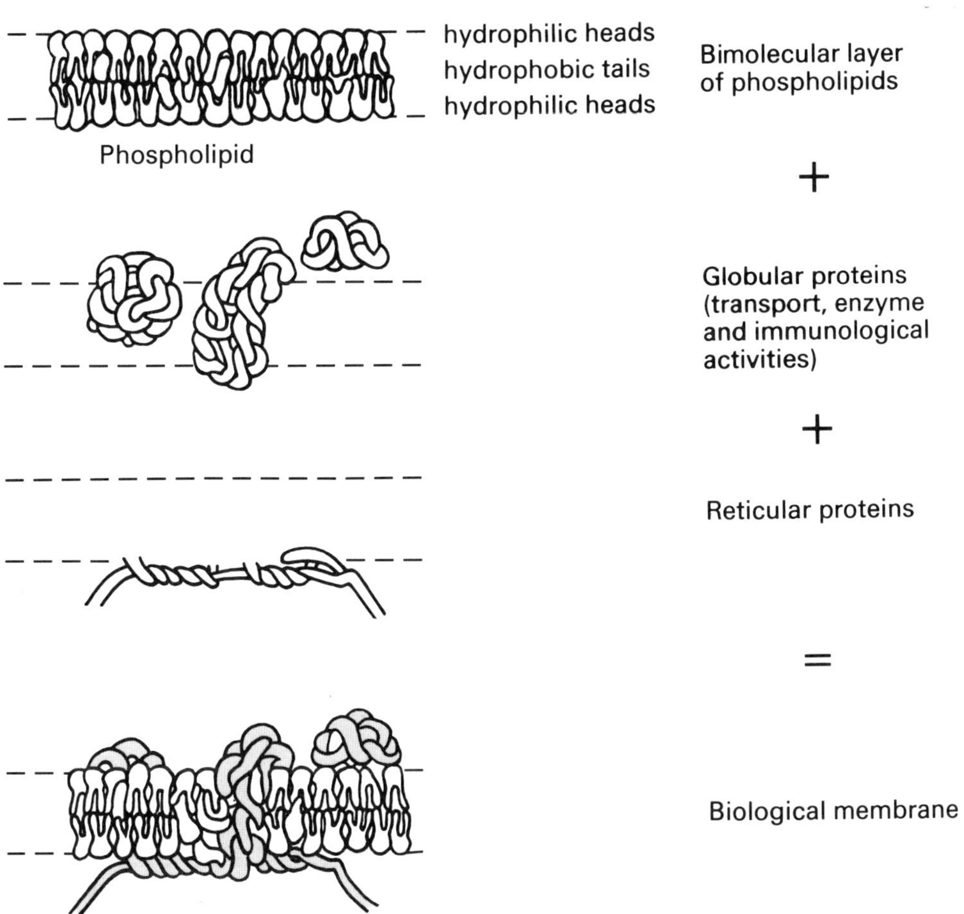

Figure 4 Highly schematic representation of the biomembrane and its structural elements.

mines the depth of its penetration into the lipid layer and whether or not it will span the entire membrane. Individual proteins can thus form specific aggregates.

A third component of the membrane structure is reticular proteins. They stabilize the specific spatial organization of the membrane and lend this mechanical support.

This membrane model represents a thermodynamically stable system, is mechanically feasible, and can explain

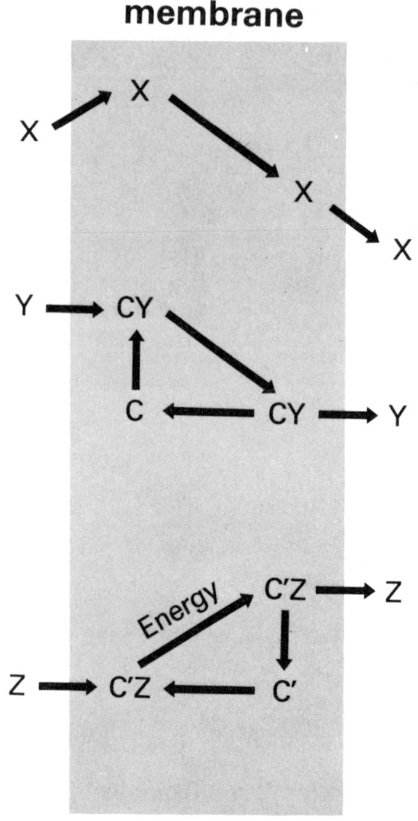

Figure 5 Simple passive diffusion (X), facilitated diffusion (Y) and active transport (Z) through a cell membrane. C and C' represent carrier molecules. All forms of transport will operate in the opposite direction if concentration conditions are reversed.

all major transport phenomena, especially carrier transport processes.

3. Membrane transfer of drugs. Drug molecules are transported through biological membranes primarily by passive diffusion. This process is based on a concentration gradient, i.e., a difference in the concentration of the substance on either side of the membrane (Figure 5). The thermal movement of the molecules makes diffusion possible.

With regard to the membrane structure, two fundamentally different possibilities emerge for passive diffusion. Water-soluble substances of very small molecular size (m.w. $<$ 100 dalton) such as urea, diffuse through water-filled pores of functional proteins. In this way, water itself can readily pass through biological membranes. Most drugs, however, are larger molecules of varying lipophilicity. Their diffusion through the membrane is facilitated by the fact that the drug "dissolves" in the lipid bilayer of the membrane. Hydrophobic interactions, (discussed in Chapter 4), between the lipophilic drug and the lipids of the membrane are therefore characteristic of such transfer processes.

Carrier-mediated transport processes play a major role for many endogenous natural substances. In this case, a distinction is made between facilitated diffusion and active transport, Figure 5. In both processes the substance is temporarily bound to a specific carrier protein. Facilitated diffusion operates only in the direction of the concentration gradient. Substances handled by an active transport are also moved against the concentration gradient. The energy necessary for active transport is ultimately derived from cell metabolism.

Although all other mechanisms cannot yet be excluded, particles or larger molecules such as proteins are commonly thought to cross membranes only by vesicular processes (termed, e.g. exocytosis, endocytosis, pinocytosis, or transcytosis). In such processes the particle is surrounded by a portion of the cell membrane, which then breaks off to totally enclose the foreign substance. This process may occur in such a way that the particle moves to the inside of the cell.

Passive diffusion is the most usual transport mechanism for drugs to cross biological membranes. The explanation for this may be twofold: first, most drugs are lipid-soluble to some degree as they contain lipophilic moieties; secondly, carrier proteins are usually quite specific for their natural substrates. Drugs do not normally resemble these substrates sufficiently in order to be transported. However, in some organs, such as kidney, liver, and intestine, carrier systems with less stringent structural requirements contribute significantly to the membrane transfer of some drugs.

C. INTERACTION OF DRUGS WITH BIOLOGICAL MACROMOLECULES

Free diffusion of a molecule to its target site is limited not only by membrane barriers but also by its interaction with endogenous macromolecules other than the target site. Such macromolecules are substantial in number and binding to them significantly influences the overall pharmacokinetic fate of a drug. Examples include plasma proteins, tissue proteins and membrane receptors.

Binding of the drug can be followed by various additional events depending on the functional properties of the endogenous reaction partner. In the simplest case, the drug is released again after a certain time, possibly at a different location within the circulating fluids. Processes that follow binding to reactive sites include transformation of the drug, transport through membranes, or chemical reaction with the ligand. These reactive sites, in particular metabolic enzymes or carriers, are involved primarily in elimination processes.

1. <u>Reversible binding to plasma and tissue proteins.</u> The extent of reversible binding depends on the binding capacity of the protein and on its affinity for the ligand. These parameters allow specific and nonspecific binding to be distinguished. Specific binding refers to a defined interaction of a drug molecule with a binding site; such

binding is characterized by a defined high affinity and limited capacity of the site for the ligand. Competitive stoichiometric displacement by other ligands (of suitable affinity and specificity) is a general requirement for specificity. On the other hand, nonspecific binding is typically linked with "low affinity" and "unlimited" capacity. Therefore competitive displacement of a nonspecifically bound drug is less likely.

In intact biological systems normally just a small fraction of all available binding sites is occupied by the drug.

Binding equilibrium is, as a rule, reached within milliseconds, since rates of association and dissociation are quite rapid. Therefore, the degree to which a drug is bound to proteins both in blood and in tissues is important for the overall equilibrium distribution and hence for the pharmacokinetic fate of a substance.

Consideration of drug binding as a reversible process reveals the dual role of binding sites, either as acceptor or as donor of a ligand. Binding has a buffer function, providing both sink conditions for free (unbound) drug and supplying drug by dissociation under suitable conditions. Hence, the drug concentration that is relevant for any other process, e.g. for elimination of the drug or for eliciting any effect, depends on the local environment where each of these processes occurs in the body. Therefore, the relevant concentration may refer to either free or total (free + bound) drug. An example is shown in Section II.C.1.

2. <u>Binding to sites that eliminate the drug.</u> Binding to reactive sites is a more complex process, as it also involves irreversible removal of the drug from the interaction site. However, these sites become important only if the chemical structure of the drug meets the specific re-

quirements for binding to and processing by the site. The probability of a particular drug molecule interacting with such a specific site decreases with increasing drug concentration in the system: the pathway is subject to becoming saturated. The process may still be quite efficient, however, and contribute significantly to disposition, if its turnover is high.

Carriers and enzymes often feature active sites with a higher affinity for their ligands than is seen for the affinity of transport proteins for the ligand. Both free and bound drug is available for interaction with the carrier or enzyme under those conditions, provided that the regional transit time is in the order of seconds, thus allowing reequilibration.

3. Implications of specific interactions. Steric properties of the ligand normally play a major role in specific interactions. Therefore, enantiomers are expected to be processed differently.

The maximum rate of reaction is determined by the number of available sites. This limiting condition will be reached, if the drug concentration becomes sufficiently high (self saturation). Other molecules that compete for the binding site, e.g. metabolites of the drug or endogenous ligands, can decrease the reaction rate (competitive inhibition). Drug interactions may become significant under these circumstances.

Variability in drug therapy is of increasing concern (see Rowland et al., 1985); individual variation frequently has a pharmacokinetic basis. It may, for example, be due to genetic or environmental effects on the abundance and efficiency of the functional proteins involved in the processing of the drug. Drugs that critically depend on such proteins in their processing show high inter- and intraindividual variability, necessitating careful therapeutic monitoring.

D. RATE-DETERMINING PROCESSES

The availability of a therapeutic agent at the site of action and, accordingly, the extent and duration of the effect can be derived in general terms from a knowledge of the rate-determining steps in absorption and disposition.

Obviously, only a minute fraction of the drug administered reaches the sites of action. The way to such sites involves a series of processes, all of which have to compete with parallel pathways of disposition. The amount of drug that is ultimately available at the site of action and the time course of that amount depend in principle on the efficiency of all the possible processes: those that lead the drug to its target and those that divert it from that path. For a particular drug, however, only a few, and sometimes only one, of these processes will be decisive. Alternative (parallel) pathways may be comparatively inefficient or slow so that they can be neglected. In addition, processes that are more efficient than the rate-determining one, the "bottleneck" in a sequence of reactions, are not manifest in the overall time course.

The processes that determine the fate of one drug may not be decisive for another one; this often holds true, even within a series of structurally related compounds. These are the reasons why structure-pharmacokinetic relationships can be successfully derived for one class of compounds but not so for others. Nonlinear relationships may be observed if more than one process contributes significantly.

1. Types of drug concentration-time curves. All three major mechanisms involved in drug kinetics, that is hemoperfusion, diffusion, and interaction with biological macromolecules, affect drug concentration in a typical manner. As a result, characteristic kinetic features are observed in concentration-time courses.

Hemoperfusion quickly evens out local differences in drug concentration within the circulatory system since the cardiac output per minute in man approximates the whole blood volume. Owing to its rapid buffering capacity, reversible binding can play a similar role throughout the organism. Consequently, local drug concentrations available for other processes are virtually constant. Thus, hemoperfusion and binding can normally influence or limit the concentration available and its rate of change, but cannot change the typical kinetic features of the other processes.

Passive diffusion can be generally described by Fick's law: the rate of diffusion is a function of the concentration difference, the surface area and the distances involved, and the characteristic factors of the nature of the biological barrier and the diffusing substance. Passive diffusion of a drug normally involves very small distances. In addition, drug concentration on the receiving site is often negligible in relation to that on the driving site (sink conditions). Therefore, within those reasonable simplifications, the rate of diffusion follows first-order kinetics, whereby the compounded influences of physiological and physico-chemical factors are collected in the proportionality constant k:

$$\text{rate of diffusion:} \quad dA/dt = -k \cdot C \tag{1}$$

This means that the rate of diffusion is proportional to the concentration at the driving site and that this concentration declines exponentially with time.

The rate of disappearance of the drug by interaction with endogenous macromolecules that eliminate the drug often depends on the stability of the complex formed. It may be expressed then by a Michaelis-Menten type equation:

$$\text{rate of disappearance:} \quad \frac{dA}{dt} = -\frac{C\ (dA/dt)\max}{(K+C)} \tag{2}$$

(dA/dt)max: the constant maximal disappearance rate;
K: affinity constant with the dimension of a concentration.

Affinity constants of drugs for macromolecules such as receptors, carriers, and enzymes are in the picomolar to micromolar range, which roughly corresponds to the range of free drug concentration at the active sites.

When C is smaller than K ($C \ll K$), that is at low substrate concentrations, the reaction becomes pseudo-first-order; the rate of disappearance becomes proportional to the drug concentration:

$$\text{rate of disappearance:} \quad dA/dt = -k \cdot C \tag{3}$$

When C is larger than K ($C \gg K$), the rate of disappearance of the drug is no longer dependent on the concentration of the drug (zero-order kinetics), but on the number of reactive sites, their activity and rate of restoration; the rate of disappearance approaches the maximum possible rate, (dA/dt)max.

2. Implications of dose-dependencies in rate-determining processes. Apparent first-order or linear kinetic conditions are also called dose-independent since the concentrations observed under comparable conditions are proportional to the dose administered. The time course of any concentration change within physiological limits is then independent of the dose administered. This is consistent with the observed pharmacokinetic fate of many drugs. Proportionality constants in this view are operational (apparent) constants which summarize the compounded influences of drug properties and characteristics of the organism. As biological descriptors they are useful for quantitative structure-pharmacokinetic relationships.

When interactions with specific macromolecules are a major determinant in the absorption and disposition of a drug, the efficiency of those processes will present limiting conditions. Above certain concentrations or dose levels the kinetics will change but no longer in proportion to the dose; total body levels of the drug can then rise dramatically with only a small increase in administered dose. In therapeutic practice this may cause problems, since the pharmacokinetic response is less predictable.

These considerations suggest that it is frequently advisable to select from several candidates that compound for further development which shows a reasonably high intrinsic activity. This compound has the best chance of following apparently dose-independent pharmacokinetics after a therapeutic dose regimen.

E. THE DESCRIPTIVE PARAMETERS OF PHARMACOKINETICS

Rates of appearance or disappearance are often described by their apparent rate constants (k), or alternatively in terms of the corresponding half-lives. The half-life is the time in which a concentration has declined to half its initial value. Under dose-independent kinetics half-lives are independent of concentration. More complex time courses are conveniently expressed in terms of a mean time which is the average time of residence of all drug molecules in a system or a subsystem.

The proportionality factor that relates a concentration observed to the amount administered, the dose, has the dimension of a volume and describes the distributive properties of a substance in the biological system. The degree of plasma protein binding, often expressed as fraction unbound of the total drug in plasma ($0<fu<1$), is of concern in the biological interpretation of distribution volumes.

The fractional amount of drug (F) that is absorbed

into the system characterizes the extent of bioavailability of that dose. An absolute bioavailability of F = 1 after oral drug administration indicates that the amount of drug that reaches the circulation from that preparation is the same on oral and on intravascular dosing. Bioavailability by many definitions covers both the extent of absorption, and also its rate. The rate aspect is conveniently described by the mean time of absorption.

The area under the plasma or blood level-time curve, AUC, is an integrated measure of concentration over time. For a systemically-acting drug it indicates how much drug is available to exert its effects. The AUC is related not only to the administered dose and the effective volumes of distribution, but also to each clearance process that eliminates the drug from the circulation.

Total drug clearance primarily reflects the rates of elimination and the volumes of distribution. Physiologically it can be interpreted in terms of plasma or blood flow and extraction efficiency of the eliminating organs discussed above. Since most elimination organs are reached by a drug in a more or less parallel fashion via the circulation, their clearances are additive. Clearance concepts also need to be considered when the drug, on its way to the general circulation, has to pass an eliminating organ; that is, when the systemically available fraction of the dose (F) is also limited by a so-called first-pass effect (cf. Figure 1).

II. PHYSIOLOGICAL CONSIDERATIONS IN PHARMACOKINETICS

The general principles of the handling of drugs by the organism will now be considered with respect to the special physiological conditions within the various organs responsible for absorption, distribution and elimination.

A. ABSORPTION

Unless given intravascularly, drugs must be absorbed in order to act systemically. A critical step for the uptake of a substance is the permeation of an epithelial layer in the gastrointestinal tract, the lung, or the skin. Very often the dosage form does not contain the drug as a solution, but rather as solid particles as in tablets, in capsules, or in suspensions. In order to penetrate the membranes, the active ingredient has to be dissolved in the aqueous environment at the site of absorption. Drugs that are sufficiently soluble in water at the pH conditions near the absorption site are most suitable. The rate of dissolution or delivery to the site of absorption as the determining factor is a major concern in biopharmaceutical considerations discussed in more detail in Chapter 9. When these processes are rapid, the type of the dosage form is less relevant for the uptake of the active ingredient.

The physiological environment in the gastrointestinal tract is quite variable. A pH value of 1 - 3 prevails in the stomach owing to the secretion of hydrochloric acid. Luminal pH values increase along the intestine up to pH 8 in the lower small intestine and the ascending colon. Transit time from the empty stomach to the colon is variable, from 3 to 8 hours on average. Particularly in the small intestine, there is a very large surface area available for absorption. In relation to a plain tube, this surface is increased 600-fold due to special morphological features, namely folds, villi and microvilli, Figure 6. At the surface of the brush border membranes in the small intestine a slightly acidic pH of 5 - 6 is maintained.

One observes with many drugs that absorption is not equally effective in the various parts of the gastrointestinal tract. In extreme cases a small region only, sometimes labeled an "absorption window", seems to be respon-

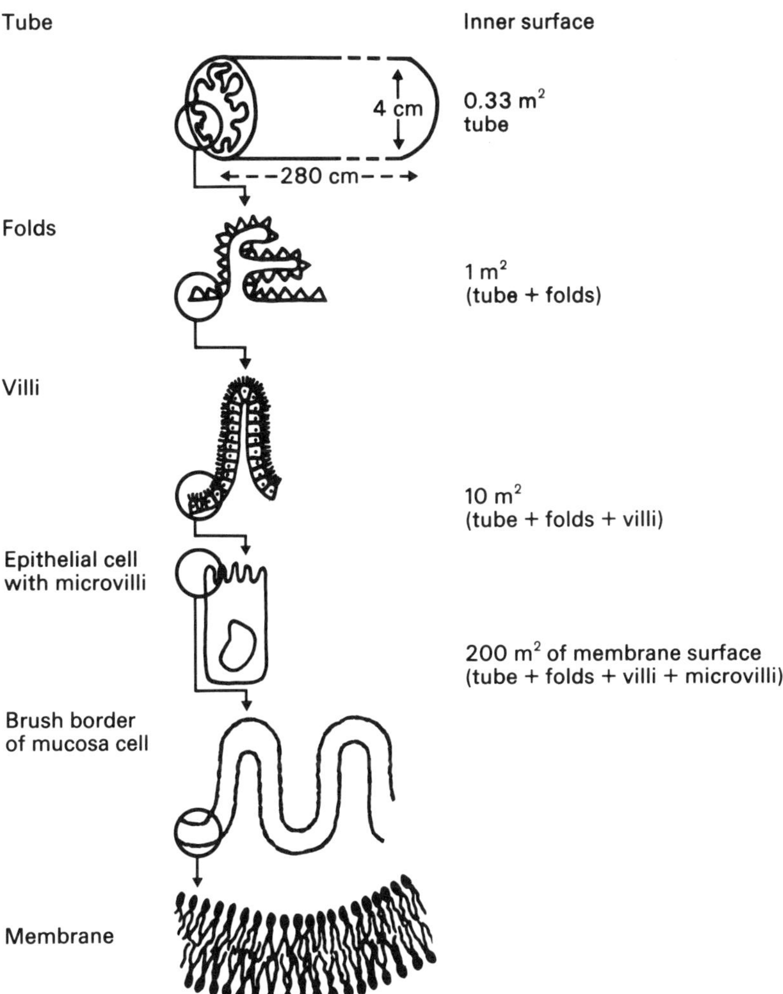

Figure 6 Morphological features that enlarge the surface area of small intestine in man.

sible for intestinal absorption of a particular drug. There is no evidence, however, that such limitations are characteristic of the mucosal cel's lining the different parts of the gastrointestinal tract or that they are due to special localized uptake mechanisms. Rather, there are other factors that may become rate-limiting and thereby contribute to the variability in absorption rate of a drug along the intestine. Examples are: gastric emptying, which governs drug transport to the absorptive sites in the intestine; the luminal pH, which influences drug solubility and ionization; the available surface area at the various sites; the accessibility of that surface, which can be limited by increasing the viscosity of intestinal contents; the rate of regional perfusion at the absorptive sites, which maintains the concentration gradient between the lumen and the capillary blood; and competitive processes of elimination by enzymatic or transport mechanisms located in the intestinal lumen or gut wall, which decrease luminal drug concentration or otherwise change the concentration gradients (presystemic first-pass effect).

For many drugs the absorption process itself follows the pH partition hypothesis: only unionized drug penetrates the cell membranes of the epithelial cells. Absorption rate thus depends on the fraction of drug molecules unionized. This in turn is a function of the pKa of the compound and the local pH near the absorptive membrane. Absorption by passive diffusion is particularly efficient in the small intestine because of the long transit time and the large area of contact. In general, the small intestine is the most active region with respect to absorption; even acidic drugs are absorbed there more effectively than in the stomach.

Active or facilitated transport mechanisms are of major importance only for foods. Some drugs may use these transport capacities in addition to passive diffusion (for review see Breimer and Speiser, 1983). Vesicular processes such as pinocytosis assume increasing interest for large entities such as peptides, proteins, liposomes, and special particles for drug targeting or delivery.

The most successful way of modifying the absorption rate with a series of molecules consists of specific variations of their hydrophobicity (for reviews see Bundgaard, 1985). Molecules for which absorption is not desirable, such as laxatives or food coloring agents, have been structurally varied by the addition of polar, hydrophilic groups to prevent their uptake. On the other hand, introduction of hydrophobic groups frequently enhances absorption.

B. DISTRIBUTION

The overall distribution of a drug into the various fluid compartments of the organism, Figure 7, is influenced by the nature of the surrounding physiological barriers and the presence of macromolecules that can modify the concentration of free drug in the various fluids.

In most cases, a drug has to pass from the circulation to a specific tissue structure to show its pharmacodynamic action. The first stage in the distribution to this tissue involves penetration of the capillary walls.

1. Penetration through the capillary wall. There are two important mechanisms for the transfer of molecules through the capillary walls. One of them, passive diffusion, is only available to lipid-soluble substances. This type of penetration occurs over the entire surface area of the blood capillaries because the membranes of the endothelial cells do not represent a barrier for lipid-soluble substances. Their diffusion rates are directly correlated with their lipid/water distribution coefficient and are

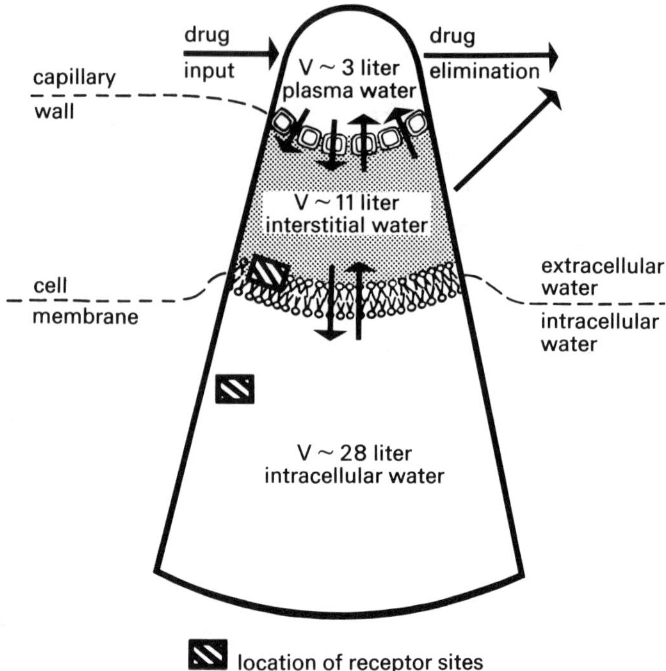

Figure 7 Physiological fluids in the human body (~70 kg) as possible distribution spaces for drugs.

many times more rapid than those of hydrophilic molecules of comparable size.

Very hydrophilic compounds are restricted to water channels between the endothelial cells. Their diffusion rate is determined mainly by molecular size. A small molecule, such as urea, with a molecular weight of 60 dalton, permeates the capillary endothelium much faster than, for example, inulin, which has a molecular weight of 5200 dalton.

The porosity of the capillary walls is different in various organs: for example a very loosely knit lining is found in the liver, whereas the capillary endothelium in the brain and central nervous system is quite tight.

Except in the brain, the capillary walls are weak barriers that allow even plasma proteins to penetrate to some extent. Without any major impediment, substances of low molecular weight are distributed almost immediately from the blood stream to the interstitial fluid which forms the outer environment of the body's cells. Therefore, the rate of appearance of those substances in most regions of the body is limited rather by the perfusion rate of the tissue than by the permeability of its capillaries.

The volume of extracellular space is the smallest possible physiological reference volume for unbound drug. Further transfer from the interstitium to the intracellular space, i.e. penetration of the cell membranes, is more restricted.

2. Penetration through the cell membrane. The main function of the cell membranes is to ensure the organizational integrity of the cell as they represent a major barrier to the diffusion of larger hydrophilic substances. Whereas natural cellular constituents such as water-soluble monosaccharides and amino acids are carried into the cells by specific transport systems, the distribution of xenobiotic hydrophilic substances is restricted primarily to the space between cells; only very small polar molecules as water can cross cell membranes. However most drugs are hydrophobic. For such substances the cell membrane is a barrier similar to a solvent layer. The drug can diffuse through this barrier at a rate dependent on its lipid/water distribution coefficient.

Most drugs contain ionizable acidic or basic groups. Since ions are in general poorly soluble in lipids, they cannot readily permeate into tissue cells. Molecules with suitable pKa values may reach the intracellular space in their unionized form.

3. Penetration through special physiological barriers. The blood-brain barrier is a particularly effective lipid

barrier. The reason for this lies in the specific structure of the cerebral capillaries: their endothelium is densely surrounded by glia cells which do not leave intercellular gaps. Therefore, a substance has to cross two cell membranes in sequence in order to enter the brain. Consequently, hydrophilic agents or ions cross into the brain only with considerable difficulty, whereas hydrophobic molecules are ideally suited for penetration. The penetration rate of drugs through the blood-brain barrier is directly proportional to their lipid/water partition coefficient and inversely proportional to their effective size (for review see Breimer and Speiser, 1983).

In many cases the presence or absence of central effects after systemic administration indicates whether or not a compound can cross the blood-brain barrier. The neurotoxicity of penicillins is clearly correlated with their hydrophobic character. The neurotoxic potential of an untested penicillin can be estimated from its hydrophobicity.

Various other physiological barriers, not yet characterized as well as the blood/brain barrier, are sometimes important for toxicology. Such barriers exist for example in the placenta and the inner eye and render these organs less accessible to hydrophilic compounds.

4. Reversible binding to endogenous macromolecules. Until now, the discussion has focussed on how biological barriers affect the distribution of drugs in the body. But distribution is also affected by reversible binding of drugs to macromolecules. The extent of binding determines the concentration of free drug, which in turn is the principal determinant of the rate of diffusion across membranes. (For a recent review using antibiotic drugs as examples, see Wise, 1986).

A relevant variable is plasma protein binding which can be determined experimentally in man. Albumin and

α_1-acid glycoprotein are the two proteins most important in drug binding.

Human serum albumin, m.w. 66 K dalton, accounts for about 60% of all plasma proteins. Its physiological functions include maintenance of the osmotic pressure of the blood and a nonspecific binding and transport capacity for numerous endogenous substrates including neutral compounds, cations and anions. This binding property of serum albumin is also seen with many drugs. The potential of the albumin reservoir can easily be estimated: assuming only one to two drug binding sites per albumin molecule, drugs can be bound completely, even at concentrations in the millimolar range. Most drugs are therapeutically active at micromolar plasma concentrations.

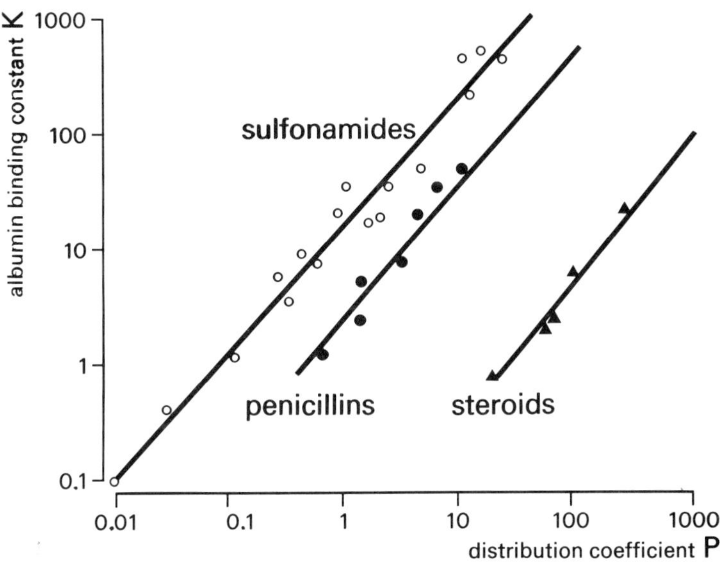

Figure 8 Schematic representation of correlations between the lipid/water partition coefficient P of various compounds belonging to different structural classes and the binding constant K to human albumin of these substances.

The relative strength of binding of drugs to albumin increases with lipophilicity, Figure 8. The fact that substances from different classes with identical distribution coefficients show different degrees of binding suggests that, in addition to the hydrophobic character, other noncovalent interactions such as electrostatic forces, hydrogen bonding, and steric effects, discussed in Chapter 4, also play a part in protein binding. Electrostatic forces are important in the binding of sulfonamides which are predominantly bound to human albumin in their anionic form.

α_1-acid glycoprotein, (m.w. 40 K dalton), has no known physiological role. Even though the number of its binding sites is one to two orders of magnitude lower than that of albumin, it plays a quantitative role in drug distribution because of its high affinity for basic drugs.

Specific and even nonspecific binding can be stereoselective. Therefore, reversible binding may be one factor responsible for differences in disposition of drug enantiomers since the degree of plasma protein binding of a compound has a potential effect on all pharmacokinetic processes.

In the process of absorption from the site of administration into the blood, binding to plasma proteins can play an important part by providing a sink and thus maintaining the concentration gradient. Additionally, the solubility of many substances in water is so low that they could not be transported in sufficient quantities in the plasma without being associated with the plasma binding proteins. Furthermore, plasma protein binding can have a pronounced effect on the rate of elimination of a substance, as discussed in the next section, and thus on the duration of action. Finally, the effect of plasma protein binding on the equilibrium distribution in the body can be appreciated from Figure 9.

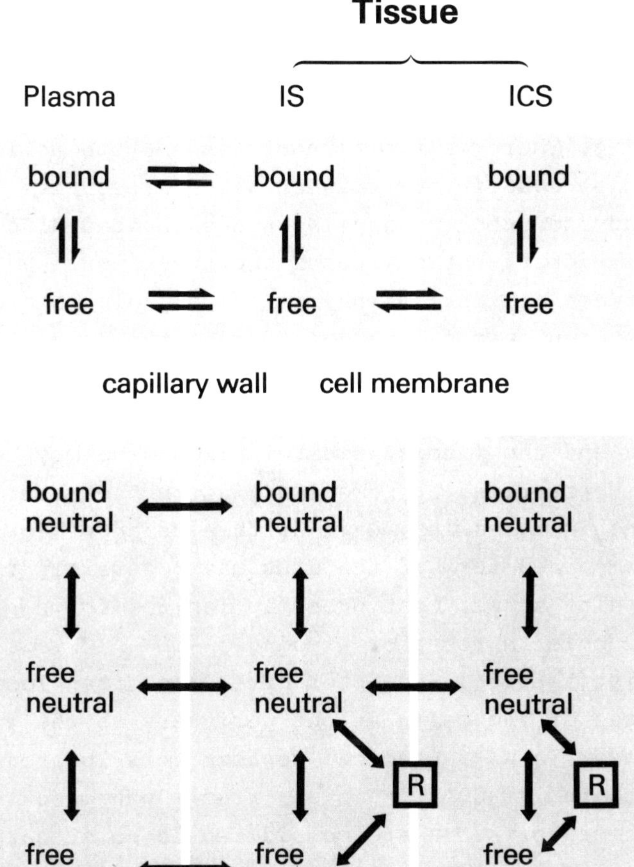

Figure 9 Equilibrium distribution of a drug within the body depends on binding to both tissue components and plasma proteins (upper panel: neutral compounds; lower panel: ionizable compounds; R: receptor, IS: interstitial space, ICS: intracellular space).

5. <u>The equilibrium distribution.</u> The rapid equilibration between free and bound drug in blood maintains a relatively constant level of active ingredient for a prolonged period of time. The same arguments apply for reversible tissue binding. There is a permanent tendency to achieve and maintain an equilibrium between all distributive states: bound and unbound, ionized and unionized drug in the respective distribution spaces shown in Figure 9. Drug transfer between the various physiological volumes occurs mainly via the unionized, unbound form of the drug. But other forms may sometimes contribute to an exchange: protein bound drug can move through the capillary wall between plasma and the interstitium; ionized drug may be transported by carriers or sometimes, as in the case of very hydrophobic molecules, even diffuse through membranes; the ionized form of the drug may reversibly interact with binding sites; most probably ionized forms play an important role in receptor binding.

6. <u>Tissue distribution.</u> Receptor sites are often located on the surface of the cells at the periphery of the extracellular space. In an anatomical context this location is attributed to the tissues. However, from a pharmacokinetic view these receptor sites are considered to be situated rather close to the circulation, since it is often the unbound drug concentration in the extravascular fluids that determines the rate and extent of the pharmacodynamic effect. Thus, anatomical boundaries of tissues do not coincide with those of kinetically defined distribution spaces. Similar arguments apply to plasma proteins, which contribute to tissue binding in the intercellular spaces, or to blood cells which often play a role equivalent to binding proteins in the intravascular space.

This consideration illustrates some of the difficulties associated with tissue distribution and reversible tissue binding. In addition, there are anatomical differences within and between tissues, and their perfusion rates are variable (as discussed in Section I.B.1). Special situations with pH gradients from tissue to tissue may influence distribution. Trapping or exclusion of a compound can occur because of pathological pH changes (e.g., inflamed tissues are more acidic) or physiological pH gradients (e.g. in glands or certain cell organelles). Data on the tissue binding of a drug are not accessible by direct measurements, since tissue homogenates or biopsy samples are of limited use only.

Nevertheless, the tissues are an important depot for many drugs. Their membranes alone constitute a reservoir for more hydrophobic drugs by simple solvent/water partitioning. Redistribution from tissues is often a rate-determining aspect in the overall elimination. For example, the long half-life of clenbuterol, Structure 44 in Chapter 6, reflects mainly its redistribution from tissue stores. Extremely lipophilic substances such as a number of insecticides show almost no tendency to diffuse back from fatty tissues to the aqueous body fluids. This is why their accumulation or sequestration in the body is of considerable toxicological importance. Tissue distribution is the main determinant in the duration of the effect of lipophilic narcotics such as thiopental. A rapid entry into tissues, including brain, mediates their rapid onset of anesthesia. This in turn is terminated by a redistribution into other, less well perfused regions of the body; small additional doses can restore the original narcotic effect.

Hydrophobicity seems to be the molecular characteristic that determines the overall distributive properties of most compounds. Other structural properties may be responsible for binding to special endogenous macromolecules: for example melanin is a pigment found in the eye that can bind to molecules with aromatic character by charge transfer interactions.

There is presently much research effort to exploit specialized carrier or metabolic systems that are involved in the specific distribution to certain tissue structures and may be more pronounced during disease states. One thereby hopes to transport drugs or drug vehicles preferentially to specific target tissues. The performance of such delivery systems in practice (for reviews see Bundgaard, 1985) remains to be established in the future.

C. EXCRETION

Excretion of a drug from the body is an important factor for its clearance, and hence for the decrease in concentration at the site of action and accordingly for the time course of the pharmacological effect. Excretion complements biotransformation in elimination of the drug: parent drug and its metabolites can be excreted simultaneously via several routes, such as urine, bile, the intestine, saliva, respiration, sweating, or lactation. In many cases, however, only renal and biliary excretion are of quantitative significance.

1. Mechanisms in renal excretion. Three major processes are involved in the renal excretion of a drug or its metabolites: glomerular filtration, tubular reabsorption, and tubular secretion. These three processes arise from the special function and structure of the nephrons, Figure 10, of which there are about one million in a single human kidney.

2. Glomerular filtration. Glomerular filtration is made possible by the pumping action of the heart. About 22% of the cardiac output goes to the kidney: approximately 10% of the cardiac output, 120 ml of plasma water per minute, is filtered rapidly through the glomeruli. The pores of the glomerular capillaries are freely permeable to substances with a molecular weight of up to approximately

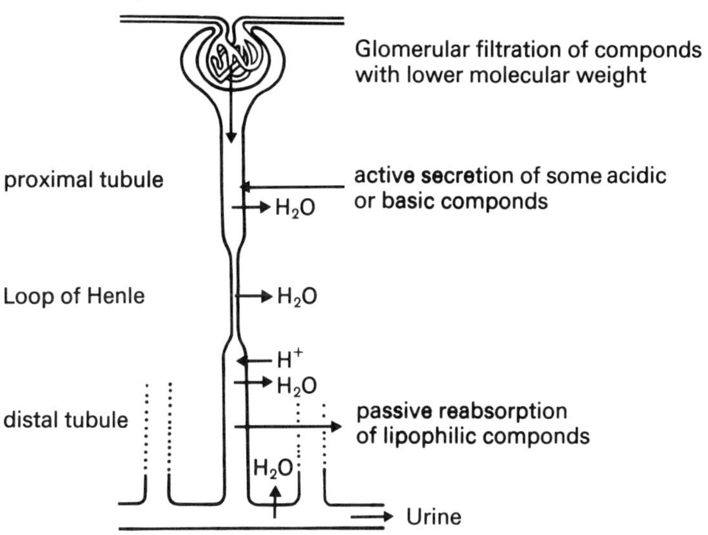

Figure 10 Schematic diagram of a nephron in a human kidney showing the main processes involved in the elimination of drugs. (For the sake of clarity, the peritubular blood capillaries have not been included in the diagram.)

10,000 dalton; larger molecules are increasingly restricted from filtration in proportion to molecular weight. Thus, the ultrafiltrate contains all substances and their metabolites that are not bound to plasma proteins. This means that the concentration of the drug in the primary urine of the proximal tubule is identical to that in the plasma water. Glomerular filtration occurs with all compounds dissolved in plasma and is only limited by reversible binding to plasma proteins. Glomerular filtration is therefore directly proportional to the filtration rate and the fraction of unbound drug.

3. Tubular reabsorption. The physiological function of the tubules is to retain substances that are lost in the filtration process but are needed to maintain the water, mineral, and energy balance of the organism. About 60% of the glomerular filtrate is reabsorbed in the proximal

tubule. By the time the fluid reaches distal regions of the nephrons, the volume has decreased even further: the final urinary output is only about 1.5 l daily from a total of almost 200 l of glomerular filtrate. This implies that substances that are not reabsorbed during urine formation, such as inulin, become highly concentrated in the tubular lumen and final urine. By contrast, lipid-soluble drugs continuously diffuse back through the tubular epithelium into the peritubular blood capillaries following the concentration gradient created by water reabsorption. Thus, tubular drug concentration is in constant equilibrium with the concentration in the plasma water. Tubular reabsorption of almost all drugs occurs by passive diffusion with the cell membranes on both sides of tubular epithelium as the diffusional barriers. The ultimate renal excretion of drugs that are extensively reabsorbed is sensitive to changes in the urinary flow. Lipid-soluble molecules are, as a rule, excreted in very small quantities via the urine. Many drugs need to be converted by biotransformation into water-soluble metabolites before they can become effectively excreted via the kidney.

In the distal regions of the nephron the urine is acidified to a maximum of about pH 4 by hydrogen ion secretion. With weak electrolytes, this lowering of the pH value leads to a change in the degree of ionization. This change is significant because generally only the nonionized form of a substance is sufficiently lipid-soluble to be reabsorbed. Weak acids and weak bases are more efficiently excreted into the urine when the urinary pH changes so that the concentration of their ionized form increases. Such pH-dependent renal excretion must be taken into account for acids with a pKa of 3 - 8 and bases with a pKa of 7 - 11.

The pH-dependence of renal excretion can be put to good use in cases of poisoning. Urinary pH can be changed purposely by administration of substances such as bicarbonates or ammonium chloride to accelerate the renal clearance of toxic agents.

Proteins and peptides that appear in the tubular fluids by glomerular filtration are reabsorbed via yet another mechanism, endocytosis. When followed by digestion, endocytic reabsorption of peptides and proteins helps the organism to recover amino acids which would have otherwise been lost. Absorptive endocytosis is a relevant elimination pathway for peptide drugs, but not for classical drug substances.

4. Tubular secretion. Peritubular capillaries receive their blood supply from the glomeruli. Blood transit time in the renal cortex is in the order of several seconds. Tubular secretion implies transport of substance from peritubular capillaries through the tubular endothelium into the lumen of the proximal tubule. Small molecules such as water and some ions may move to the lumen through intercellular gaps, but drug-like molecules have to cross the antiluminal and the luminal membranes of the tubular cells. Since the concentration gradient for this process is unfavorable, this movement cannot occur by passive diffusion. At least one of the membrane transfers must be achieved by active transport. The rate of active secretion follows the laws of enzyme kinetics and can often be reasonably described by a Michaelis-Menten type equation.

After administration of therapeutic doses, the plasma concentrations of most drugs that are actively secreted are considerably below the affinity constant of the carrier. In the case of a pseudo-first-order reaction, the secretion rate is proportional to the concentration of unbound drug in plasma. Both free and bound drug, however, can be removed by renal secretion at that rate, since the transit time through the capillaries is long enough to allow the replacement of secreted drug by a fast reequilibration from the plasma pool of bound substance.

Active secretion has been detected for a number of organic cations and anions; several independent active transport systems (for review see Breimer and Speiser, 1983) are involved. These carriers are usually not especially stringent in terms of structural requirements apart from the charge of the ion to be transported. However, they sometimes handle stereoisomers at different rates. Two or more ions can compete for transport by the same carrier. For example, probenecid blocks the active transport of penicillins at the capillary side and thus markedly prolongs the activity of these antibiotics in the body. Differences in the efficiency of two anion carrier systems, the one at the luminal side and the other at the antiluminal side, seem to be responsible for the accumulation of cephalosporins within the tubular cell with consequent nephrotoxicity.

5. <u>The rate of renal elimination.</u> The rate of renal excretion is the net result of filtration, secretion and reabsorption. Filtration always occurs and is strictly a first-order process. Its efficiency is limited directly by binding, and indirectly by passive reabsorption of already excreted drug. Reabsorption may or may not occur depending on the lipophilicity and the pKa of the drug, and on the average pH gradient from tubular fluids to plasma. Reabsorption normally follows first-order kinetics as long as the urine flow is reasonably constant. Active secretion requires the presence of certain structural features in the drug molecule. The rate of active secretion is often, but not necessarily, proportional to the drug concentration in the blood. Renal perfusion can however become rate-limiting. Besides excretion, occasionally metabolic transformation is also a part of renal elimination.

6. Physiological basis of hepatic elimination. Hepatic elimination is important for most drugs. The ability of the liver to eliminate drugs effectively from the circulation is based on two factors: its almost unique complement of most enzymes for metabolic transformation and its capability of secreting drugs and their metabolites into the bile. The major implications of biotransformation for the pharmacological and toxicological profile of drugs are treated in Section III. Only the physiological and kinetic aspects of hepatic clearance are dealt with here.

Hepatic extraction of drugs from the circulation again may best be considered in the light of the clearance concept shown in Figure 3. In man the hepatic blood supply accounts for approximately 27% of the cardiac output, approximately 1.4 l/min. Three quarters of this enters the liver via the portal vein and the remainder arrives via the hepatic artery. The processes that determine the clearing capacity of the liver may be best understood in terms of their physiological nature and location. There are strong limitations to studying these processes experimentally in man so that, in spite of great progress in this field, many aspects remain speculative.

The functional unit of the liver is schematically depicted in Figure 11. The blood supply from branches of the hepatic artery and of the portal vein is mixed in an anastomosis. Only sinusoids connect the blood-donating vessels across the functional unit to a branch of the central vein that returns the blood to the venous circulation. Small molecules pass the sinusoid from the periphery to the center within 5 to 10 seconds.

The liver sinusoid is a unique capillary. Its endothelial lining has wide pores to the interstitial volume that borders on the parenchymal cells. The exchange of fluids between the sinusoidal space and the interstitial space is thus quite efficient. It is further enhanced by the mechanical action of blood cells. Diffusion of substances from the interstitial fluids through the parenchymal cell membrane is facilitated by the enlargement of the cell surface by microvilli. Parenchymal cells, hepatocytes, are functional cells that represent four fifths of the total liver cell volume. All major enzymes for biotransformation are localized in the hepatocytes, either in the cytosol or integrated into membranes of special organelles such as the endoplasmic reticulum. Hepatocytes seem to be arranged in long cords along the sinusoids, generally in a layer two cells thick. Cellular enzymic activities are not evenly distributed, but often gradually increase or decrease on the way to the collecting vein.

The hepatocytes have a polarity indicated by the features of their membrane surfaces: the sinusoidal pole which accounts for 65% of the membrane surface, the intercell contact region with a smooth surface, and the canal-

CV: branch of central vein

BC: bile canaliculi

S: sinusoid

PV: branch of portal vein

HA: branch of hepatic artery

BD: bile duct

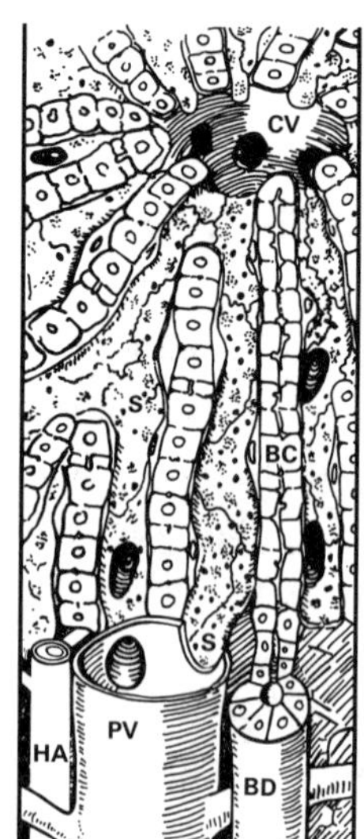

Figure 11 Microcirculatory complex in the liver.

icular pole constituting the lining of the bile canaliculi which account for 15%. The bile canaliculi, the origin of the biliary tree, drain to branches of the bile duct, which ultimately empties into the duodenum, part of the small intestine.

7. Kinetic aspects in hepatic elimination. At least six processes, schematically outlined as processes *a* - *f* in Figure 12, determine the fate of drug molecules in the liver. Both the rate of these processes and the equilibrium conditions are important. Depending on its lipophilicity, unbound drug can diffuse from the sinusoid through the hepatocyte membrane into the cell (*c*). Very frequently cellular concentrations exceed those in the plasma. Various explanations for these observations are given (for a review on hepatic transport processes see Breimer and Speiser, 1983). There are indications that bidirectional transport of some anions into the hepatocytes occurs by an active process, while for other ions, co-transport together with counterions has been suggested. There may also be an albumin receptor on the hepatocyte membrane (*g*) to facilitate permeation of neutral substances.

Protein-bound and often erythrocyte-bound molecules are cleared from the sinusoid because of reequilibration (*a* and *b*). Rapid removal of the free drug from the cytosol creates sink conditions in the hepatocyte and eliminates the need for carrier transport. Sink conditions would also result from reversible binding to parenchymal proteins (*e*), the interaction with enzymes responsible for biotransformation (*d*), or biliary excretion (*f*). The latter may occur via different mechanisms, including active transport systems, and is frequently a rate-limiting step in hepato-biliary transport.

The concentration of drugs in bile may be lower but may also be much higher than in the plasma of the sinusoids. Therefore biliary clearance is quite variable.

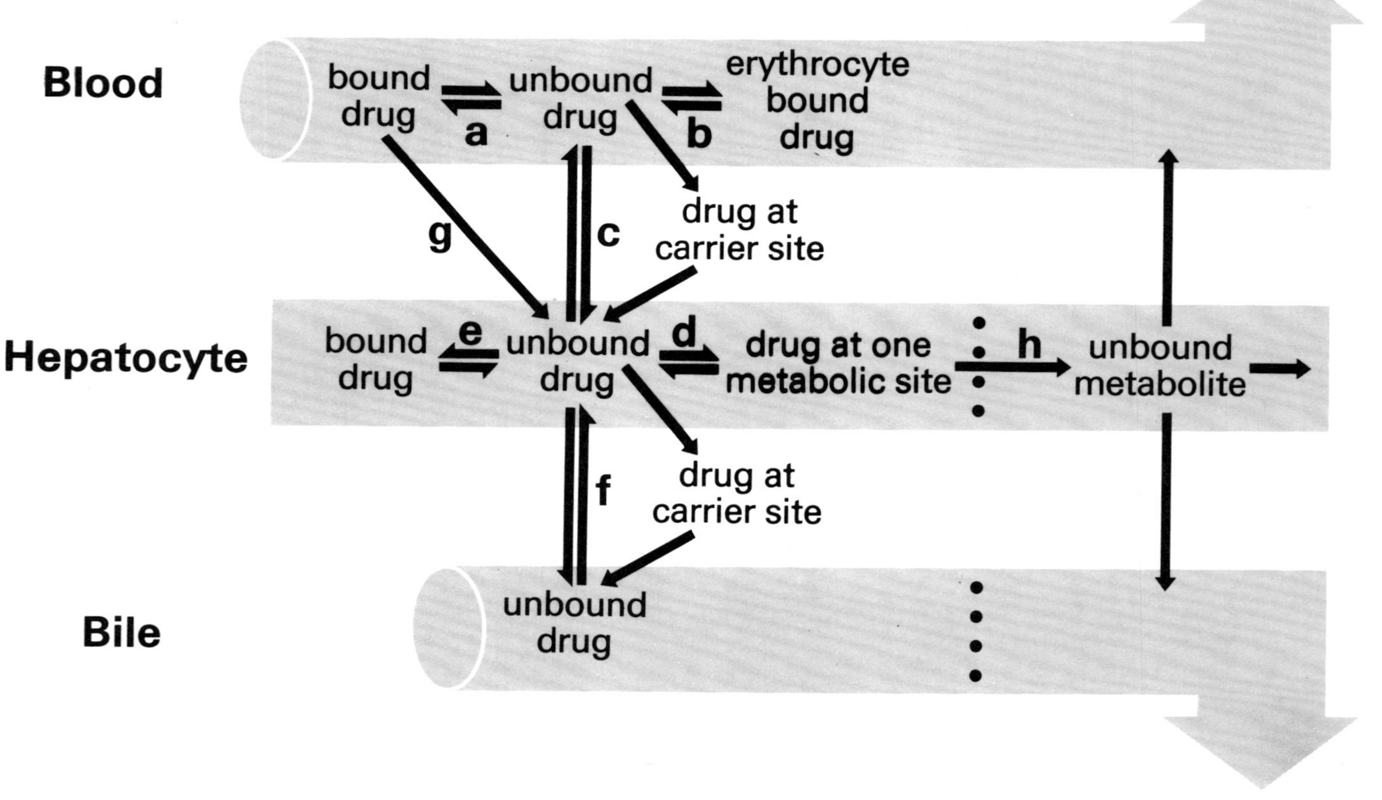

Figure 12 The complexity of possible major routes for the distribution of pharmaceutical substances within the liver, highly schematic (explanation of symbols in text).

Structural requirements for extensive biliary excretion in man are a certain polarity in the molecule and a molecular weight exceeding 450 dalton. The threshold value for biliary excretion in terms of molecular size differs in different species, and is poorly defined. Most drugs can be cleared into bile only after metabolic transformation to the conjugates that fulfil the requirements stated above.

With proteins or liposomes, hepatic uptake can occur by absorptive endocytosis. In addition, special constituents of the reticulo-endothelial system, macrophages, located in the sinusoids, remove and degrade such particles as red blood cells or liposomes.

Kinetically, hepatic clearance processes may be very complex and are not well understood, especially for drugs with high hepatic extraction. For drug-like substances almost any of the processes discussed may determine hepatic clearance. The liver also represents a first-pass eliminating organ for all oral dosage forms, since the venous outflow of the intestine is collected in the portal vein. Even some of the venous outflow from the rectum drains into the portal vein. Thus, high hepatic clearance of a drug inevitably decreases its absolute bioavailability after oral or rectal administration. The hepatic first-pass effect is often saturable when higher oral doses are administered. Systemic plasma concentrations then increase more than in proportion to the dose.

Drug molecules excreted into the bile may be reabsorbed in the intestinal tract. Metabolites that are excreted via the bile may be reconverted to the drug by the microflora of the gut or by enzymes in the gut mucosa and thus contribute to enterohepatic recycling. If such processes occur, the transit time of the drug in the body can be substantially prolonged. Absorptive materials such as activated charcoal administered, for example, in the case

of drug poisoning can compete effectively with reabsorption and thus enhance the fecal excretion of toxic agents.
8. Various other clearing organs. Even though the majority of classical pharmaceutical substances are predominantly disposed of by kidney and liver, in some special cases other organs require some attention.

One of these is the lung, which receives 100% of the cardiac output. On i.v. administration drugs reach the lung prior to any other eliminating organ, the pulmonary first-pass effect. The lung can remove drugs from blood not only by metabolic transformation (an elimination aspect), but also through binding (a distribution aspect). This makes the pharmacokinetic interpretation in the case of drugs with appreciable pulmonary clearance quite difficult (for review see Bend et al., 1985).

Drugs can also be excreted into the saliva. Measuring the corresponding excretion rate is a useful noninvasive method of monitoring plasma concentrations of free drug. The fraction of drug that is excreted via the saliva may be reabsorbed in the intestine.

The stomach, the intestinal contents or the mucosal environment may act as trapping volumes for drugs with suitable pKa via pH partition effects. In addition, the capability of the small intestine to secrete drugs has been underestimated for a long time. Transport systems for anions, cations and neutral compounds are present in the small intestine. These systems resemble the ones found in the renal tubule. However, in man the quantitative contribution of intestinal secretion in relation to the other excretory pathways remains to be clarified for many drugs.
9. Total clearance. In conclusion, systemically-acting classical drugs and their metabolites are excreted mainly via renal and hepatic routes. The two routes are complementary to each other, at least in the range of molecular

weights of most currently used drugs. Impairment of one of these excretory pathways can be compensated for, to a certain degree by the other (for review see Fleck and Bräunlich, 1984).

Usually drugs are excreted in the form of metabolites. Consequently, total clearance that refers to the parent drug only and is the net result of its clearances by all organs is often dominated by metabolic clearances. Metabolite clearances, however, are determined by excretion.

For newer active principles such as peptides or recombinant proteins, renal and/or hepatic uptake and subsequent degradation are the major elimination pathways. For removal and degradation of newer carrier systems such as liposomes and red blood cells, quite specialized systems in the body will have to be exploited.

D. BIOLOGICAL SYSTEMS TO STUDY PHARMACOKINETICS

Investigations on pharmacokinetics imply studies of the drug in the intact organism. The principles of physiology are the same in all mammalian species. Animal models are thus highly predictive for man. Comparative results from various species can be interpreted with increasing success using an interspecies pharmacokinetic scale (for reviews see Breimer and Speiser, 1987).

In vitro or in situ preparations of whole organs allow the study of certain aspects of pharmacokinetics without interference from competing processes (for reviews see Breimer and Speiser, 1987). Cell cultures, isolated enzymes or carrier preparations are of value in special investigations concerned with mechanistic aspects. These systems are particularly suitable for predicting metabolic elimination pathways and even their relative rates. In addition, various in vitro tests are very valuable in predicting the pharmacokinetic properties of drugs. Plasma protein binding and erythrocyte distribution studies are excellent predictors of the overall distribution of a compound in the organism (Hinderling, 1987). This is also true for human placenta preparations apart from their special value for the risk assessment of transfer of drug from the mother to the fetus.

These test systems are well suited to explain the influence that structural characteristics of a drug, such as lipophilicity, pKa values, and molecular size, exert on its pharmacokinetic behaviour. But in general, for a conclusive pharmacokinetic evaluation no alternative to the intact organism is foreseen. Investigations must be done at all dose levels and in all species which have been used in the assessment of efficacy and safety of a drug. Also, all relevant administration forms must be investigated in these experiments. The full time range of administration in toxicological investigations has to be complemented by pharmacokinetic studies.

III. BIOTRANSFORMATION OF XENOBIOTICS

Almost every chemical compound, whether of natural or synthetic origin, is transformed in the body by metabolic processes. Biotransformation refers in particular to the metabolic conversion of substances of nonbiological origin, the so-called xenobiotics; (for review see Testa and Jenner, 1976).

Biotransformation is an important factor in drug therapy because it contributes to the clearance of biologically-active metabolites from the body; and it may produce molecules, metabolites, whose pharmacological and toxicological properties differ from those of the parent drug.

The second aspect can be used to liberate the actual drug from precursors, prodrugs, that may not be biologically active themselves.

Generally, a distinction is made between two phases of biotransformation. Phase-I reactions include hydrolytic, oxidative, and reductive processes. In the phase-II reactions, the products of these processes or the parent drug itself are conjugated with hydrophilic endogenous substances. The resulting conjugates are often considerably more water-soluble and therefore more easily eliminated from the body than the is parent drug or its phase-I metabolites.

The biotransformation enzymes are old systems. For example, the cytochrome P-450 dependent monooxygenases are common to mammals and nonmammals including insects. There is no reason to assume that there are special enzymes involved in the biotransformation of drugs. Metabolically active enzymes have evolved to cope with certain requirements:

- to respond to toxic plant constituents such as alkaloids, a defense mechanism.
- to supply the organism with energy and physiological important compounds.

Biotransformational enzymes are often responsible for the defensive aspect. This is supported by the multiplicity of many drug-metabolizing enzymes and the fact that their levels increase, are induced by, repeated exposure to the xenobiotic. When a drug shows a close structural relationship to an endogenous compound, such as a steroid, catecholamine, amino acid, or fatty acid, it will most likely interact with the natural metabolic processes for that compound. In this case the efficiency with which most endogenous substrates are handled in the body becomes important: endogenous substrates are transported by specific mechanisms; they are metabolized by enzymes that have a very high specificity and high turnover rates, which are well regulated; there are specific storage mechanisms for endogenous compounds.

As a result, minimal structural variations within drugs derived from endogenous substrates, e.g. ß-2-adrenergics, can change the type of biotransformation thereby altering the metabolic fate and the pharmacokinetic behaviour dramatically. Examples of ß-adrenergics are shown in Chapter 6, Structures 41 - 44.

A. ENZYMES INVOLVED IN PHASE-I BIOTRANSFORMATIONS OF XENOBIOTICS

Almost all of the essential enzyme systems for biotransformation are found in the liver cell (for review see Jacoby, 1980). Some occur in the cytoplasm as soluble proteins, but most are bound to membrane structures of the cell. Some enzymes are located in the mitochondria. However, the endoplasmic reticulum, a cavity system of branching and interweaving tubes, contains most of the metabolic functions of the cell: most of the essential phase-I and phase-II reactions of drugs are controlled by enzymes associated with this structure.

1. Cytochrome P-450 dependent monooxygenases. The most important enzymes for metabolizing drugs are those in the class of cytochrome P-450 dependent monooxygenases, also known as mixed function oxidases. About 15 isoenzymes are known in the rat. Cytochrome P-450, the oxygen-activating component of the enzyme system, takes its name from the spectral change of the reduced form that occurs after addition of carbon monoxide. The NADPH-O_2-dependent generation of the active oxygen intermediate is common to all isoenzymes. The overall reaction is summarized in Figure 13. The oxidation of a drug (R-H) requires a cytochrome P-450, reduced NADPH, and molecular oxygen.

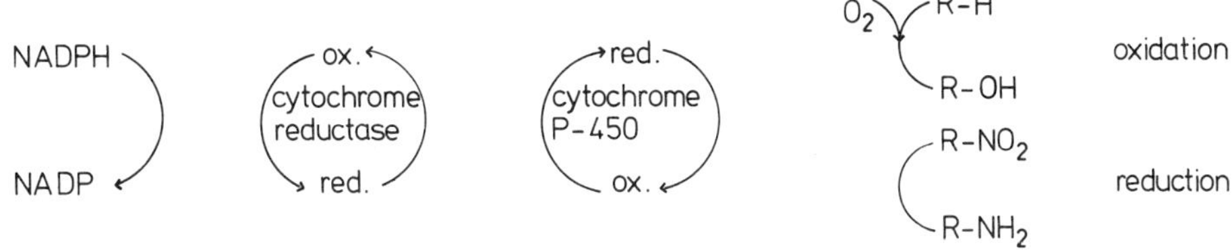

Figure 13 Oxidation and reduction of the drug (RH) by cytochrome P-450.

Cytochromes P-450 are located in the endoplasmic reticulum of different organs, the most important ones being the liver, kidney, and lung (for review see De Montanello, 1986).

The multiplicity of enzymes is the key to the observation that, on the one hand, an incredible variety of compounds are metabolized, whereas, on the other hand, a high degree of specificity to substrates is seen. Cytochromes P-450 oxidize a variety of structurally different aliphatic and aromatic hydrocarbons, alcohols, amines, amides, ethers and sulfides. Oxidation of C-, N- and S-atoms has been observed. These enzymes are also involved in the oxidation of lipophilic endogenous compounds such as steroids and fatty acids. Under certain circumstances P-450s mediate reductions.

It is worth noting that the reaction rates of the cytochrome P-450 monooxygenases are very low compared with those of other enzymatic reactions. The turnover rate is only 1-10 nmole per nmole P-450 per minute. On the other hand, the concentration of the enzyme in the liver is relatively high.

Specific forms of cytochrome P-450 can be induced up to 50-fold by drugs such as sulmazole or phenobarbital, food constituents, pesticides, and environmental compounds such as polychlorinated biphenyls or tetrachlorodibenzodioxins. Autoinduction of P-450 enzymes by a drug after multiple doses may result in reduced plasma levels and pharmacologic effect. Since the extent of induction varies from individual to individual, the variability of the response to the medication increases. Furthermore, the rate of metabolism of other drugs may also be increased. Other drug interaction problems may occur if a drug is a potent inhibitor of cytochrome P-450 and therefore of biotransformation, as is the case with cimetidine.

2. Cytochrome P-450 independent oxidating enzymes. In addition to the mixed function oxidases, there are many enzymes present in the mitochondrial and soluble fractions of tissues, such as the liver, that catalyze oxidation and reduction of foreign compounds. The following list, even though it is not complete, illustrates the complexity of the enzymatic basis of biotransformation.

Monoamine oxidase catalyzes the oxidative deamination of primary, secondary, and tertiary amines. It is located in the mitochondria and is found particularly at the sites of formation and storage of the biogenic amines dopamine, norepinephrine, epinephrine, and serotonin.

Alcohol dehydrogenase catalyzes the oxidation of primary or secondary aliphatic or aromatic alcohols to aldehydes or ketones. It requires NAD^+ as a co-substrate. It also catalyzes the reverse reaction, the reduction of aldehydes to alcohols.

Xanthine oxidase is a complex iron-sulphur enzyme that contains molybdenum and FAD. Xanthine oxidase is present in the cytosol and plays an important role in the oxidation of endogenous purines and of structurally related drugs. The oxygen atom incorporated into the substrates is ultimately derived from water and not from molecular oxygen, as in processes catalyzed by cytochrome.

Aldehyde oxidase, which oxidizes aldehydes to carboxylic acids, appears to be similar to xanthine oxidase. Aldehyde oxidase is located mainly in the cytosol of the liver.

Apparently small structural variations in a molecule can cause a change in the enzyme by which it is metabolized. As an example, the drug AR-L 115 (1) is metabolized by cytochrome P-450 isoenzymes; however, UN-LZ 32 (2), is a substrate of xanthine oxidase. This metabolic switching can result in considerable differences in the pharmacokinetics or toxicity of two structurally related compounds.

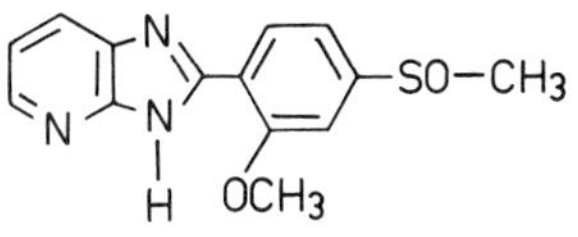

1

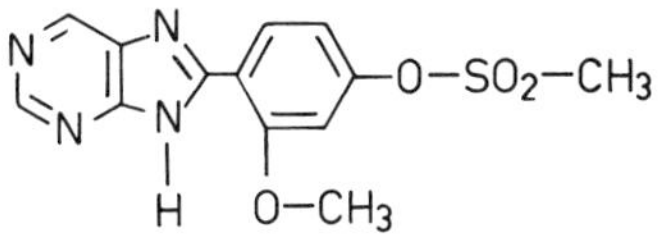

2

Often the products formed by one enzyme are further transformed by one or more of the other enzymes described above.

3. Hydrolases. These are among the most important enzymes involved in phase-I biotransformation. Various soluble esterases and amidases, mostly with low substrate specificity, are found in intestine, blood, and the cytoplasm of cells of many tissues and organs.

The epoxide hydrolases hydrolyze epoxides to diols. They are always associated with the cytochrome P-450 system and are inducible in the same way. However, the extent of induction appears to be lower, only up to 5-fold.

B. PHASE-I REACTIONS

1. Oxidation of carbon atoms of aliphatic chains. The methyl groups attached to aromatic, heterocyclic, or aliphatic moieties are often oxidized to the corresponding alcohols, aldehydes, and finally the carboxylic acids. If there is a substituent in the vicinal position, the alcohol is usually the final product. Relatively long aliphatic side chains of molecules are readily oxidized, the preferred attack then being at the penultimate and ultimate carbon atoms.

2. Hydroxylation of carbon atoms of saturated rings. For example the secretolytic drug bromhexine (3) is hydroxylated in the alicyclic ring and N-dealkylated to yield the drug ambroxol (4) and other regio- and stereoisomeric hydroxylation products.

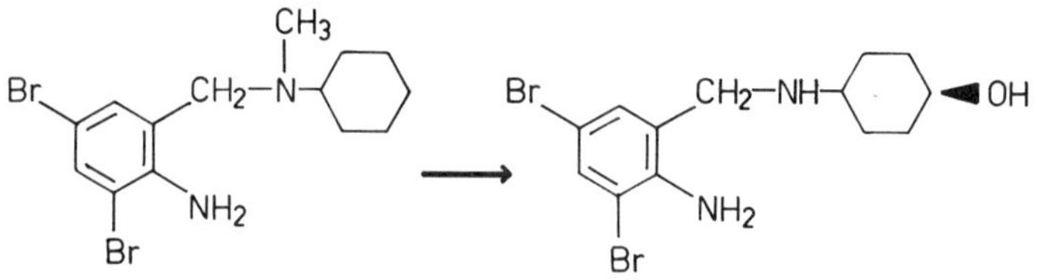

3 → 4

3. Oxidation of aromatic compounds. Two mechanisms have been proposed for the oxidation of aromatic C-H bonds. The direct insertion of an oxygen in a C-H bond to give a phenol is observed in some cases. The more important mechanism, however, is the epoxidation of an aromatic ring by cytochrome P-450. These epoxides are reactive species that undergo four possible enzymatic or nonenzymatic biotransformations, Figure 14: first, epoxide hydrolases may catalyze hydrolysis; the resulting diols are usually chemically less reactive and therefore less toxic than the epoxides. Second, nucleophilic attack by glutathione to form

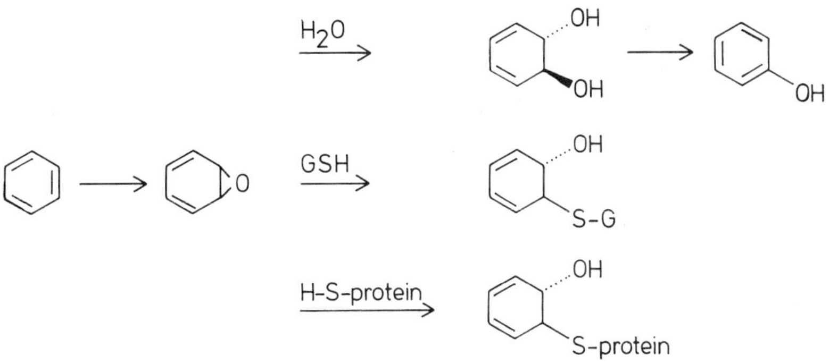

Figure 14 Pathways of reaction for epoxides.

thioether conjugates may occur; this reaction is discussed under phase-II mechanisms. Third, epoxides can directly isomerize nonenzymatically to phenols. Fourth, epoxides may bind covalently to proteins or DNA. This may form immunologically reactive products or may result in mutagenic lesions, respectively.

An example of such an aromatic oxidation is provided by diphenylhydantoin (5, 6). An interesting case (8) where

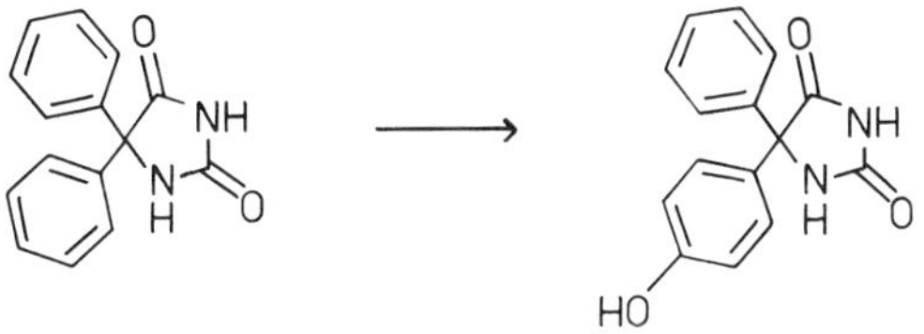

5

6

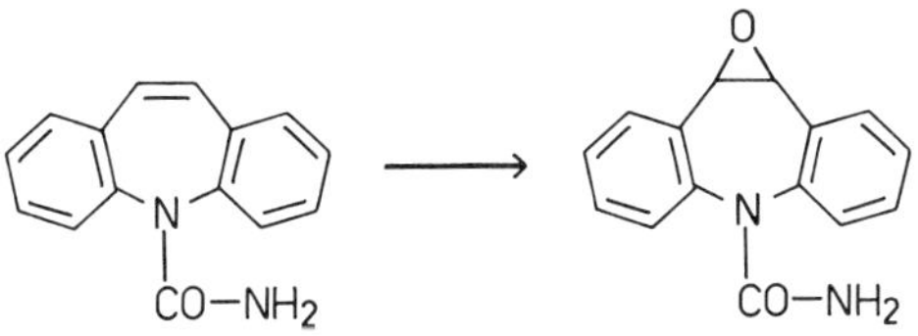

7

8

the epoxides are stable enough to be isolated is the metabolic product of the tricyclic drug carbamazepine (7).

4. Dealkylation of short chain alkyl ethers. The initial reaction is the oxidation by cytochrome P-450 monooxygenases at the carbon atom connected to the oxygen to form a hemiacetal intermediate. The hemiacetal then decomposes to an alcohol or phenol and an aldehyde respectively. In the example depicted, pharmacologically active phenacetin (9), is converted via an intermediate (10) to paracetamol (11), another pharmacologically active compound.

In a series of homologous alkyl ethers the dealkylation rate often remains nearly constant up to butoxy groups and then decreases. With long aliphatic chains, oxidation of the penultimate carbon becomes more important.

NH—CO—CH3 NH—CO—CH3 NH—CO—CH3

O—CH2—CH3 → O—CH—CH3 (OH) → OH

9 10 11

5. Dealkylation and deamination of amines. The cleavage of carbon-nitrogen bonds is one of the most important chemical reactions in drug metabolism. Secondary and tertiary amines are dealkylated by various P-450 monooxygenases and flavin-containing monooxygenases. Tertiary amines are generally more rapidly metabolized than are secondary amines. Short alkyl groups are more easily attacked than longer chains. The conversion of imipramine (12) to desimipramine (13) is a well known example of this process.

6. N-oxidation of amines. Tertiary amines are often oxidized to the corresponding N-oxides, a transformation that

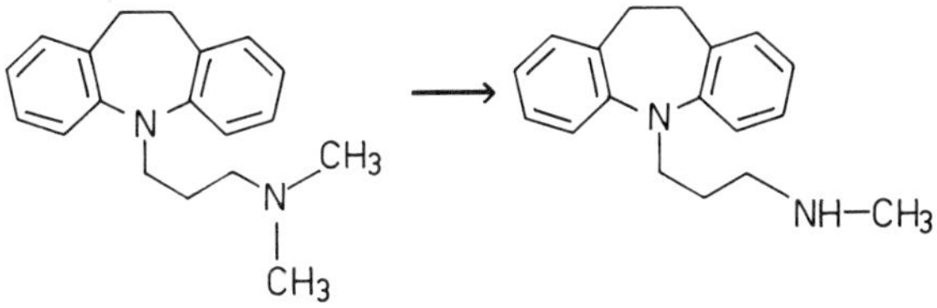

12 13

decreases the lipophilicity of a drug. One example is the formation of imipramine N-oxide (14).

N-oxidation converts primary and secondary amines and amides into hydroxylamines and nitroso compounds. For example amphetamine is converted to N-hydroxy amphetamine; phenacetin (9) yields N-hydroxy phenacetin (15), which is at least partially responsible for the toxic effects of phenacetin.

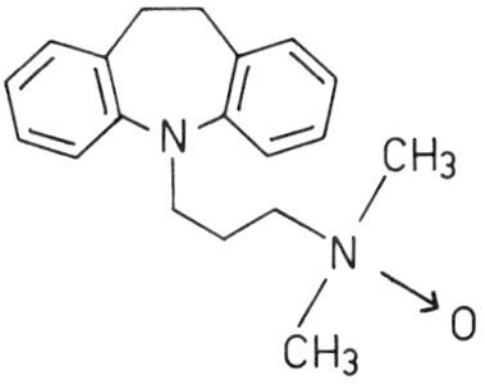

14 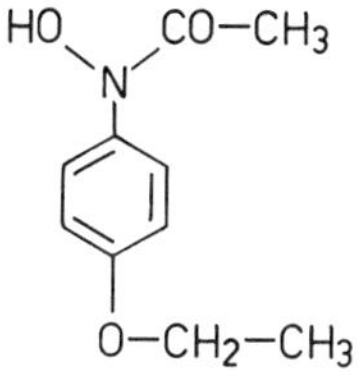 15

7. Oxidation of thioethers. The oxidation of thioethers to sulfoxides and sulfones is significant in the biotransformation of molecules such as chlorpromazine (16,17). The oxidation to sulfoxides is reversible. Sulfones, however, are metabolically stable products.

8. Reductions. Reductions are, under normal circumstances, biotransformation pathways of minor importance. Table 1 shows the chemical groups that are prone to reduction.

Several enzymes are involved in reductions: Cytochrome P-450, cytochrome c reductase and, alcohol dehydro-

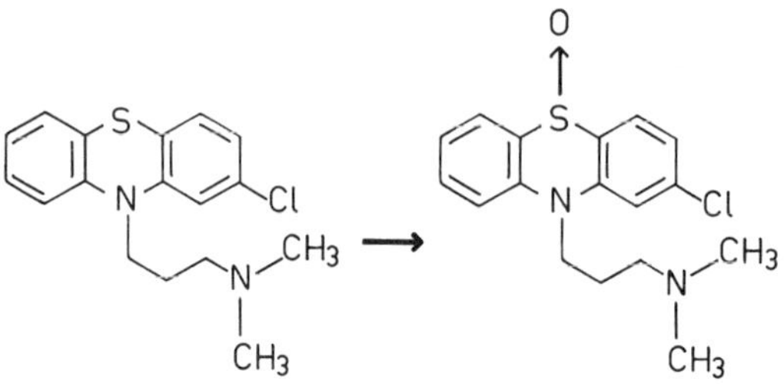

16 17

genase. Finally, bacteria in the large intestine are able to reduce a number of compounds.

The reduction of the nitro group to a nitroso group suggests toxicological risks because nitroso compounds are often mutagenic. The reduction of a sulfoxide to a sulfide increases the lipophilicity of the compound dramatically. This in turn changes the distribution of a drug and may increase its penetration into the brain.

Examples of drugs that are metabolized by reduction are chloral hydrate and sulfinpyrazone (18). The latter is a platelet aggregation inhibitor. It is reduced to the sulfide (19), which is pharmacologically more potent.

9. Hydrolysis. Hydrolytic processes are very important phase-I reactions. The hydrolysis of epoxides has been discussed above. The ubiquitous presence of esterases (for review see Williams, 1985) has been taken advantage of in the prodrug approach to drug design discussed in Chapter 6.

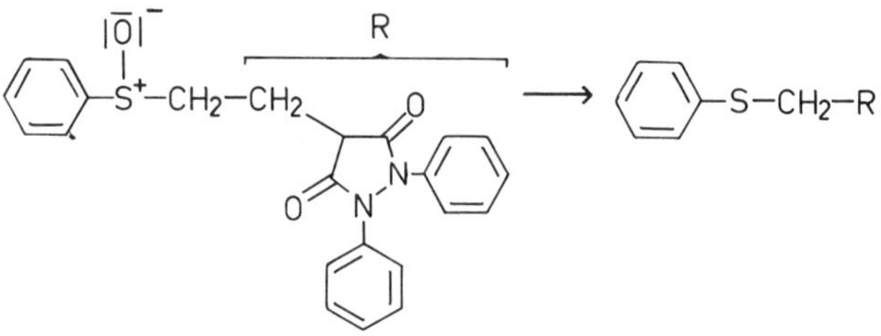

18 19

Table 1 Metabolic pathways of reduction.

Functional groups	Example
Aldehydes	$R_1{-}CHO \longrightarrow R{-}CH_2OH$
Ketones	$R_1{-}CO{-}R_2 \longrightarrow R_1{-}CHOH{-}R_2$
Alkenes	$R_2C{=}CR_2 \longrightarrow H{-}CR_2{-}CR_2{-}H$
Nitro groups	$R{-}NO_2 \longrightarrow R{-}NO \longrightarrow R{-}NHOH \longrightarrow R{-}NH_2$
Azo groups	$R_1{-}N = N{-}R_2 \longrightarrow R_1{-}NH{-}NH{-}R_2 \longrightarrow R_1{-}NH_2 + R_2{-}NH_2$
Sulfoxides	$R_1{-}S({=}O){-}R_2 \longrightarrow R_1{-}S{-}R_2$

Some esters are hydrolyzed by a wide variety of esterases, one example being acetylsalicylic acid. However, other esters are only attacked by a subgroup of these enzymes. An example is pethidine (20) which, in man, is hydrolyzed to the acid (21) mainly by microsomal esterases of the liver. In the case of pivaloylampicillin (22), which was synthesized to improve the oral bioavailability of ampicillin, the acetal ester is hydrolyzed in the mucosa and in the plasma of the portal vein so that only ampicillin (23) reaches the systemic circulation.

Amides are generally much less susceptible to hydrolysis than esters. Procainamide (24), an amide derivative of p-aminobenzoic acid (25), for example, is markedly more resistant to hydrolytic processes than is the corresponding ester procaine (26).

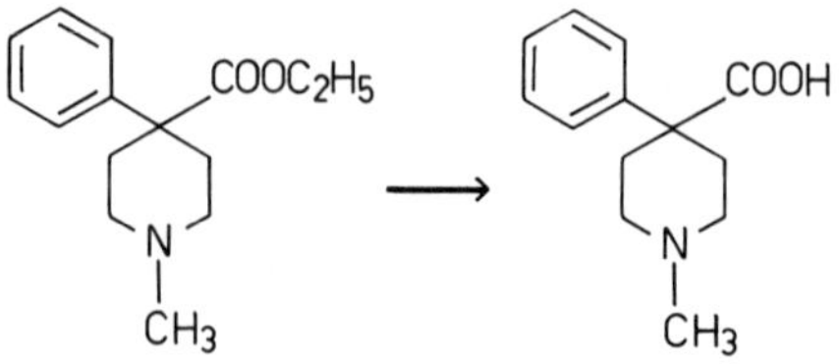

$COOC_2H_5$ $COOH$ N CH_3 N CH_3

20 21

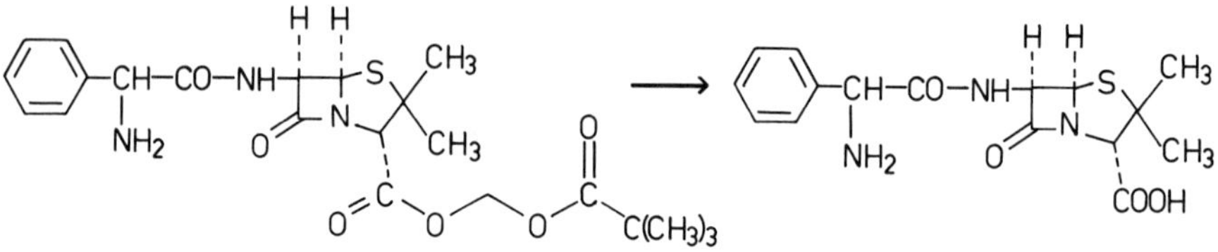

H H H H
$CH-CO-NH$ S CH_3 NH_2 O N CH_3 O C O O C $C(CH_3)_3$

$CH-CO-NH$ S CH_3 NH_2 O N CH_3 $COOH$

22 23

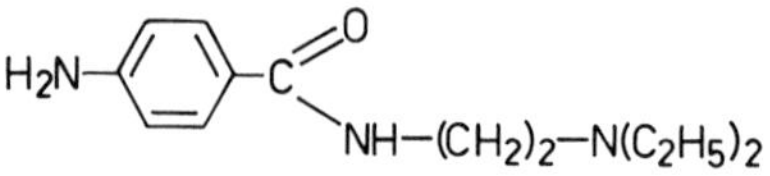

$H_2N-C_6H_4-C(=O)-NH-(CH_2)_2-N(C_2H_5)_2$

24

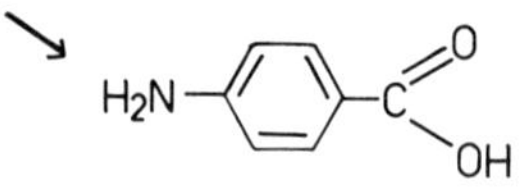

$H_2N-C_6H_4-C(=O)-OH$

25

$H_2N-C_6H_4-C(=O)-O-(CH_2)_2-N(C_2H_5)_2$

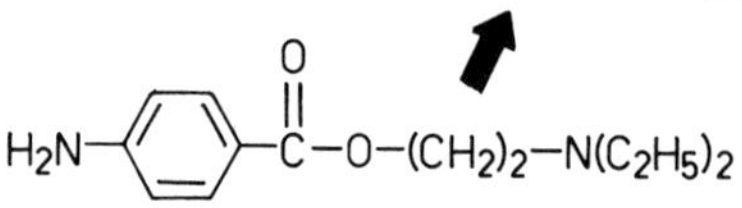

26

The hydrolysis of amide groups is an important pathway for the cleavage of N-containing heterocyclic compounds such as hydantoins and barbiturates. Aromatic amides are frequently metabolized to the corresponding amines, e.g. phenacetin (9) to phenetidine (27). This metabolite is thought to be responsible for the kidney toxicity of phenacetin.

Glucuronides and sulfates, which are formed in phase-II reactions, can be hydrolyzed in the intestinal tract. Enzymes of the intestinal wall and the intestinal flora may be involved. This will be discussed in Section III.D, below.

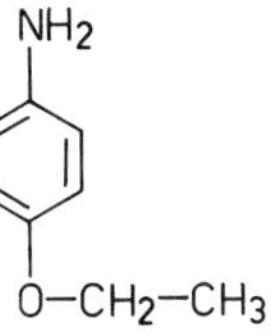

NH_2

$O-CH_2-CH_3$

27

C. PHASE-II REACTIONS

Phase-II reactions, also specified as conjugation reactions, are a group of synthetic reactions in which a foreign compound or a metabolite thereof is covalently linked with an endogenous molecule or grouping to give a characteristic product known as a conjugate. The most important conjugation reactions are listed in Table 2.

Phase-I biotransformation reactions usually reduce the hydrophobicity of the molecule. The reactions of phase-II normally increase the hydrophilicity of the molecule even more dramatically. This increase in hydrophilicity facilitates the elimination of the product into the urine or bile.

Table 2 Metabolic conjugation reactions (the abbreviations are explained in the text).

Reaction of a	conjugating agent	with functional groups
A Reactions involving activated conjugating agents		
Glucuronidation	UDP-glucuronic acid	-OH, -COOH, $-NH_2$
Sulfation	PAPS	-OH, $-NH_2$, -SH
Glucose conjugation	UDP-glucose	-OH, -COOH, -SH, $-NH_2$
Methylation	S-Adenosyl-methionine	-OH, - NH_2, -SH
Acetylation	Acetyl-CoA	-OH, $-NH_2$
B Reactions involving activated substrates		
Glutathione-conjugation	Glutathione	Epoxides Activated halides
Amino acid-conjugation	Glycine, Taurine Glutamine, Ornithine	-COOH

The formation of a conjugate is usually a two-step reaction. The first step involves the activation of the endogenous substrate or the drug. In the second step, phase-I metabolites and endogenous molecules are coupled by transferases. The synthesis of the activated endogenous compound requires energy. Therefore, in contrast to phase-I biotransformation, phase-II reactions require energy to be supplied by the body. As a result, phase-II reactions often compete with the metabolism of endogenous compounds. This happens especially under extreme conditions such as immaturity, starvation, and advanced liver disease.

It is important to appreciate that there are pronounced differences in conjugation reactions from species to species. This will be discussed further in Section III.E.

1. Glucuronidation. For most drugs and many endogenous compounds, glucuronidation is a dominant route of elimination. Glucose-1-phosphate, which has its origin in carbohydrate metabolism, is activated by uridine triphosphate (UTP) and oxidized by a dehydrogenase to form uridine diphospho-α-D-glucuronic acid (UDPGA).
Finally, transferases catalyze its reactions with a wide variety of nucleophilic functional groups (Table 3). The enzymes that activate glucose-1-phosphate occur in the cytoplasm of all tissues. Transferase activity, however, is concentrated in the endoplasmic reticulum of the liver, the kidney, the lungs, and intestinal mucosa. Transferase activity is inducible by a variety of xenobiotics.

2. Sulfation. An important process is the conjugation of phenols, alcohols, and aromatic amines with sulfuric acid. In the first step the sulfate ion is converted by two molecules of adenosine triphosphate to "active sulfate", 3'-phosphoadenosine-5'-phosphosulfate (PAPS). The sulfate group is then transferred by sulfokinases to the substrates (Table 4).

Sulfation often competes with glucuronidation. In man the ratio of sulfate to glucuronide excretion decreases with increasing drug doses. There are two reasons for this phenomenon: transferases involved in sulfation often have a higher affinity but lower capacity for conjugation compared with those involved in glucuronidation. Secondly, the availability of sulfate itself limits the extent of sulfation.

Table 3 Glucuronide formation with various xenobiotics.

	Functional group	Example	Structure
OH	Alcoholic (prim., sec., tert.)	Oxazepam	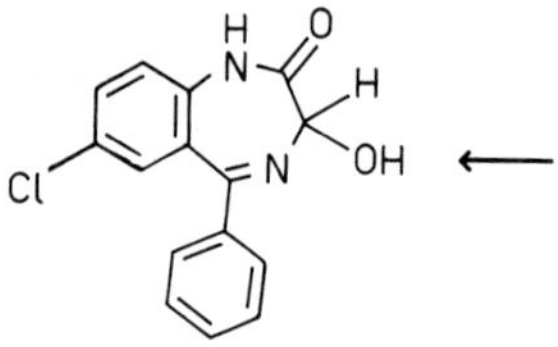H O N H N OH Cl
COOH	Carboxyl- (aromatic, aliphatic)	Salicylic acid	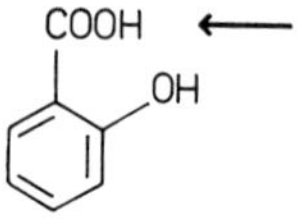COOH OH
NH	Aromatic amino-	Aniline	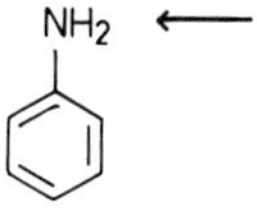NH_2
N	Tertiary amino-	Cyproheptadine	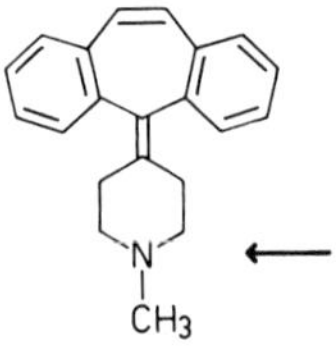N CH_3
CH	activated C-H bond	Phenylbutazone	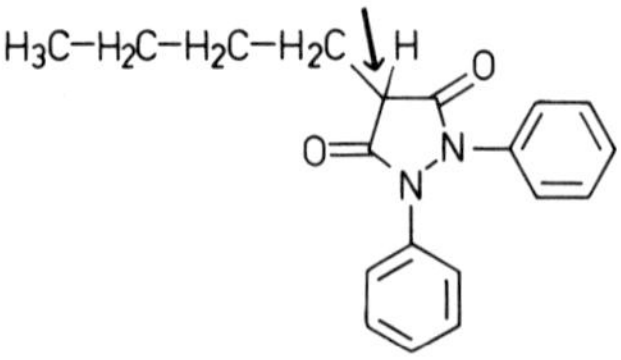$H_3C-H_2C-H_2C-H_2C$ H O O N N

The formed sulfates, like glucuronides, are more hydrophilic than the parent molecules. However, sulfates are more often subject to further biotransformation, as in the case of steroids. Sulfates can also act as activated substances, as the sulfate moiety is an excellent leaving group. Examples are N-hydroxy-2-acetyl-amino-fluorene (second structure in Table 4), and N-hydroxy phenacetin (15). Thus conjugates, especially sulfates, may have pharmacological and toxicological significance.

3. Acetylation. As a result of normal metabolic processes, there is an abundance of activated acetic acid, acetyl coenzyme A, in the mitochondria of all cells. However, acetylation is a major pathway of conjugation

Table 4 Sulfate conjugation with various xenobiotics.

Functional group	Example	Structure
Phenolic-	Isoprenaline	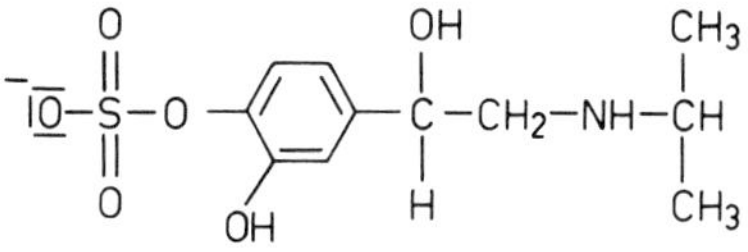
N-Hydroxy-	N-Hydroxy-2-acetyl-aminofluorene	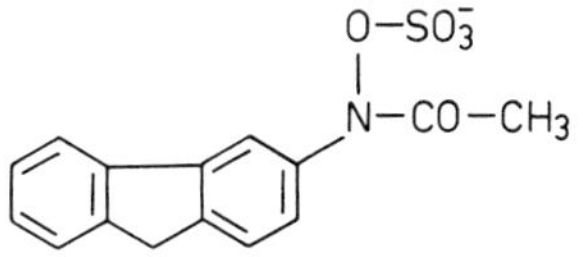
Amino-	Tiaramide	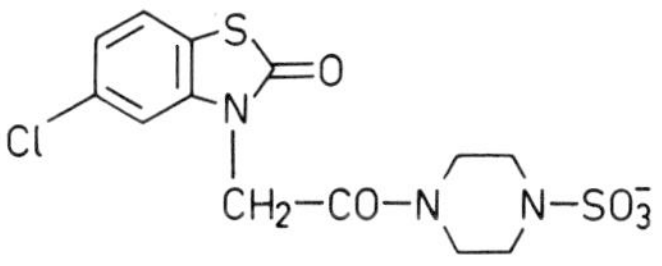

only for xenobiotics that contain NH_2-groups and not for those that contain OH-groups. Examples are aliphatic amines such as histamine, aromatic amines, sulfonamides such as sulfanilamide, and hydrazides such as isoniazid 28,29. As a rule, the product of such a reaction is no more hydrophilic than the original compound. Therefore, it is not surprising that acetylation does not result in an enhanced elimination of drugs from the body.

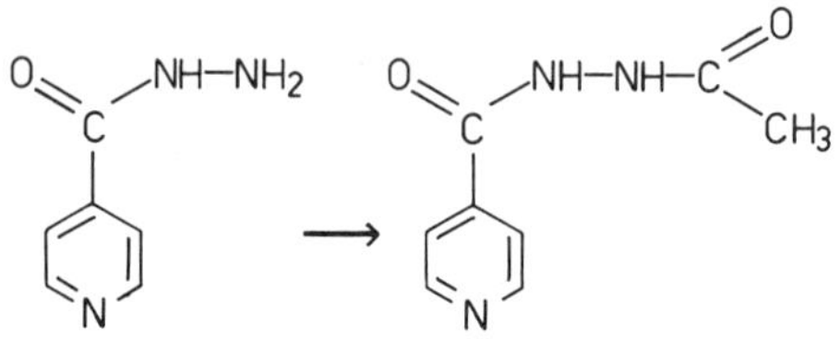

28 → 29

The polymorphism of this reaction in the population has been well investigated and correlated with differences in drug toxicity. Aspects of this are discussed in more detail in Section III.E.

4. Methylation. O-methylation plays an important part in the metabolic transformation of endogenous catechols. It is of minor importance for drugs except catechols. As an example isoproterenol (Structure 43, Chapter 6) is methylated by C-O-methyltransferase. Known methyl group donors are S-adenosylmethionine and 5-methyl-tetrahydro-

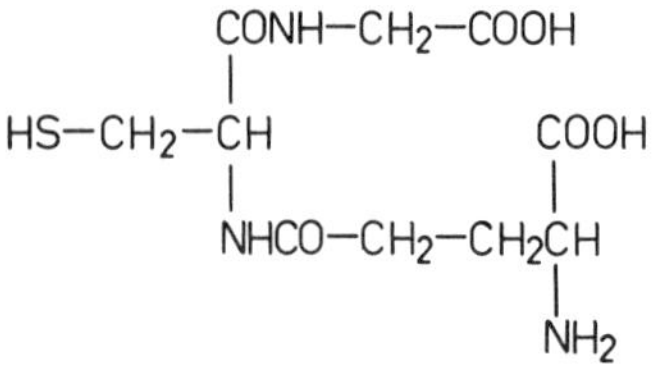

30

folic acid. Especially in liver, lung and brain there are enzymes capable of catalyzing this reaction.

5. Glutathione conjugation. The tripeptide glutathione (GSH), γ-glutamyl-cysteinyl-glycine (30) is important in xenobiotic metabolism due to its reactive thiol group, its high polarity, and its high intracellular concentrations. In the rat liver its concentration of ~8 mM exceeds even that of glucose or ATP (~5 mM).

The highly nucleophilic sulfhydryl group can react enzymatically and even nonenzymatically with many electrophils such as epoxides, activated double bounds, activated halides and quinones. The reactions are catalyzed by glutathione transferases, which are predominantly localized in the cytosol. Examples of drugs that are excreted as glutathione conjugates are paracetamol (Figure 15) and the cardiotonic drug amrinone (31,32). Paracetamol is thought to be activated to a reactive quinone-imine, which

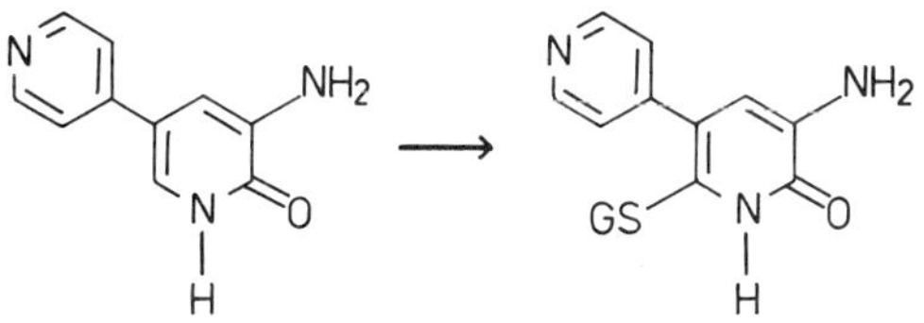

31 → 32

then reacts with glutathione. The quinone-imine intermediate is responsible for the acute liver toxicity of the drug after very high doses.

Owing to their high molecular weight, glutathione adducts are eliminated exclusively via bile. More often the glutathione conjugates are further transformed by enzymes that cleave the glutamyl and glycyl residue. Finally N-acetylation occurs, mainly in the kidney, to yield mercapturic acids. The GSH-adduct of paracetamol is shown in Figure 15.

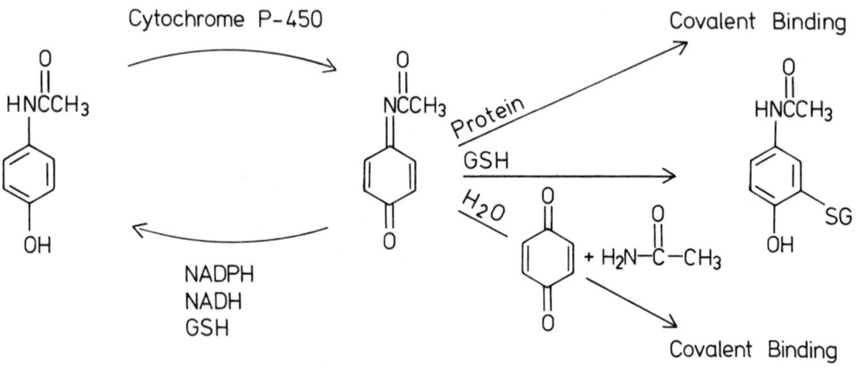

Figure 15 The bioactivation of paracetamol leading to GSH-adducts.

Due to their high polarity and instability, only a few glutathione conjugates of drugs have so far been isolated and fully characterized by nuclear magnetic resonance and mass spectrometry. However, the appearance of glutathione adducts and degradation products thereof is always an indication of reactive intermediates occurring during biotransformation and should be considered a warning during further drug development. This will be discussed in more detail in Section IV.B.

6. Conjugation with amino acids. For aromatic carboxylic acids the formation of an amide with amino acids is a very important reaction. In man, the conjugates of glycine and of glutamic acid are particularly abundant. Other species use a number of different amino acids and even peptides. Salicylic acid (33) offers an example of conjugation with glycine (34).

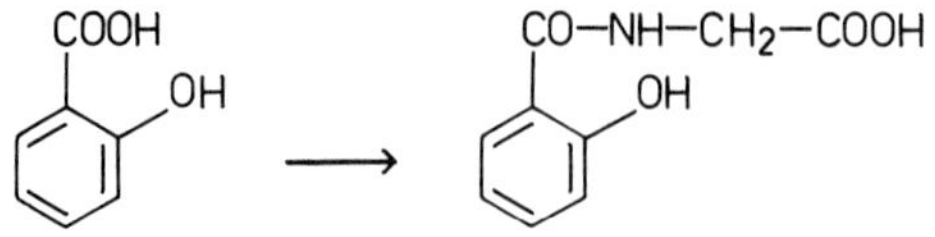

33 34

Various other conjugation reactions are reported in the literature. Few phosphate conjugates of xenobiotics have been identified, a fact which is surprising, if we consider the fundamental importance of activated phosphates in the body.

7. Novel biotransformation reactions. As a result of modern analytical procedures, an increasing number of biotransformation pathways have been described that differ from the "classical" ones described above. These "novel" routes include various rearrangements of the molecular structure of drugs and some unusual conjugations. Their biological significance is at present often unclear.

D. EXTRAHEPATIC BIOTRANSFORMATION OF XENOBIOTICS

Although the liver is the most important organ for biotransformation, there are other organs that perform this activity. The skin, blood, lung, kidney, brain, placenta, and gut are especially significant. Extrahepatic biotransformation is sometimes underestimated in pharmacokinetic research. This is especially true for the lung as a drug metabolizing organ. Metabolism by the intestinal mucosa and the gastrointestinal flora is significant for the biotransformation and the pharmacokinetics of many drugs. The reason for this is apparent if one considers that the bile, which can contain biotransformation products and the unchanged drug, is emptied into the intestinal tract. Then the drug or its metabolites may become subject to the following processes:

- excretion via the feces;
- further metabolic transformation during intestinal passage;
- reabsorption into the portal circulation (enterohepatic circulation).

Most metabolic pathways increase the hydrophilicity of drugs thus making them less prone to intestinal reabsorption by passive diffusion. However, glucuronides, sulfates, and even mercapturic acids may be cleavaged by intestinal microorganisms. Extensive enterohepatic circulation increases the half-life of a drug. Enterohepatic circulation of indomethacin correlates with its ability to cause intestinal damage.

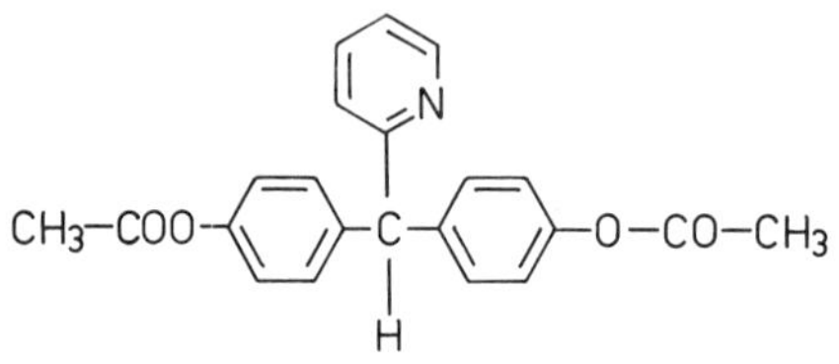

CH_3—COO—C₆H₄—C(H)(2-pyridyl)—C₆H₄—O—CO—CH_3

35

HO—C₆H₄—C(H)(2-pyridyl)—C₆H₄—OH

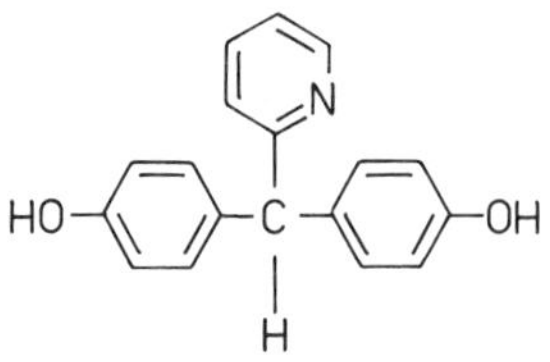

36

Intestinal microorganisms can perform a wide variety of other metabolic reactions: hydrolysis of esters and amides; decarboxylation; dehalogenation; reduction of nitro, azo, N-oxide and hydroxylamine groups; and reduction of aldehydes and ketones. For example, bacteria of the large intestine hydrolyze the ester of the laxative bisacodyl (35) to the diphenol (36). The latter is the active principle.

E. SPECIES DIFFERENCES AND GENETIC FACTORS IN DRUG METABOLISM

1. Species differences. Different animal species may vary markedly in their responses to a drug (Caldwell, 1981). This is a serious point of concern during drug discovery and development because studies in experimental animals are required before any testing is done in man and an important part of the safety evaluation of a new drug is done in animals. Species-specific differences in drug metabolism arise for one or more of the following reasons: competing reactions, metabolic defects and the occurrence of unusual metabolic reactions in some species.

For the ideal testing of a drug, an animal species should be chosen which is similar to man in four respects:

- the rates and routes of metabolism,
- the rates and routes of excretion,
- the pharmacokinetic profile which is influenced by metabolism and excretion,
- the pharmacological response.

Since in man, early studies of a compound comprise pharmacokinetics and metabolism, similarity is based initially on the overall pharmacokinetics and metabolism. The situation changes if, during further drug development, serious interspecies differences in a pharmacological or toxicological response are observed. If this difference is related to the presence of a metabolite, risk evaluation requires its concentration to be measured over a wide range of doses of the drug in all species, including man.

A classic example of species-specific differences due to competing reactions is the biotransformation of 2-acetyl- amino-fluorene, 2-AAF, the second structure in Table 4. 2-AAF is a potent carcinogen in a variety of species. The carcinogenicity of 2-AAF depends on the ratio between the toxifying pathway N-hydroxylation and the detoxifying pathways, mainly aromatic hydroxylations. This ratio differs from species to species. In the rat, rabbit and dog the degree of N-hydroxylation and subsequent conjugation is considerable. Thus 2-AAF is a carcinogen in these species. In vitro assays with human microsomes produced this N-hydroxylated metabolite, too. So we assume that 2-AAF is carcinogenic in man. In the guinea pig, which does not form this type of metabolites, 2-AAF is non carcinogenic.

Further examples of species differences concerning pharmacological activity are discussed in Chapter 6, Structures 47 - 50.

2. Defective reactions. Metabolic pathways that occur in man but not in animals are summarized in Table 5. In some cases the missing pathway is the primary route of detoxification of a compound. Thus, species which lack the pathway are not appropriate species for the assessment of the toxicological risk of the compound in man.

Although primate species are usually more similar to man with respect to biotransformation than non-primate species, generally, it is not possible to predict the

Table 5 Defective reactions of metabolism in some species.

Reaction	Species
N-hydroxylation of aliphatic amines	rat
glucuronide formation	cat
sulfate formation	pig (not minipig!)
arylamine acetylation	dog
mercapturic acid formation	guinea pig

appropriate "manlike species" (to what man?) from the chemical structure of a drug. The prediction is further complicated because the rate of biotransformation may also be affected by sex and age (Vesell, 1983).

3. Genetic differences in man. There are some genetically determined enzyme deficiencies in humans that influence drug metabolism (Clark, 1985). Marker substrates for genetic polymorphism in man are isoniazid for acetylation, debrisoquin for an aromatic hydroxylation, mephenytoin for another hydroxylation pathway, and phenacetin for de-ethylation.

The genetic differences vary between ethnic groups. 95-100% of Eskimos are rapid acetylators; only 18% of Egyptians, however. The plasma half-life of isoniazid is 3 times longer in slow acetylators. If the dose of this drug is not adjusted for this metabolic difference, serious toxic effects can be observed in the slow acetylators. Therefore new drugs have to be tested to determine whether or not they are substrates of known genetically variable enzymes and whether such enzymes contribute in a rate-limiting manner to the biotransformation of the new drug candidate.

F. KINETIC ASPECTS AND IN VITRO MODELS OF DRUG METABOLISM

The kinetics of biotransformation are complex. There are mainly three principal factors whose interplay determines the overall kinetics: the individual metabolic transformation by the enzyme system, transport processes to the metabolic site, and competing metabolic and elimination pathways.

Metabolic transformation itself is a multi-step process in which three steps are always involved: binding of the drug to the enzyme; chemical modification of the drug; and desorption of the metabolite from the enzyme.

Often an important determinant of the binding to the enzyme is the lipophilicity of the drug. Not surprisingly, binding to an enzyme often shows an additional dependence on the other physico-chemical properties described in Chapter 4.

The chemical modification step depends strongly on the steric and electronic properties of the drug.

Finally, for desorption, inverse dependencies on the factors governing enzyme binding can be expected. As lipophilicity is usually decreased by metabolism, desorption is facilitated.

It is currently impossible to predict from the structure of the drug molecule the rates of all metabolic processes. Predictions are possible only if, within a defined class of substrates, the structure-activity relationships of the rate of metabolism by a certain type of enzyme has been examined. Thus, the relationship between physicochemical properties and the in vitro hydrolysis rate has been successfully applied in designing prodrugs that undergo rapid enzymatic hydrolysis in blood. An example is esters of penicillins. In contrast, such relationships are apparently less likely to be found for transformations by cytochrome P-450. Dramatic changes in reaction rate of compounds are often observed following minor structural variations. This observation is understandable if one considers that different isoenzymes, each with a distinct structural specificity, may be involved. Therefore, application of quantitative structure-activity relationship analysis (Chapter 4; Austel and Kutter, 1983) is more appropriate where specific questions such as the structure of the active site of one biotransformation enzyme are concerned, or where an explanation of unusual pharmacokinetic behavior is required.

In vitro models with microsomal enzymes and model substrates such as ethoxycoumarin are used to assess enzyme induction, activation, and inhibition. Hepatocytes are a tool to study active processes involved in the transport of conjugates through the membranes of the liver cell.

An example where in vivo kinetics could be explained by in vitro studies is provided by the drug 8-methoxsalen (8-MOP). This drug suffers from a saturable first-pass effect in rat and man, even at low doses (Figure 16). Moreover the drug induces its own biotransformation after multiple administration in rat. Both effects could be explained by the metabolic enzymes involved: 8-MOP is a substrate of a cytochrome subclass named P-448; it has a very high affinity for this special enzyme. P-448, however, is only a minor constituent of the total cytochrome and is therefore readily saturated by 8-MOP, even at low doses. Furthermore, 8-MOP is an inducer of the cytochromes on multiple administration.

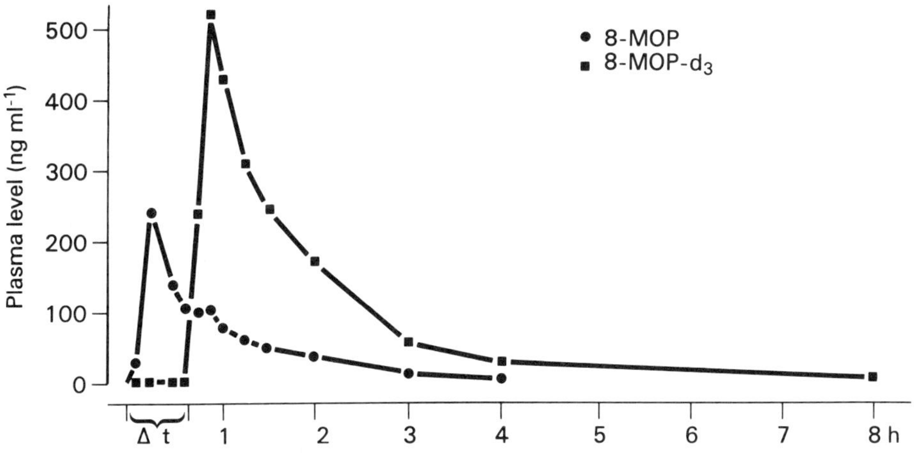

Figure 16 Oral administration of 20 mg 8-MOP (at t = 0) and of 20 mg deuterated 8-MOP (within a time interval Δt) to a human volunteer (quantification by GC/MS).

There is convincing evidence that drugs that contain planar, polycyclic aromatic systems, and do not exceed certain molecular dimensions are metabolized by the high affinity biotransformation enzyme cytochrome P-448, the levels of which are induced by the drug.

IV. THE ROLE OF PHARMACOKINETICS AND BIOTRANSFORMATION IN THE DEVELOPMENT OF BETTER AND SAFER DRUGS

Investigations on pharmacokinetics and drug metabolism have proved their value for therapies based on small molecules. The role of such investigations has been to aid interpretation of the safety assessment of potential drugs by explaining certain pharmacologically and/or toxicologically important observations. Many different drugs have been studied in this way as part of the drug approval process. Additionally, this information has led to the discovery and design of prodrugs.

From the accumulated body of knowledge one can now make educated guesses of the major overall rate-limiting processes. Pharmacokinetics are mainly concerned with the overall fate of the drug in the organism. Where the pharmacokinetics are governed by distribution and excretion, predictions can be based on quite valid rules considering lipophilicity and pKa. If the fate of a drug is governed by biotransformation, the present rules are less reliable. Acceptable estimates can, however, be made as long as major biotransformation routes are involved.

The toxicological and pharmacological profile of a drug may, however, often depend on minute quantities of drug-derived products such as reactive intermediates or active metabolites. Predictions of their toxicological profile and their quantity generally cannot be made with any degree of certainty. At best, one can identify structural elements that, on the basis of prior knowledge, may have greater potential to cause such effects than other structures.

A. EMPIRICAL RULES FOR THE SELECTION OF DRUG CANDIDATES

Pharmacokinetic and metabolic data can be used to select from several candidates those drugs most likely to be safe. In order to promote a consistent therapeutic effect, in general, one would select those compounds that show the least interindividual variability in their pharmacokinetics and metabolism. Another selection criterion is the absence of structural elements that may cause metabolically-mediated problems.

A few empirical rules for such an evaluation of drug candidates are summarized in the following. The rules apply to systemically-acting compounds excluding prodrugs and may readily be modified for different situations.

1. <u>Drugs with apparently linear pharmacokinetics are preferable!</u> When the pharmacokinetic response is proportional to the dose, the implication is that additional doses can be handled effectively by all pathways. Unpredictable responses due to capacity-limited processes are less likely. A high intrinsic pharmacological activity usually helps meet that objective since a lower total amount of compound is administered. These considerations lead to the following suggestions for the design or selection of potential new drugs:

Drug molecules should lend themselves to efficient and predictable pathways of elimination, the soft-drug concept! There are only a few processes with a virtually unlimited capacity for elimination. Among these are the metabolic processes of hydrolysis and glucuronidation. Glomerular filtration is also unlimited, but it is not effective with lipophilic compounds as they are subsequently reabsorbed. Because diseases, genetic predisposition, enzyme induction, and drug-drug interactions often affect only one disposition pathway, any possible alternative metabolic pathway provides additional safety.

Only drugs with appropriate physical properties for absorption should be used! A certain minimum lipophilicity is required for passive diffusion through membrane barriers. On the other hand, a minimum water-solubility is essential for transport to the absorptive site. For each class of drugs only a certain lipophilicity range is optimal. Even the most advanced delivery systems can only work if these basic requirements are met.

Compounds with a hepatic extraction ratio close to unity should be avoided! Because of the first-pass effect, such drugs are not suitable for oral administration, which is generally the most safe and efficient route. Saturation of the first-pass effect at high doses is likely to occur. Consequently, all pharmacokinetic and pharmacodynamic responses may show high variability after oral administration of such drugs.

Compounds which are bound excessively to plasma proteins should be avoided! The level of the free drug often determines the pharmacologic effect. When the fraction unbound is less than 10%, variations up to severalfold in the concentration of free drug in the plasma can occur when bound drug is competitively displaced by metabolites, other drugs, or endogenous or exogenous compounds from food or disease processes.

2. Chemically reactive structures should be avoided because they may cause toxicological problems! All reactive compounds have the potential to modify endogenous molecules and thus interfere with their physiological function. There is also a danger of depleting the natural scavenger systems (Gilette, 1986). The exceptions to this

Table 6 Stuctural elements with a high risk of metabolic toxification.

Structural element	Example
Electron-rich π-systems (e.g. furans)	4-ipomeanol
Quinons and quinonimines, or their precursors (catechols, aminophenols)	practolol
Aromatic methylene-dioxy groups	safrole
Aromatic nitro groups	nitrefazole
Primary aromatic amines	2-naphtylamine
Sulfhydryl groups	captopril

prohibition are therapies in which biological functions are purposely blocked such as in cancer therapy.

In this context one should consider reactive groups in the parent molecule as well as moieties that are activated metabolically. Features belonging to the first group are strong electrophils such as activated esters or alkylating agents. Examples of the second group can be found in Table 6 and in Sections III.B and C. In order to avoid problems due to metabolic activation, one might consider the active metabolite theory of the soft-drug concept. It suggests that the drug of choice should be the active metabolite in the highest oxidized state.

Problems may also be circumvented by selecting drug candidates that contain "safe" structural features, such

Table 7 The role of pharmacokinetic and metabolic studies in various stages during drug development.

Stage of development	Role of pharmacokinetic and metabolic studies
Selection of drug candidates for development:	Consideration of the pharmacokinetic profile desired in connection with known biotransformation processes; explorative studies.
Preclinical development:	Design and interpretation of pharmacological and toxicological investigations also with respect to species differences.
Safety and activity evaluation in man:	Establishing dose regimens; identification of metabolites and evaluation of their contribution to the biological profile of the drug.
Broad clinical application:	Studies in special patient groups at potential risk (age, disease, metabolic disorders, comedications) to adjust dose regimens.
Appearance of unexpected adverse effects at any stage:	Design of relevant, specific test models for the adverse effects in order to select another drug candidate in this class.

as those found in many drugs that have not given rise to concern with respect to toxicological and biotransformational problems. This approach is supported by a study of 855 unrelated xenobiotics in the Ames test to give an indication of the potential genotoxic properties of the compounds (McMahon et al, 1979). The reactive types of compounds such as aromatic nitro compounds and nitrosamines showed a high incidence, 75%, of positive effects. Of the other 783 compounds, only 16% produced positive effects; of these, a considerable fraction (35%) were benzimidazoles. All pyridines, thiazoles and pyrimidines were negative.

B. BIOTRANSFORMATION AND PHARMACOKINETIC STUDIES AS INTEGRAL PARTS OF DRUG DEVELOPMENT

Metabolic and pharmacokinetic investigations concerning the safety of a drug have to be performed according to official regulations. Such investigations are not merely a formal routine task. On the contrary, intelligent contributions from metabolic and kinetic studies are essential to minimize drug development costs and to ensure that the most promising compound in a series is selected. One objective is to fulfil the pre-clinical requirements for testing a drug candidate in human volunteers as early as possible.

1. The integrated drug development program. This strategy requires thoughtful and well-timed integration of drug metabolism and kinetic studies into the drug development program, and hence a joint cooperative effort with pharmacology, toxicology, pharmaceutical development and clinical pharmacology. Similar approaches have to be pursued in later phases of drug development when pharmacokinetic and metabolic studies must complement clinical studies, Table 7.

Only such interdisciplinary cooperation can cover all aspects of the desired therapeutic indication together with the special considerations in the individual class of drug under investigation. In the course of such cooperation, pharmacokinetics and drug metabolism studies can provide invaluable information that is not only essential for selecting and evaluating drug candidates but also can help other disciplines to pursue their aims within the search for better and safer drugs, Table 7.

The impact of previous pharmacokinetic and drug melabolism studies on the design of potential drugs has already been alluded to. Further developments in this direction will include the establishment of data bases of pharmacokinetic processes and known drug metabolic transformations together with their relative importance in different species. These data bases must then be integrated into practicable computer programs that can predict the fate of chemical compounds in the organism. The result of such efforts would be to help to design more informative and conclusive experimental studies, thus saving development time and resources including animals.

When a class of compounds with a promising activity profile has been discovered, the next step is to select candidates for structural optimization (see Chapter 6) or for development. Explorative investigations in drug metabolism and pharmacokinetics can answer key questions at that stage:

- Will the compound be absorbed?
- Is the parent compound or rather one of its metabolites more likely to be responsible for pharmacological activity?
- Is predominately one metabolite formed?
- Are there metabolites with very long half-lives?

Recent improvements in analytical instrumentation provide rapid semiquantitative methods to provide answers to these questions. High performance liquid chromatography, photodiode array detection and all newer mass-spectroscopy techniques, each with improved data storage and data handling facilities, are often useful. Already at this stage in vitro and cell culture techniques will increasingly allow the collection of data of major relevance to man.

When a compound is already selected for development as a potential new drug, pharmacokinetic and drug metabolism studies play their traditional role, aiding in the interpretation of the pharmacological and toxicological findings. At the start of the detailed pharmacological and toxicological evaluation of a potential new drug, experimental dosage formulations must be devised and tested to ensure the validity of the whole biological test program. This is especially important when the drug candidate is highly insoluble. Inconsistency between the results of various pharmacological or toxicological models such as in vitro/in vivo discrepancies must be interpreted in the light of their possible pharmacokinetic basis. At this stage, the selection of the most appropriate species for drug safety studies is considered. In situations where pharmacological or toxicological findings are species-dependent, a problem-oriented investigation on metabolism and kinetics is required. Studies over a wide range of doses and over a prolonged period of time can give clues as to possible explanations for species differences and their relevance to man. Improved evaluation of animal data by suitable computer programs together with a better interpretation of the kinetic results obtained in various species will help in their extrapolation to man.

When a drug candidate has made its way to man, the understanding of the complete human pharmacokinetic and pharmacological profile is sought, including estimates of risks of undesired effects. Clinical pharmacology allows the time course of the various biological effects of the compound to be followed and quantified. Such a description of effect kinetics is becoming a valuable tool. There are many examples of drugs that give rise to active metabolites. They may have either a similar (e.g., acetylsalicylic acid/salicylic acid), a modified (imipramine/desimipramine), or an antagonistic (trazodone/m-chlorphenylpiperazine) activity profile when compared with the parent drug (Garattini, 1985). Similarly, an undesired or toxic effect of the parent drug may be different from that of its metabolite (phenacetin/paracetamol). The biological consequences of stereoisomerism are of increasing concern. Until now most drugs have been administered as racemates, even though in many cases only one of the isomers seems to be responsible for the effects (for review see Walle and Walle, 1986).

In the final stage of the development of a drug candidate, its efficacy and safety has to be proven in the target patient population. Pharmacokinetic and drug metabolism monitoring is a part of such studies. Patient groups at special risk, (those with renal, hepatic or circulatory diseases; elderly or multimorbid people and neonates and children), may need to be studied to devise appropriate dose adjustments. Pharmacogenetic differences in drug metabolism may affect the response to a drug. Its possible pharmacokinetic basis has to be established (Clark, 1985). Population kinetics is becoming increasingly important and will greatly broaden our possibilities for dealing with individual differences in the pharmacokinetics and metabolism of drugs (Sheiner and Benet, 1985). Population kinetics is based on multivariate analysis of, for example, an unselected patient population with respect to the kinetic fate or the pharmacodynamics of a drug. Such analyses allow one to pinpoint special risk factors or predispositions within a particular patient group that may necessitate special attention in therapy.

Quite frequently, disappointing results with one drug candidate from an otherwise promising class of compounds may be obtained at some stage of development. It is completely wrong at this point to discard the whole class. Very often compounds can be identified in which the desired effect is separate from the less favorable one. However, to proceed in a cost-effective manner and provide the medicinal chemists with the information they need to separate these effects further, one needs to establish sensitive models to selectively screen many related compounds. One example may highlight this strategy: N-substituted azoles are potent inhibitors of yeast cytochrome P-450. Due to their interference with ergosterol biosynthesis they have antimycotic properties: the desired effect. At higher concentrations they also affect the mammalian P-450 system interfering, for example, with corticosterone biosynthesis: the undesired effect. Therefore, appropriate test systems for in vitro evaluation are yeast cultures and mammalian enzyme preparations. The best selectivity was found with itraconazole.

2. Early recognition of caveats concerning efficacy and safety. It is unusual for results from pharmacokinetic and metabolic studies alone to be taken to decide upon the further development of a drug or a whole class of drugs. Pharmacokinetic and drug metabolism studies do, however, play a significant role as indicators of efficacy and safety: they give certain warnings that must be considered during further biological testing.

One type of warning comes from one or more of the following observations: pharmacokinetics and metabolism of the drug change dramatically after chronic administration; effects differ substantially between various routes of administration; or pharmacokinetics vary with the dose. The causes of such problems can be saturable metabolic or transport processes, saturable hepatic first-pass elimination, first-pass effects in various other organs, depletion of natural scavenger systems, enzyme induction, inhibition of elimination pathways by metabolites or even functional impairment of eliminating organs by the drug and/or its metabolites.

A second type of warning comes from indications that drug metabolism involves reactive intermediates: in studies with radioactively-labeled drug one may observe accumulation of radioactivity in certain organs or organelles, long persistance of radioactivity, covalent binding to plasma proteins or tissue constituents, and incomplete recovery of adminstered radioactivity. Unusual distributions in the biological system or its substructures as well as long persistence of the labeled material necessitates special attention in histological studies during toxicological evaluation as well as in general pharmacological evaluation. Irreversible binding of radioactivity to cell constituents may indicate problems in terms of an allergenic, genotoxic, or mutagenic potential, or point to an organ-specific toxicological response. Dark or colored urine and bile, or glutathione adducts as metabolites are another strong indication of highly reactive intermediates in biotransformation (Gilette, 1986).

Such observations alone should not necessarily cause the development of a promising drug candidate to be curtailed, but should rather direct the subsequent research program to bringing these findings into a general perspective. These caveats influence the depth of subsequent investigation in some fields and the degree of thoroughness in the studies. That is by no means to say that, for example, metabolites have to be routinely traced down to minute quantities.

On the contrary, there is still a common guideline that has to be applied throughout the whole investigational program. It is given by the scientific and ethical principle of all life sciences:

> An experiment is justified only when all possible results to be expected will have clear consequences for the project and make a real contribution to the objectives, in this case, for the discovery of safer and better drugs.

3. Conclusion. Investigations in pharmacokinetics and biotransformation are of value, beyond fulfilling the guidelines given by the authorities. These investigations are essential

- for the selection of the optimum compound from a series of candidates,
- for the detection and quantification of potential risks of administering the compounds,
- and for the design of optimum dosage forms and dosing schedules.

There is a steady progress in the investigation of pharmacokinetics and biotransformation, in particular, from improvements in analytical methodology, of in vitro techniques, and of computer software and hardware. In addition, there is an increased understanding of the common biological principles that are relevant at all stages of preclinical and clinical drug development. This progress has compensated for some of the time delays in drug development which must be taken into account in order to make new drugs better and safer than existing ones.

"Over all, a sequentially designed and integrated drug development program considering pharmacokinetics in relation to the other disciplines involved, will have the greatest potential for success in the development and use of safe and effective new drugs with reproducible onset and duration of activity."
(S.A. Kaplan, 1983)

REFERENCES

Austel, V. and E. Kutter, Absorption, Distribution and Metabolism of Drugs, in "Quantitative Structure-Activity Relationship of Drugs", (J.G. Topliss, ed.), Medicinal Chemistry, Vol. 19, Academic Press, New York, 1983, pp. 437 -496.

Bend, J.R., C.J. Serabjit-Singh and R.M. Philpot, The Pulmonary Uptake, Accumulation, and Metabolism of Xenobiotics, Ann. Rev. Pharmacol. Toxicol., 25, 97-125 (1985).

Breimer, D.D. and P. Speiser (eds.), "Topics in Pharmaceutical Sciences 1983", Elsevier, Amsterdam, 1983.

Breimer, D.D. and P. Speiser (eds.), "Topics in Pharmaceutical Sciences 1987", Elsevier, Amsterdam, 1987.

Bundgaard, H. (ed.), "Design of Prodrugs", Elsevier, Amsterdam, 1985.

Caldwell J., The Current Status of Attempts to Predict Species Differences in Drug Metabolism, Drug metabolism reviews, 12, 221 - 237 (1981).

Clark, D.W.J., Genetically Determined Variability in Acetylation and Oxidation: Therapeutic Implications, Drugs, 29, 342-375 (1985).

Dost, F.H., "Der Blutspiegel: Kinetik der Konzentrationsabläufe in der Kreislaufflüssigkeit", Thieme, Leipzig, 1953.

De Montanello, P.R.O. (ed.), "Cytochrome P-450: Structure, Mechanism and Biochemistry", Plenum Press, New York, 1986.

Fleck, C. and H. Bräunlich, Methods in Testing Interrelationships between Excretion of Drugs via Urine and Bile, Pharmac. Therap., 25, 1-22 (1984).

Garattini, S., Active Drug Metabolites: An Overview of their Relevance in Clinical Pharmacokinetics, Clin. Pharmacokin., 10, 216-227 (1985).

Gilette, J.R., Significance of Covalent Binding of Chemically Reactive Metabolites of Foreign Compounds to Proteins and Lipids, in "Biological Reactive Intermediates III" (J.J. Kocsis, D.J. Jollow, C.M. Witmer, J.O. Nelson and R. Snyder, eds.), Plenum Press, New York 1986, pp. 63-82.

Hinderling, P.H., Drug Distribution in the Body: In Vitro Prediction and Physiological Interpretation, J. Pharmacokin. Biopharm., in press (1987).

Jacoby, W.B. (ed.), "Enzymatic Basis of Detoxication Vol I + II", Academic Press, New York, 1980.

Kaplan, S.A., Pharmacokinetics Research: More than a Simple Disposition Profile of a New Chemical Entity, Trends Pharmacol. Sci., 4, 372 - 375 (1983).

McMahon, R.E., J.C. Cline, and C.Z. Thompson, Assay of 855 Test Chemicals in Ten Tester Strains Using a New Modification of the Ames Test for Bacterial Mutagens, Cancer Research, 39, 682 - 693 (1979).

Rowland, M., L.B. Sheiner, and J.-L. Steimer (eds.), "Variability in Drug Therapy: Description, Estimation and Control", Raven Press, New York, 1985.

Rowland, M. and T.N. Tozer, "Clinical Pharmacokinetics", Lea & Febinger, Philadelphia, 1980.

Sheiner, L.B. and L.Z. Benet, Premarketing Observational Studies of Population Pharmacokinetics of New Drugs, Clin. Pharmacol. Ther., 38, 481-487 (1985).

Testa, B., and P. Jenner, "Drug Metabolism: Chemical and Biochemical Aspects", Marcel Decker, New York, 1976.

Vesell, E.S., Why Individuals Vary in their Response to Drugs, in "Drug Metabolism and Disposition" (J.W. Lamble, ed.), Elsevier , Amsterdam 1983, pp. 143-147.

Walle, T. and U.K. Walle, Pharmacokinetic Parameters Obtained with Racemates (part in a series of reviews on chirality), Trends Pharmacol. Sci., 7, 155-158 (1986).

Williams, F.M., Clinical Significance of Esterases in Man, Clin. Pharmacokin., 10, 392-403 (1985).

Wise, R., The Clinical Relevance of Protein Binding and Tissue Concentrations in Antimicrobial Therapy, Clin. Pharmacokin., 11, 470-482 (1986).

4

Theoretical Basis of Medicinal Chemistry: Structure-Activity Relationships and Three-Dimensional Structures of Small and Macromolecules

Yvonne Connolly Martin

Computer-Assisted Molecular Design Project
Pharmaceutical Products Division
Abbott Laboratories
Abbott Park, Illinois

A question that has intrigued generations of chemists is what features of a particular small molecule are responsible for its biological activity. This question is important to some because of its significance in understanding biology and to others because it may lead to a better and safer drug. Many types of data and theories may be used to probe this question (Jolles and Woolridge, 1984), but most start with some hypothesis as to the general types of forces that are involved in molecule-molecule interactions. The classic answer to the question involves molecular dissection called structure-activity analysis. QSAR, quantitative structure-activity relationships, takes the analysis one step further by representing each molecule of a series in terms of a set of physical chemical descriptors and examining the correlation of potency with the value of one or more of these physical properties. Target biomolecule binding site mapping considers the three-dimensional properties of biologically active molecules to propose a three-dimensional map of the shape and chemical properties of the target biomolecule. Finally, macromolecular structural methods provide atomic level resolution of the details of small molecule-macromolecule interactions. Each of these topics will be considered in turn in this chapter.

I. FORCES WITH WHICH A BIOLOGICAL MACROMOLECULE BINDS A SMALL MOLECULE

A. IMPORTANCE OF UNDERSTANDING INTERMOLECULAR INTERACTIONS

A frequent objective in making better and safer drugs is to design molecules that are improvements over existing compounds because they bind more strongly with the target and less strongly with other molecules of the body. Thus one needs to develop an understanding for the basis for the strength of such interactions and how changing the structure of the molecules might change the strength of such interactions (Cantor and Schimmel, 1980).

The structural formulae in the previous and following chapters are designed to show in two dimensions what atoms in the structure are connected to what other atoms and whether a bond between atoms is single, double, aromatic, or triple. However, molecules are three-dimensional. Every spatial arrangement of the nucleii of these atoms describes one conformation of the molecule. Usually a molecule has several bonds about which groups rotate with respect to each other; this produces many different conformations. Figure 1 shows some of the low-energy conformations of dopamine.

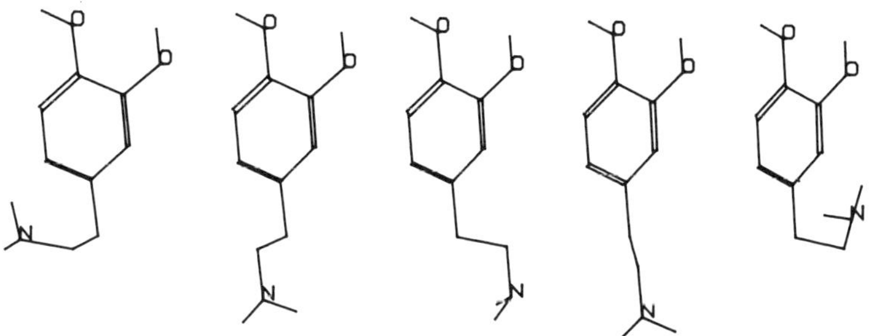

Figure 1 Some of the conformations of dopamine.

Which conformation a molecule adopts when bound depends on the intermolecular forces it experiences. Thus molecules are three-dimensional objects of constantly changing shape.

Yet another level of complexity of molecular structure originates from the fact that when two molecules come into contact, it is their electron clouds that interact, not their nuclei. The nuclei serve as (the constantly moving) positively charged skeleton to which the (also constantly moving, but at a much faster rate) negatively charged electrons are held. Those regions of the space near a molecule where there are excess electrons are electrically negative whereas regions deficient in electrons are electrically positive. The regions of one molecule with a net negative charge tend to interact with the positive charge regions of another. The types of atoms and bonds in the molecule determine the regions of electron excess and deficiency in space.

The strength of drug-macromolecule binding is generally very specific and very strong. From protein crystallography we have learned that both the strength and specificity of such binding arises from the accumulation of many weak forces between the drug and the macromolecule. For example, the cofactor NADPH is bound to dihydrofolate reductase by a total of 30 weak interactions (Goodford, 1984). The close contacts between the cofactor and the protein are seen in Figure 2. This accumulation of weak forces is more than additive because of the "chelate effect". The chelate effect arises because any intermolecular interaction must use energy to restrict the motion of molecules. However, once the molecular motion is restricted with one interaction, then further

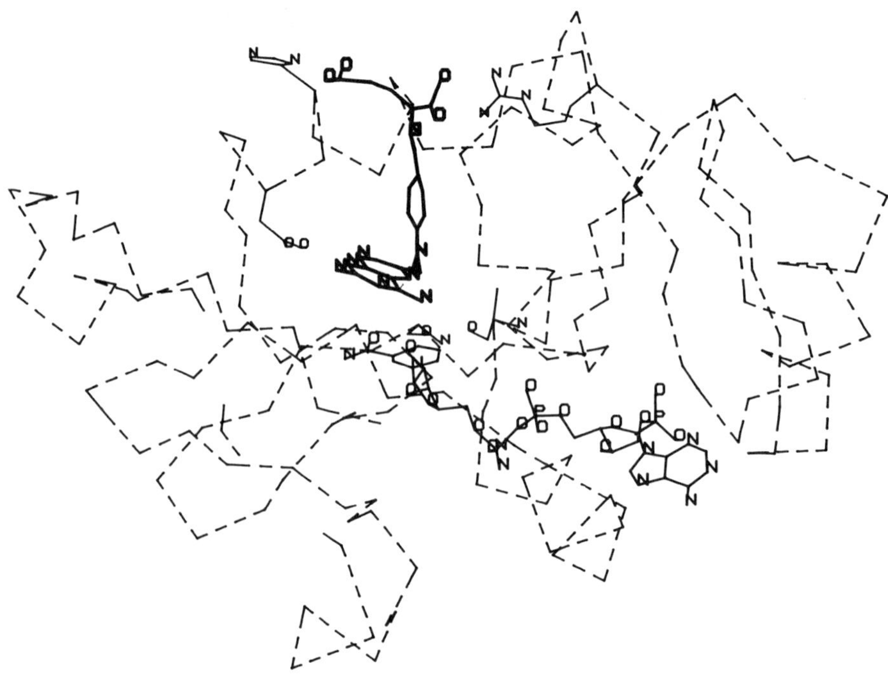

Figure 2 A trace of the α-carbons of the enzyme dihydrofolate reductase (dashed lines), the inhibitor methotrexate (bold lines), some of the groups of the enzyme that interact strongly with the inhibitor, and the cofactor NADP. The coordinates are from the Brookhaven Data Base file 3DFR of the work of D.J. Filman, D.A. Matthews, J.T. Bolin, and J. Kraut.

interactions reinforce the strength of the binding with no additional energy cost.

Intermolecular forces differ from each other in the relative energy that they can ideally contribute to an interaction, to the rate of fall-off of the interaction energy as the distance between the interacting groups is changed, and to the effect of solvent on the strength of interaction.

Although it is artificial to separate the total energy of interaction into discrete types of forces, we do so because it helps us understand the physics of these interactions and to predict the effect of changes of chemical structure on the strength of the interaction. In addition, this separation has allowed theoretical chemists to devise empirical potential energy equations that reproduce experimental data to a first order of approximation.

B. DESCRIPTION OF INTERMOLECULAR FORCES BETWEEN MOLECULES

1. Electrostatic interactions. The force that is easiest to understand is the coulombic attraction between a positively and a negatively charged ion and the replusion between two positively or two negatively charged ions. Partial atomic charges also give rise to coulombic attraction or repulsion between groups within one molecule or different molecules.

If one imagines the binding site for a drug to be a pocket in a protein, that pocket would be lined with atoms bearing partial atomic charges, some positive and some negative. The drug would tend to orient itself within the pocket to maximize electrostatic attraction and minimize electrostatic repulsion. Because of this large number of subtle positive and negative inter-

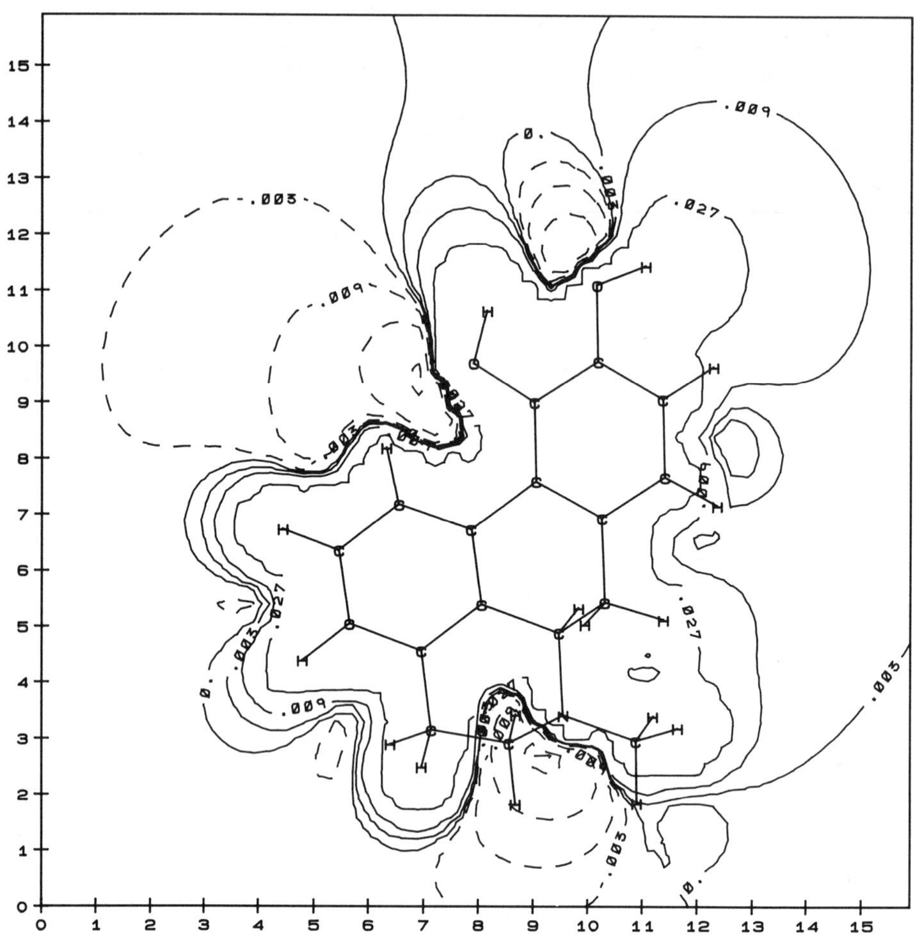

Figure 3 A contour plot of electostatic potential of apomorphine in the plane of the aromatic ring. The solid lines are positive contours and the dashed lines, negative contours. The most positive contours are not shown because they arise from steric repulsion to the nuclei.

actions, electrostatics play a prominent role in molecular recognition (Perutz, 1978).

For example, Figure 2 shows the neutralization of the charge of one carboxylate group of the methotrexate by a guanidinium group of the enzyme and the neutralization of the charge on another carboxylate group by an imidazole of the enzyme.

Electrostatic interactions are much stronger in vacuum or non-polar media than in water because water molecules themselves have partial atomic charges and so interact with the other molecules of interest. Never-the-less, electrostatic interactions are very strong. Furthermore, they fall off more slowly with distance than do other interactions. Hence, they contribute substantially to the strength of drug-receptor interactions.

Because electrostatics are so important, theoretical chemists have devised means to calculate and display this property of a molecule. Figure 3 shows a contour diagram of the energy of interaction of an H^+ ion, the electrostatic potential, with apomorphine. Notice that the oxygen atoms are attractive to an H^+ whereas the hydrogen atoms attached to the oxygens are strongly repulsive to an H^+. Another popular representation is to calculate the electrostatic potential at the molecular surface and then to use color or shading to differentiate the different energy levels, Figure 4.

2. Dispersion interactions. When one molecule comes close to another, any temporary or permanent unequal distribution of electrons in the first tends to induce a change in the distribution of electrons in the latter. This phenomenon is labeled a dispersion, dipole-induced dipole, or induced dipole-induced dipole interaction. This is the force that keeps benzene or hexane a liquid.

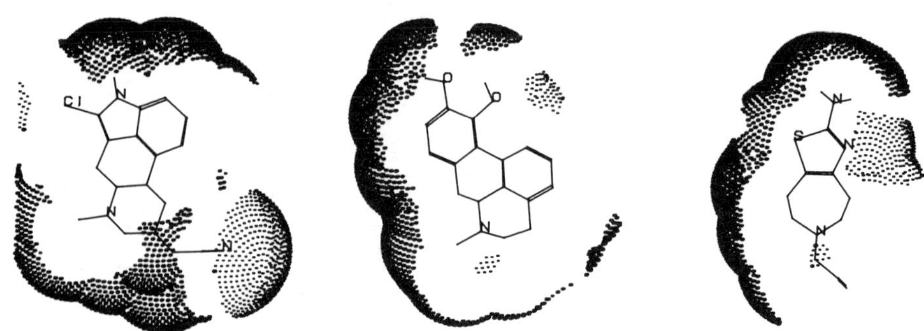

Figure 4 A representation of the electrostatic potential of three dopaminergic compounds, lergotrile, apomorphine, and B HT-920. The surfaces are located at the van der Waals radius plus 1.4 Å, roughly where the nucleus of an interacting atom would be located. The regions of potential >2.0 Kcal/mole are shown in heavy dots and <-2.0 Kcal/mole in fine dots.

Dispersion interactions require that the molecules be rather close: they fall off as the sixth power of distance. Since every atom in a molecule participates in dispersion interactions, if there is a good complementarity between the shape of the drug and its binding pocket, then dispersion binding can contribute substantially to the strength of the interaction.

The best way to show the sites of optimum dispersion interaction of a molecule is to show the molecular surface. Figure 5 shows the close correspondence between the surface of the binding pocket on dihydrofolate reductase and the surface of methotrexate. Figures 2 and 5 show that the cofactor (NADPH) makes part of the binding sites for the inhibitor.

3. <u>Steric repulsions</u>. The electron cloud about an atomic nucleus establishes a certain radius which represents the closest approach of another molecule. These Van der Waals radii of atoms are established from experimental observations of closeness of intermolecular approach in crystal structures.

Steric repulsion increases as the ninth or twelfth power of interatomic distance. Thus steric repulsion can be a major determinant of lack of fit of a particular compound into a binding site and, conversely, it is a key factor in molecular recognition.

The molecular surface shows the closest allowed approach of other atoms to a molecule.

4. Hydrogen bonds. The hydrogen atom is of a special character since it has only one electron. When hydrogen is bonded to any atom, but particularly when it is bonded to electronegative atoms such as oxygen or nitrogen, the binding electrons are pulled toward the electronegative atom. This leaves the hydrogen with a quite substantial partial positive charge. Such a

Figure 5 A clipped view of the surface of the binding pocket in dihydrofolate reductase for methotrexate (fine dots) and the surface of methotrexate in heavy dots.

hydrogen atom is thus ideally suited to participate in a strong electrostatic interaction with electronegative atoms such as fluorine, oxygen, or nitrogen atoms in the same molecule or a different one. This special non-covalent association of a hydrogen atom with an electronegative atom is called a hydrogen bond.

Hydrogen bonds are strongest when the hydrogen atom is on the line between the atom to which it is bonded and the one with which it interacts. The hydrogen atom penetrates the Van der Waals radius of the interacting atom. Figure 2 shows several of the hydrogen bonds between the nitrogens of the aromatic ring of methotrexate and groups on dihydrofolate reductase. The most beautiful example of hydrogen bonds is ice in which each water molecule donates and accepts two hydrogen bonds.

Because water is such a good hydrogen-bond donor and acceptor, little energy is gained by forming a hydrogen bond between two molecules dissolved in water. However, once two molecules are close together, by the chelate effect hydrogen bonding can contribute to the strength of the interaction.

Hydrogen bonds contribute to molecular recognition because of their directional character and because most atoms of a molecule cannot participate in a hydrogen bond.

5. Hydrophobic interactions. The strong hydrogen-bonding network of water gives rise to a more subtle type of force called a hydrophobic bond (Tanford, 1980). A hydrophobic bond forms where two relatively non-charged (non-polar) molecules are squeezed out of water solution: the driving force is the tendency of water to associate with itself in a constantly changing hydrogen-bonding network. Non-polar molecules increase the energy of the system by freezing the water around

themselves. Only a small percentage of the driving force of a hydrophobic bond is the attraction of the non-polar molecules for each other.

The strength of a hydrophobic bond is correlated with the surface area of the non-polar part of the molecule, in other words with the energy required to remove water from a volume to replace it with a non-interacting molecule. Hydrophobic bonds describe the fact that oil and water don't mix. They are essential to the maintenance of the three-dimensional structure of membranes and proteins.

II. STRUCTURE-ACTIVITY RELATIONSHIPS (SAR's) IN TERMS OF SUB-STRUCTURAL FEATURES

One of the cornerstones of modern organic chemistry is the power of structure-property relationships. For this analysis the chemist synthesizes and measures the property of interest in a set of compounds that systematically vary the structure of the original "lead" molecule. Structure-biological activity relationships (SAR) are usually very subtle. Until the 1960's this was the only method for analogue design (Wolff, 1980).

In this approach to SAR's, molecules are described in terms of the sub-structures of which they are composed. Respective SAR's then simply specify the sub-structures associated with the desired activity. Similar relationships can be formulated for toxicity, undesirable side effects, or metabolic properties. In order to derive such SAR's one prepares many analogues of the lead in which the perceived sub-structures are systematically modified or deleted. For example, if the lead contains a basic nitrogen one may choose to investigate the biological properties of analogues in which the nitrogen has been replaced by an oxygen, sulfur, or carbon atom. Similarly, methyls, hydroxyls, or halogens may be replaced by hydrogens. One would in this manner arrive at a notion of the minimum structural features associated with activity and if activity and toxicity are parallel or divergent.

In addition, SAR's may also pertain to positions on the molecule at which substitutions may be made without loss of activity. One needs to identify these regions of bulk tolerance because it is there that many groups will be tested to arrive at a better and safer drug.

Typically more than one sub-structure is required for biological activity. Accordingly, one may want to derive SAR's for the dependence of potency on the spacing between these groups. Thus, analogues might differ in the length of an aliphatic chain between required groups or in the substitution pattern of an aromatic or aliphatic ring.

Yet another point of interest is whether one group of atoms in the lead can be replaced by another group of atoms that are considered to be biologically equivalent, bioisosteric. For example, a methylene in a chain might be replaced by an ethereal oxygen or sulfur atom; a carboxylic acid group by a sulfonic acid or a tetrazole; or a benzene ring by a thiophene. Chapter 6 describes the use of bioisosteres in more detail.

For example, Table 1 shows several early analogues of the monoamine oxidase (MAO) inhibitor pargyline 1 (Martin, et al., 1975). Compounds 2, 3, and 8 show that the phenyl is required for maximum potency; 4 and 8 that the (N) methyl is not required; 5 and 6 that the benzylic carbon is not required; and 13 that the acetylenic hydrogen is not required. Furthermore, the inactivity of 12 suggests that 8 is the minimum essential structure for inhibition of MAO. Accord-

Table 1 Example of structure-activity relationships revealed through systematic modification of the drug pargyline

$$X - Y - \underset{\underset{R^1}{|}}{N} - R^2 - Z - R^3$$

Cmpd. No.	X	Y	R^1	R^2	Z	R^3	pI_{50}
1	∅	CH_2	CH_3	CH_2	C≡C	H	6.0
2	c-C_6H_{11}	CH_2	CH_3	CH_2	C≡C	H	5.4
3	H	CH_2	CH_3	CH_2	C≡C	H	3.7
4	∅	CH_2	H	CH_2	C≡C	H	5.1
5	∅	$(CH_2)_5$	H	CH_2	C≡C	H	5.3
6	∅	$C(CH_3)_2$	CH_3	CH_2	C≡C	H	5.2
7	-	∅	H	CH_2	C≡C	H	<3.0
8	-	H	H	CH_2	C≡C	H	4.0
9	∅	CH_2	C_2H_5	CH_2	C≡C	H	3.1
10	∅	CH_2	CH_3CH_2	$CH(CH_3)$	C≡C	H	<3.0
11	∅	CH_2	CH_3	CH_2	C≡N	-	3.6
12	∅	CH_2	CH_3	CH_2	CH=CH	H	<3.0
13	∅	CH_2	CH_3	CH_2	C≡C	CH_3	5.0
14	2-OEt-∅	CH_2	CH_3	CH_2	C≡C	H	7.5

ingly, subsequent synthesis of analogues concentrated on the modification of groups a and b. This resulted in a thirty-fold enhancement of the potency of 1, compound 14.

SAR's also lead to hypotheses as to the molecular mode of action of compounds. For example, the requirement for the triple bond suggested that these compounds irreversibly inhibit MAO. This was verified by the observation that 1 could not be washed away from the enzyme. On the other hand, the dramatic effect of changing from 1 to 9 suggested that the compounds act initially as alternate substrates for MAO. This suggestion was also experimentally verified.

The outcome of an SAR analysis is a notion of the pharmacophore, the minimum structure required for the compound to possess the particular biological property. Compound 8 is the pharmacophore in the example. The correctness of the hypothetical pharmacophore depends on the design of the series. As with any hypothesis, a pharmacophore can never be proved, only disproved. It is useful to the extent that it suggests new compounds for synthesis, new biological tests, or a new way to interpret the SAR.

Because the practice of deriving an SAR is so empirical, it meets with varying success depending on the creativity and energy of the practitioner. The hard work in the case of pargyline came from the decision of the medicinal chemist to test, as inhibitors of monoamine oxidase, as many amines as possible. Special emphasis was placed on unsaturated or electron-rich compounds. The lead, 8, was only marginally active. However, the creativity of the medicinal chemist led to the incorporated the benzyl group of a substrate, benzylamine, into the molecule. The result was the first potent non-hydrazide monoamine oxidase inhibitor,

4. Its N-methyl analogue became the drug pargyline, 1. It is much safer than the previously used hydrazides.

The empirical nature of SAR analysis and the corresponding large number of compounds required to find a better and safer drug led to the derisive term "molecular manipulation" as a description of the medicinal chemistry until the 1960's. However, one must remember that such methods led to the major drug therapies against bacterial infection as well as cardiovascular and mental disease; a dramatic extension of the life-expectancy in cancer; the safe induction of anesthesia; and reliable, inexpensive, convenient, non-intrusive, and safe birth control. The accomplishments of chemists armed with such crude methods suggest that, when armed with much finer tools, the rate and perfection of the resulting therapies should be overwhelming.

Informative though such SAR analysis may be, it has several serious limitations. The most overwhelming is that the number of analogues needed is more than any one group can reasonably accomplish. There are always additional analogues to prepare. Furthermore, different workers would define the structure differently and so synthesize different analogues. If they are your competitors, they might find a better compound. A second limitation is that there is no easy way to summarize the information gained since each compound supplies independent information. To the more theoretically minded, a third and very serious limitation is that it does not explain the observed potency differences between compounds in terms of atomic-level details of the kinetics and thermodynamics of the drug-biomolecule interaction. Firm predictions of activity of an untested compound can be done reliably only when one has such an understanding.

III. QSAR: CORRELATION BETWEEN PHYSICAL PROPERTIES AND BIOLOGICAL POTENCY

The first step toward the revolution in medicinal chemistry came with the realization in the 1960's by Hansch and Fujita that one should examine not structure-activity, but rather physical property-activity relationships (Hansch, et al., 1963). The problem is that any one structural change alters many properties. How does one decide which properties are relevant to biological potency? To do so each chemical structure is transformed into a list of physical properties: the relationship between these physical properties and biological properties is examined in the computer (Martin, 1978; Seydel and Schaper, 1979; Franke, 1980; Martin, 1981; Topliss, 1983; Fujita, 1984). This method is named linear free energy or QSAR (Quantitative Structure Activity Relationships).

QSAR is based on Hammett's demonstration in the 1930's of the parallel influence of a particular substituent on the chemical properties of a variable parent molecule. For example, a para chloro group lowers the pKa of three different kinds of acids: benzoic acid, (15); phenol, (16); and benzene sulfonic acid, (17). Thus one can use the measured effect of a remote substituent on one chemical reaction to predict the effect of the same substituent on a different but related reaction. Once several analogues have been studied in the two reactions, a correlation line can be drawn and the effect of yet another substituent on the second reaction can be predicted from its effect on the first.

A. QUANTITATIVE DESCRIPTORS OF PHYSICAL PROPERTIES

Scientists have devised linear free energy descriptors for the effect of substituents on the ability of a molecule to participate in each of the types of interactions described above (Martin, 1978; Hansch and Leo, 1979).

1. Descriptors of the effects of remote substituents on the tendency of a common group to participate in electrostatic interactions. This is the Hammett σ constant. It is defined by the logarithm of the equilibrium constant for the ionization of a substituted benzoic acid in water at 25°C compared to that for the unsubstituted benzoic acid. The σ value for -H is defined to be 0.0; hence, the sigma values of other substituents may be defined by the following:

$$\log K_X = \log K_H + \sigma$$

$$\sigma = \log K_X - \log K_H = \log \frac{K_X}{K_H} \quad (1)$$

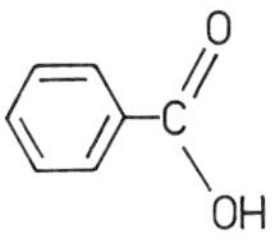

15 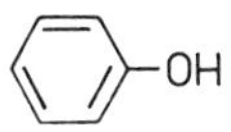 16 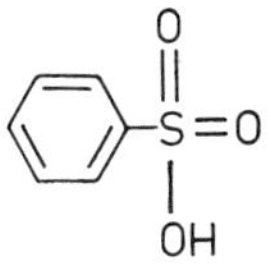 17

Table 2 lists the σ values of some common substituents.

It has been shown that the substituent effect on a very large number of chemical properties is well correlated with σ:

$$\log K'_X - \log K'_H = \rho\sigma \quad (2)$$

In Equation 2 the ρ value describes the sensitivity of the reaction to the electron-attracting or withdrawing properties of substituents. For benzoic acids at 25°C, ρ=1.0 by definition. When one methylene group is inserted between the phenyl and the carboxylate, the ρ value falls to 0.5; with two it falls to 0.2. Thus ρ

is a quantitative measure of the electronic involvement between a site of substitution and a constant reaction site.

Many chemical reactions have been correlated with σ: acid-base pK_a's of carboxylic acids, amines, oximes, and phenols; oxidation-reduction potentials; rates of esterification of acids and hydrolysis of esters and amides; rates of aromatic substitution; rates of alkylation of amines; and ^{1}H, ^{13}C, and ^{19}F magnetic resonance chemical shifts.

However, it has been observed that normal σ constants do not describe the effects of substituents on ionization of phenols. In particular, the acid strengthening effect of the para nitro group is underestimated because of the direct resonance interaction between the nitro and phenolate groups, 18. One may solve this problem is by using special σ^- constants for the effect of substituents on reactions in which there

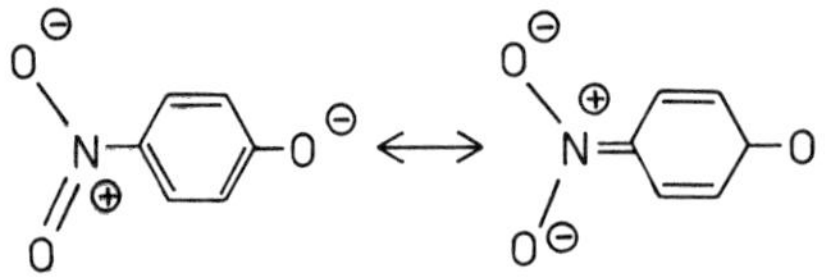

18

can be direct conjugation with an electron excessive site. By analogy, σ^+ constants are defined for reactions in which an electron deficient site is in direct conjugation with the substituent, and σ^* for substituents on an aliphatic chain. Alternatively, scientists have partitioned σ into σ_I, which describes the inductive-field effect of the substituent, and, σ_R, which describes the resonance effect. Table 2 lists some of these σ values also.

σ, σ^+, and σ^- constants are additive: each substituent makes an independent contribution to the

Table 2 σ Constants for typical substituents (Hansch and Leo, 1979)

Group	σ_m	σ_p	σ^+_p	σ^-_p	σ^*
H	0.00	0.00	0.00	0.00	0.49
Br	0.39	0.23	0.15	0.28	2.80
Cl	0.37	0.23	0.11	0.27	2.94
Me	-0.07	-0.17	-0.31	-0.15	0.00
NH_2	-0.16	-0.66	-1.30	-0.15	----
NO_2	0.71	0.78	0.79	1.24	----

electronic nature of the constant part of the molecule. Accordingly, when the series contains multiple sites of substitution, the σ constants can be simply summed. However since the proportion of inductive and resonance influence is not constant by position, one cannot simply sum σ_I and σ_R constants over all positions.

One may also use measured electronic QSAR descriptors such as acid dissociation constants (pKa's), NMR chemical shifts, or oxidation-reduction potentials. Aside from the work of measuring such properties, the use of measured descriptors limits the predictive value of the resulting equations. However, they may provide important information for selection of appropriate σ values. A very simple example of the correlation between pKa and potency is shown with the antibacterial sulfonamides, 19, Figure 6.

$H_2N-C_6H_4-SO_2NHR$

19

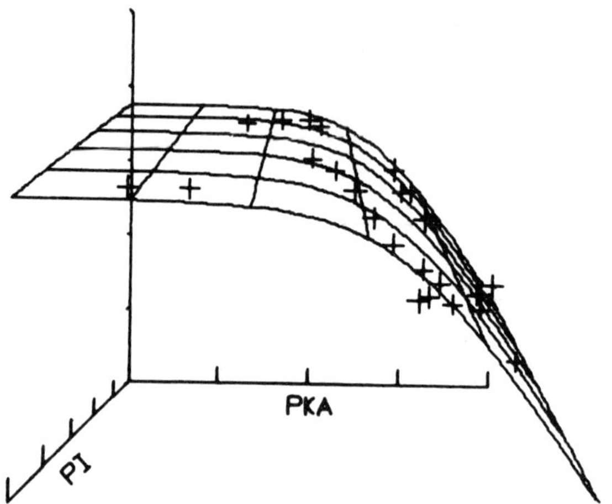

Figure 6 The antibacterial potency of sulfonamides as a function of π and pKa. The theoretical curve was generated by a fit to a compartment-model based equation:

$$\log(1/C) = -\log([H+]/Ka + 1) - 0.19\,\pi - 0.12.$$

$$R^2 = 0.85,\ s = 0.18,\ n = 24.$$

The methods and results are discussed in Martin, 1980.

2. Descriptors of substituent effects on the tendency of molecules to participate in dispersion interactions. In QSAR terms the molar refractivity, MR, of a molecule is used. Tabulated values are available; some are listed in Table 3. The values are additive at one position, so computation is straightforward.

The major limitation to the use of molar refractivity as a descriptor of the tendency of a molecule to participate in dispersion interactions is that MR is frequently highly correlated with the size of the substituent. If there is a steric interference with the drug-biomolecule interactions, the correlation equations may show a negative correlation between potency and MR.

Molar refractivity values are not typically summed over all positions, but rather separate variables are used for each position.

3. Descriptors of substituent effects on the hydrophobicity of molecules. Modern QSAR became possible only when Hansch and Fujita realized that a particular substituent makes a rather constant contribution to the logarithm of the oil-water partition coefficient. By analogy with the definition of σ constants, Hansch and Fujita defined the π value:

$$\pi_X = \log \frac{P_X}{P_H} \tag{3}$$

P_X and P_H refer to the octanol water partition coefficient of the neutral form of the substituted and unsubstituted molecules, respectively. Typical π values are listed in Table 3. As expected, ionization lowers partition coefficient by approximately one-thousand fold.

Table 3 Hydrophobic, steric and dispersion constants for typical substituents (Hansch and Leo, 1979)

Group	π_{aro}	π_{ali}	E_s	B_1	B_4	L	MR
H	0.00	0.00	0.00	1.00	1.00	2.06	1.03
Cl	0.71	0.39	-0.97	1.80	1.80	3.52	6.03
Me	0.56	0.50	-1.24	1.52	2.04	3.00	5.65
Et	1.02	----	-1.31	1.52	2.97	4.11	10.30
tBu	1.58	----	-2.78	2.59	2.97	4.11	19.62
OH	-0.67	-1.12	-0.55	1.35	1.93	2.74	2.85
NH_2	-1.23	-1.19	-0.61	1.50	1.84	2.93	5.42
NO_2	-0.28	-0.85	-2.52	1.70	2.44	3.44	7.36

If a molecule contains several functional groups, they may interact with each other and solvent to violate the simple additivity rules for calculation of log P. Thus it is not always possible to calculate the octanol-water logP of a molecule unless that of a related molecule is known. However, there is a computer program, CLOGP, that calculates logP from a set of empirical rules devised by Leo from examination of thousands of measured values (Leo and Weininger, 1987). Much effort of theoretical and physical organic chemists is still used for the investigation of molecular properties that predict hydrophobicity and solvation.

As an illustration of the significance of log P for the biological properties of compounds, Figure 7 shows the correlation of the rate of buccal absorption of aliphatic acids with log P.

4. Descriptors of substituent size. The earliest linear free energy descriptor of substituent size is the E_s constant defined as the logarithm of the relative rate of the acid-catalyzed hydrolysis of acyl-substituted methyl acetate:

$$E_{s,X} = \log k_{XCO_2Me} - \log k_{MeCO_2Me} \tag{4}$$

Since E_s values of spherical substituents are correlated with their atomic radii, one may calculate the E_s value of other substituents. A more sophisticated description of substituent size is included in the constants of Verloop and Tipker who reported values for the length (L), minimum width (B_1), and maximum width (B_4) of many substituents. Table 3 lists the values of these constants for some common substituents. Austel and Charton have shown that steric effects can also be described by simple counts of the number of heavy atoms or branches in a substituent. More elaborate, yet simple, graph theoretical descriptions of molecules

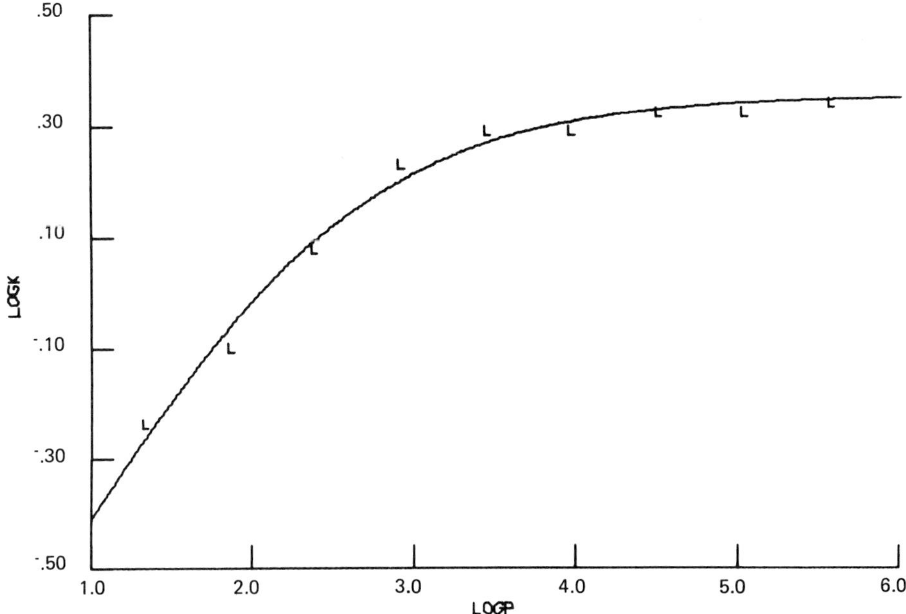

Figure 7 The rate of absorption of compounds from the buccal cavity of the mouth as a function of the partition coefficient of the compound. The theoretical curve was generated by a fit to a compartment-model based equation:

$$0.56 \log(k) = -0.89 + 0.56 \log P - \log(1 + 0.058(P)^{0.56})$$

$R^2 = 0.99$, $s = 0.027$, $n = 9$. See the legend to Figure 6 for the reference.

have also been proposed by Simon and colleagues and by Hall and Kier. Such description might include positive steric effects (dispersion interactions) as well.

The accurate linear free energy description of steric effects for biological reactions has not been totally satisfactory. One reason for this failure is that bulky substituents may both diminish productive alignment of the drug and target biomolecule and also change the relative percentage of the various conformations of the drug, only one of which is the biologically active. An additional complication is that steric repulsion increases roughly with the twelfth power of distance between atoms and that there is typically a very close fit between the drug and the relatively inflexible target. The result is that steric repulsion is extremely sensitive to the precise size and shape of the molecules and that a minor change in the size of a substituent may move one from an active to an inactive molecule. It is because of these deficiencies that much attention has recently been paid to computer calculation of conformation to be discussed in Part III of this chapter.

5. Indicator or all-or-none descriptors. The power of SAR analysis suggests that one also include variables for the presence or absence of specific functional groups. For example, if a series of amines included only primary and N,N-dimethyl analogues, such an indicator variable could be used to describe this difference. Indicator variables are also used to describe potential sites for drug-receptor interaction that are not described by σ, MR, or E_s. For example, a hydrogen-bond donor group at a particular position in the molecule may appear to increase or decrease potency. Such properties are included in the data matrix with values of 1.0 (present) or 0.0 (absent).

The maximum use of indicator variables is to use no physical property descriptors, but to correlate potency with the presence or absence of particular functional groups at different positions. This is called the Free-Wilson method. To use it one must be sure that each descriptor applies to more than one molecule and that no two descriptors are always present simultaneously.

6. Descriptors of the biological potency of the molecules. In accord with the derivation of the

descriptors of the physical properties of the molecules, the biological properties are also expressed whenever possible by some value that is related to the logarithm of the relative rate or equilibrium constant for interaction with the biological target. For in vitro measurements, one would typically have actual dissociation constants for the compounds. The situation is more complex with in vivo assays. For such cases, one typically chooses to correlate with the negative logarithm of the dose required to produce some predetermined response, for example the negative logarithm of the molar ED_{50}. For simplicity in the equations and discussion that follows, the biological parameter is denoted as log(1/C).

B. PRINCIPAL METHODS OF DERIVING STRUCTURE-ACTIVITY RELATIONSHIPS (Martin, 1978)

1. Tables. Typically one would list the compounds from most potent to least potent. The listing of the associated physical properties might be enough to reveal a relationship between a physical and a biological property.

2. Plots. Examples of such plots are shown in Figures 6 and 7. Plots can show both the data and the function fit to reveal the closeness between observed and predicted values. Plots, however, do not provide statistical information on the agreement between observed and calculated values. If several properties appear to be related to potency, how does one distinguish between them?

3. Regression analysis. This is the process of least squares fitting of a function to a set of data. Several properties may contribute to relative potency: multiple regression can easily handle such cases. In addition, the statistical evaluation of the fit is part of the analysis.

The regression analysis establishes the values of a, b, c, ρ, and e in the Hansch equation, Equation 5:

$$\log 1/C = a + b(\log P) - c(\log P)^2 + \rho\sigma + eE_s \quad (5)$$

For any data set the value of any one or more of the fitted coefficients may be not statistically significantly different from zero. In a typical regression analysis each combination of physical properties is used to predict the biological activity. For each combination one calculates the significance level of each coefficient and that of the overall equation. Only equations in which each term is significant are considered further. The goodness of fit of the equation to the data is measured by two statistics, R^2 and s. The R^2 is the fraction of the variance in the data that is explained by the equation. Hence, an R^2 of 1.00 indicates a perfect fit of the data to the equation whereas one of 0.50 indicates that only 50% of the variance in the data is explained by the equation. A complementary statistic is s, the standard deviation of the observed values from those predicted by the equation. The data are overfit if an equation has an s value smaller than the standard deviation of duplicate determinations of biological potency. The QSAR equaions are shown in the legends of Figures 6 and 7.

4. Discriminant analysis. Often the bioactivity is a classification such as active *vs* inactive. Discriminant analysis statistically evaluates which combination of physical properties assigns compounds into the correct activity classes. The resulting discriminant function may then be used to predict the classification of new compounds. For example, discriminant analysis of the inhibition of monoamine oxidase by monomethoxy and monohydroxy analogues of **20** suggested that active compounds are characterized by a small R

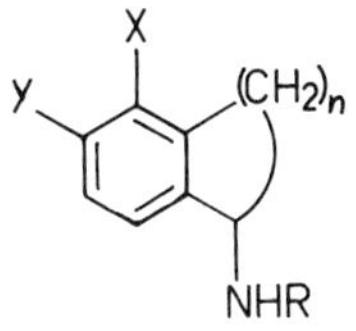

$$\text{20}$$

substituent and X = H and Y = OH or OMe (Martin, et al., 1974). The function correctly classified 19/20 of the compounds. Thus discriminant analysis is an extension of regression analysis.

5. Principal components, partial least squares, SIMCA, and factor analysis. These methods are used to study the relationships between physical or biological properties of a set of molecules. They investigate the question, in the correlation matrix of n descriptors, how many independent properties are there? If there are fewer independent properties than dexcriptors, then the set may not be well designed. If the analysis yields different answers for the whole data set than for only active compunds, it suggests that a certain combination of the properties is associated with biological activity. The method of partial least squares, PLS, extracts successive linear combinations of physical properties that best predict bioactivity (Wold, et al., 1984).

6. Cluster analysis. Cluster analysis is a non-statistical method that is used to study the relationship between compounds based on their physical or biological properties. Clusters are formed on the basis of the distances between compounds in a space formed by the physical or biological properties. Each compound is a point in this space. Distance may be calculated several different ways but a common measure is the Euclidean distance, or the root mean square of the sum

of squared differences in each property. Various clustering programs then group compounds that are close to each other into clusters of supposedly similar members.

One application of cluster analysis is to prepare a data matrix of all compounds that one may possibly synthesize, run the cluster analysis, and then actually synthesize only one compound from each unique cluster.

C. ROLE OF QSAR IN THE DESIGN OF BETTER AND SAFER DRUGS

QSAR is the only drug design method currently available that allows one to make a quantitative prediction of the potency of a new analogue. There are literally thousands of QSAR equations reported in the literature, at least a hundred examples of the correct prediction of the biological potency of a molecule before its synthesis, and two compounds on the market that were designed by QSAR (Martin, 1981; Topliss, 1983; Fujita, 1984). Thus such methods have a place in modern drug research.

QSAR may be used for the design of a series to follow up a lead. With the synthesis of a set of analogues with uncorrelated combinations of electronic, hydrophobic, and steric properties, more information is gained per compound. One may also by such means probe a new region on the receptor and thus start of a whole new series. We would have missed a series of high ceiling diuretics, 21, if we had continued to synthesize hydrophobic analogues similar to the first set of compounds.

If one finds a QSAR of predictive value, it may happen that the optimum compound has already been made and that none is appropriate for advanced testing. Thus QSAR may help one to decide to stop synthesis in a series. For example, in our explorations of analogues

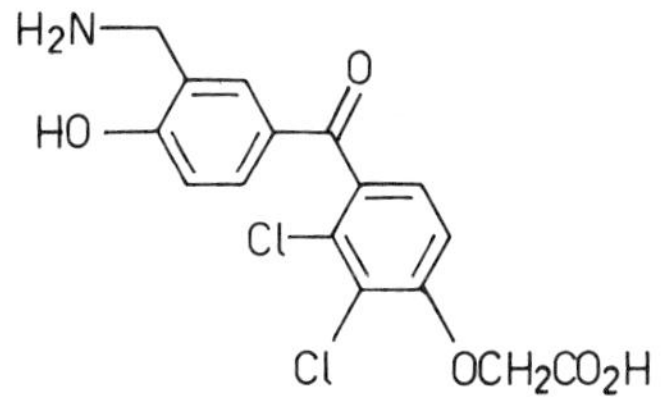

21

of erythromycin, 22, the QSAR analysis suggested that we could not enhance in vitro potency by variation of substitution at the 2', 4", 9, 11 or 12 positions. In spite of work in many laboratories around the world, the in vitro potency of erythromycin has been enhanced at most three-fold.

Finally, a QSAR may provide testable hypotheses of the details of the mechanism of action of the molecules and thus complement receptor mapping techniques.

Various technical mistakes can lead to a flawed QSAR. The two most common are the examination of too many physical properties as possible predictors and the deletion of outliers from an apparent relationship.

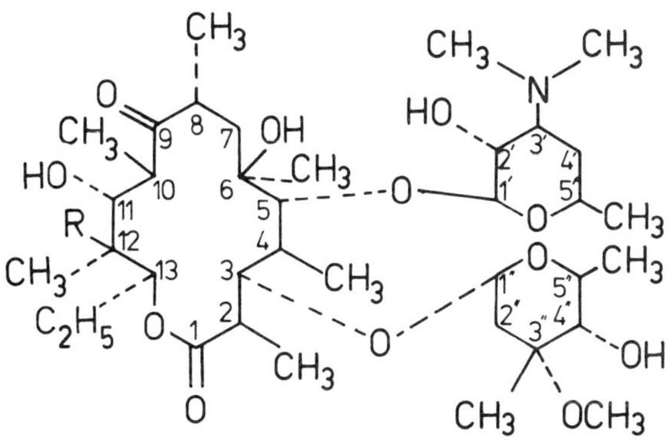

22

Either action increases the apparent fit of the data to the relationship but sacrifices the predictive value of the relationship.

However, QSAR techniques also suffer from a fundamental flaw. The interaction between a drug and a biomolecule is an interaction in three-dimensional space. Thus, any method that does not explicitly consider three-dimensions will ultimately reach the limit of its predictive power.

IV. THE ANALYSIS OF THREE-DIMENSIONAL PROPERTY-ACTIVITY RELATIONSHIPS TO MAP BINDING SITES ON BIOMACROMOLECULES

Since the binding site on a macromolecule for a small molecule is three-dimensional, a powerful probe of the structure of such a binding site is the three-dimensional strücture of the small molecules that are recognized by it (Olson and Christofferson, 1979; Gund, et al., 1980; Humblet and Marshall, 1981; Hopfinger, 1985). Academic medicinal chemists have synthesized conformationally restricted analogues to probe receptor structure since the 1960's. However, interest in these methods increased when two industrial groups reported novel compounds designed from such considerations using mainly crystal structures and mechanical models (Humber et al., 1979; Olson et al., 1981). However, only since 1982 have the molecular graphics and computational chemistry tools needed to do this easily been available to industrial chemists.

A. GENERAL STRATEGY FOR MAPPING BINDING SITES

The central concept of binding site mapping (called receptor mapping sometimes) is that any small molecule, ligand, that binds to a protein must both physically fit into the binding site and it must have no chemical properties that prevent binding. When tight binding is seen, it is expected that the regional chemical properties of the ligand are complementary to those of the binding site. Since from radioligand binding it is known that one ligand may displace another from its binding site, for binding site mapping it is proposed that all members of a family of related molecules interact with the same region on the biomacromolecule. Thus any region in space open to a substituent on the bound conformation of any active analogue must be open to a substituent on a proposed analogue in its bound conformation (Humblet and Marshall, 1981; Martin and Danaher, 1988).

Consequently one's strategy is to discover the atoms necessary for recognition by the target biomolecule, to propose the bound conformation of each active ligand, to superimpose all active ligands in such a way as to maximize their chemical and shape resemblance, and to calculate the shape of the union of all active molecules. This volume encloses the minimum volume available to a new ligand. The chemical criteria used for superposition lead to proposal of complementary chemical properties in the biomolecule. Compounds will be inactive if they cannot assume an active conformation or if when doing so they collide with the biomolecule. Thus inactive molecules provide valuable information for the binding site map.

Such mapping is most appropriately applied to in vitro assays of potency such as radioligand binding or inhibition of an enzyme. Since the data are interpreted in terms of structure, the assays should be designed to measure interaction of ligands with only one macromolecular structure.

Since the basic tools for binding site mapping are so new, the techniques will undoubtedly change and new ones emerge as a result of experience, technical advances in three-dimensional structure analysis of macromolecules, and greatly enhanced computer power. While we sense the dawn of a new era of medicinal chemistry, we do not know its details. Certainly the new era will not discard the previous approaches of structure-activity analysis and QSAR, but rather it will streamline such work and use it within a larger context.

Binding site mapping helps the medicinal chemist design new molecules with greater specificity by using maps of several related receptors or enzymes. The resulting compounds are expected to be better and safer because they will interact with greater affinity with their target and with much less affinity with other biomolecules.

Binding site mapping also allows the chemist to prepare molecules that are both more likely to be active and more structurally novel than conventional approaches. Thus there is a better pool from which to select a clinical candidate. This results in better and safer drugs.

B. MOLECULAR GRAPHICS

To do binding site mapping one must be able to conveniently display and manipulate three-dimensional molecular structures. This capability was provided by the development of affordable color computer display screens that allow one to rotate the molecule, set of molecules, or parts of a molecule in real time, to zoom up on parts of interest, and to see it in stereo (Langridge, et al., 1981). One usually can blink or dash certain parts of the display, change the color of molecules and their surfaces, move one part of the display with respect to the other, etc.

Molecular graphics software also allows one to do chemical operations such as to build a molecule from parts; to rotate about bonds; to calculate distances between atoms and relative energies of different conformations; to show only the molecular backbone or the backbone plus a surface; to superimpose molecules over user-defined pairs of atoms; to display chemical properties calculated by quantum mechanics; and to calculate and display the union, intersection, and difference between regions in space occupied by different sets of molecules.

The advantage of molecular graphics over traditional molecular models are: that the computer model is precise and will not be changed by gravity or a curious co-worker, and if it is changed, the two versions can be compared; that it can be used to superimpose many molecules; that it is coupled to theoretical chemistry capabilities; and that chemical properties can be displayed overlaid on the molecular backbone.

C. ENERGY REFINEMENT OF A THREE-DIMENSIONAL STRUCTURE

Molecular graphics would be of no advantage for drug design if one could not also accurately calculate the structure and relative energies of several conformations of a molecule, as well as its simple chemical properties. Theoretical chemistry methods to do this were already mature and waiting for the required computer power and real-time color graphics to be widely used.

One usually starts a calculation with a provisional three-dimensional structure of the molecule built from an X-ray structure of a related molecule or from standard bond angles and bond lengths such as one builds a hand-held molecular model. Two main methods have been used for the refinement of the structure; quantum mechanical and potential energy. Today, potential energy calculations are the most commonly applied, with quantum mechanical methods used mainly to derive information required for the parameterization of the potential energy functions.

Different potential energy methods differ in the details, but in essence they describe a molecule as a set of balls (nuclei) connected by springs (bonds) (Burkert and Allinger, 1982). Energy is required to move each fragment of the molecule from its ideal value and an energy minimum represents the best compromise. The bond lengths are described by stretching functions, the bond angles by bending functions, the rotation about bonds by torsional potential functions, steric attraction and repulsion by distance functions, hydrogen bonds (often) by a special function, and electrostatic attraction and repulsion by a Coulombic function.

Empirical parameters describe the dependence of energy on deviation from ideal values as well as the ideal values themselves. They are derived from many types of experimental information or ab initio quantum chemical calculations. In essence, potential energy methods supply a composite of all experimental and theoretical information as it applies to the structure of the molecule under investigation. The differences between the various methods lie in the types of molecules used to parameterize the functions, in the exact form of the functions themselves, and in the type of minimizer used to find the minimum energy conformation. Structure refinement is simply the minimization of energy as a function of the coordinates of the molecule. For a molecule the size of apomorphine (27) minimization may take approximately 10 minutes. Hence it is possible to examine many conformations.

Molecular dynamics calculations are used to explore atomic motion. They use the same type of energy function but instead of minimizing the energy they move to a new conformation by solving Newton's laws of motion. The results are often shown as a movie.

If parameters are not available for potential energy calculations on the molecule of interest, then quantum mechanical calculations are sometimes used. Semi-empirical methods such as CNDO/2 or AM1 may give quite good results. Such a calculation may take as long as 25 hours on a super-mini computer to optimize the structure of a drug-sized molecule, or an hour or so to calculate the energy of a single structure. Ab initio calculations would take at least five times longer.

D. DISCOVERY OF THE BOUND CONFORMATION OF LIGANDS AND THE SHAPE OF THE BINDING SITE

1. Identification of the key atoms required for binding. The first hypothesis as to the pharmacophore is usually derived from classic SAR. If loss of certain atoms leads to no substantial loss in potency, then the deleted atoms are not part of the pharmacophore. For example, in the case of the dopaminergic aminotetralins, 23 and 24 bind strongly to the D_2 receptor and are full agonists whereas 25 is much less potent. Thus the meta hydroxyl is part of the pharmacophore whereas the para is not.

The pharmacophore atoms need not be identical in all biologically related compounds, rather they must have similar chemical properties. An example of this is that a number of ergoline derivatives 26 are also dopaminergic. It is proposed, and structure-activity studies support this, that the indolic -NH is the hydroxyl equivalent. Such hypotheses can be examined by molecular graphics displays of the results of quantum chemical or potential energy calculations. Figure 11 shows the regions around three dopaminergic compounds at which a carboxylate oxygen is calculated to preferentially bind. In addition, one could compare for all active molecules displays of the type shown in Figures 3 and 4 to arrive at hypotheses of the chemical properties required for activity.

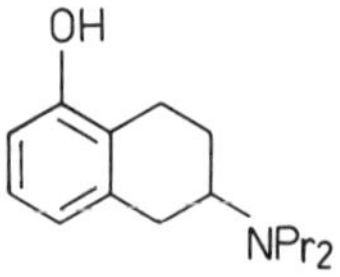

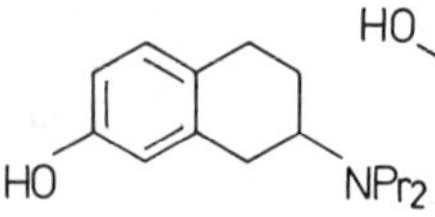

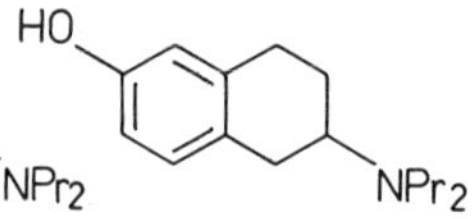

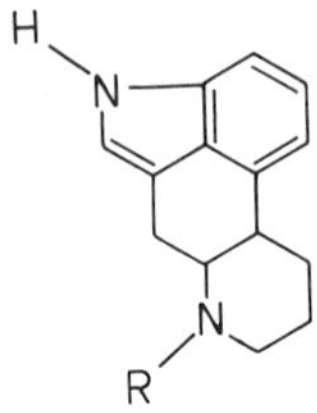

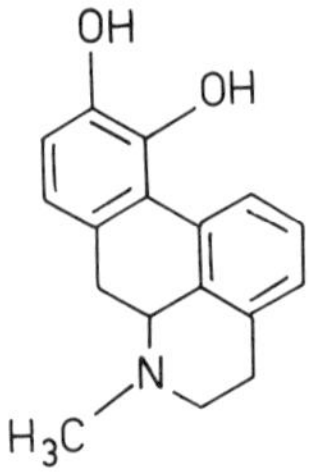

2. Sources of the receptor-bound conformation or distances between the key atoms. Figure 2 shows that single-crystal X-ray diffraction of the ligand-macromolecule complex reveals the bound conformation. Sometimes one may propose the active conformation from

the X-ray structures of a related ligand when bound to a related enzyme as in the design of the converting enzyme inhibitor captopril (Petrillo and Ondetti, 1982).

Experimental observation of the bound conformation of a ligand may also be approached by nuclear magnetic resonance, NMR. If the conformational and tautomeric equilibration of the ligand is slow compared to the rate of its binding to the macromolecule, the structure depleted from solution upon binding is probably close to that when bound (Fesik, 1988). For example, in solution the sugar KDO exists as two slowly interconvertible pyranoses, Figure 8. That in the center of the figure is depleted when the enzyme KDO-CMP synthetase adds a CMP group to the OH at C1. This led to the synthesis of the stable analogue shown at the right of the figure. It is an inhibitor because it binds to the enzyme, but lacks the essential OH for further processing.

The bound conformation may also be fairly well established if the ligand is not flexible. For example, apomorphine, 27, is a potent D_2 agonist. The only conformational flexibility in this molecule is

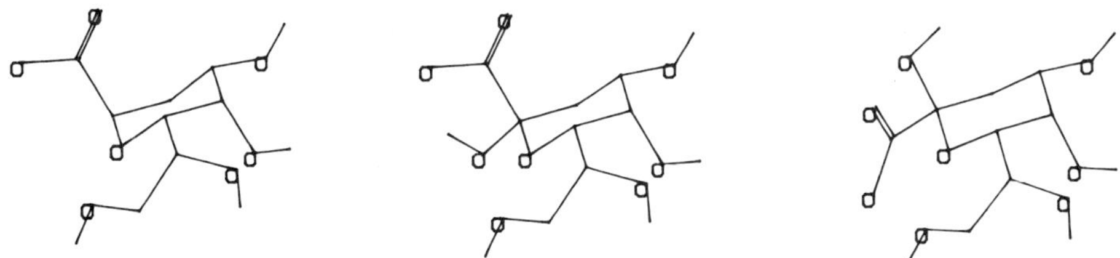

Figure 8 The two anomers of KDO that interconvert, but slowly are shown at left and center. That in the center is depleted from solution by KDO-CMP synthetase. The analogue at the right of the figure is an extremely potent inhibitor of KDO-CMP synthesis.

rotation of the phenolic groups, inversion of the N, and ring puckering shown in Figure 9. The design and synthesis of such rigid and semi-rigid analogues is one of the key strategies in receptor mapping. A rigid analogue itself might result in a better and safer drug if rigidification makes it no longer recognized by undesired receptors or if it prevents metabolic degradation.

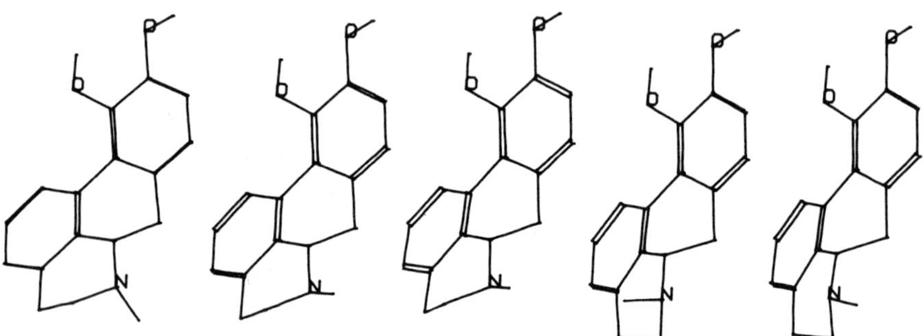

Figure 9 The various conformations of apomorphine. The crystal structure is shown in the center of the figure.

If none of the above definitive experiments is possible, then one must search by some exhaustive technique over all conformations of all active compounds to discover if only one of the arrangements of the key atoms is found in all compounds (Humblet and Marshall, 1981). If so, then the conformation leading to it is the proposed active conformation. If several possible bound conformations are found, then synthesis and testing rigid analogues of each may help settle the question.

3. Comparison of the chemical and shape properties of molecules to derive the binding site map. For the correct superposition of molecules, one must super-

impose the chemical properties of the molecules. In the earliest publications, molecules were superimposed so that corresponding ligand atoms were overlapped. A more accurate approach is to superimpose imaginary target atoms that have been placed at the appropriate position near the ligand. For example, one might place an O^- of a proposed carboxylate 2.8 Å away from a hydrogen bonding hydrogen atom of the drug molecule in line with the electronegative drug atom. Potential energy calculations may be used to locate such proposed binding sites around a ligand. One would evaluate the energy of interaction between various probe atoms representative of various typical binding groups found in proteins and the small molecules of interest. Figure 10 shows the favorable binding sites around the three D_2 agonists.

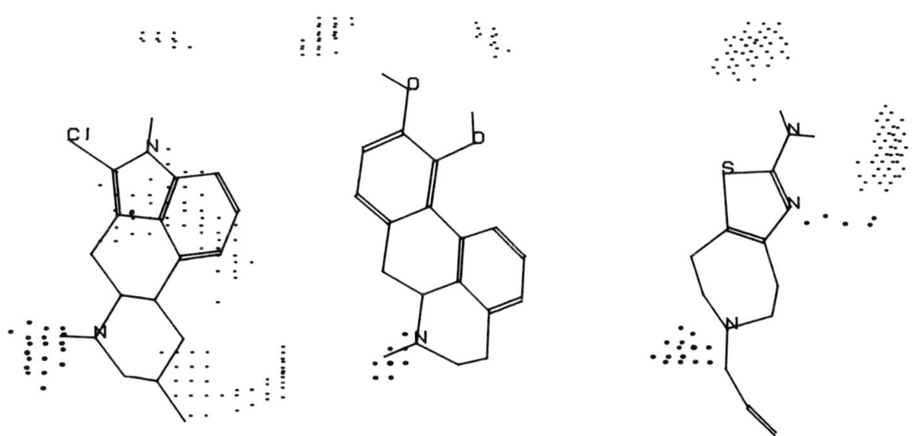

Figure 10 A plot of the potential energy interaction of an O^- (fine dots) and an NH_3^+ (heavy dots) with three dopamine agonists. Notice that all three have strong hydrogen bond donor sites at the top of the figure and hydrogen bond acceptor sites at the position of the lone pair on the nitrogen atom.

Once the active molecules are superimposed, the surface that encloses all active molecules is calculated. It is the hypothetical boundary of the binding site. Inactive molecules frequently protrude outside these boundaries. Their new regions in space define forbidden regions, or regions required by the target biomolecule. The final map of the binding site

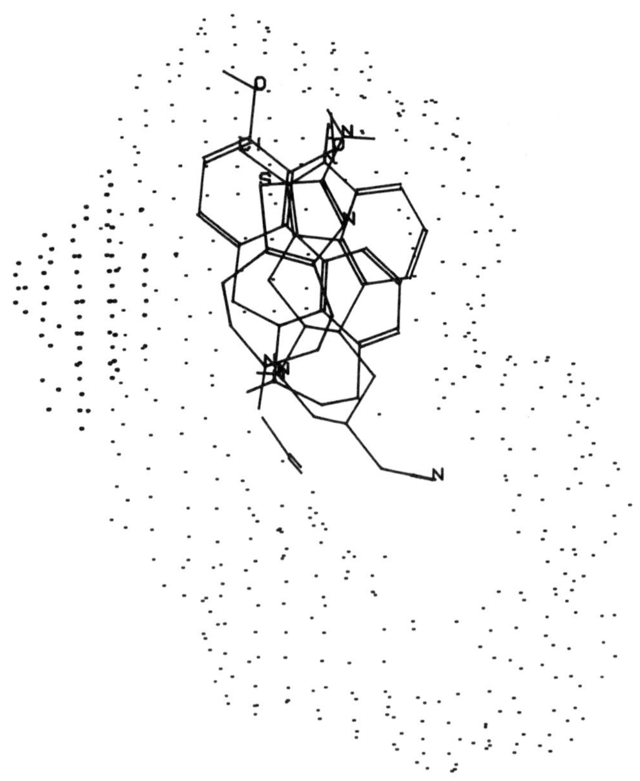

Figure 11 A clipped view of the three dopamine agonists superimposed in the proposed receptor map for D-2 agonists. The surface that encloses all potent agonists is shown in fine dots and the new volume occupied by inactive molecules is shown in heavy dots.

includes the boundary of space examined, the proposed positions of key atoms in the biomolecule, and new regions in space occupied by inactive molecules. Figure 11 shows our proposed receptor map for D_2 agonists and Figure 12 a comparison of receptor maps for two receptors that bind dopamine.

A binding site map can be used to propose the active conformation, enantiomer, or superposition of molecules not yet in the map. Qualitative structure-activity relationships plus the enantiomeric selectiv-

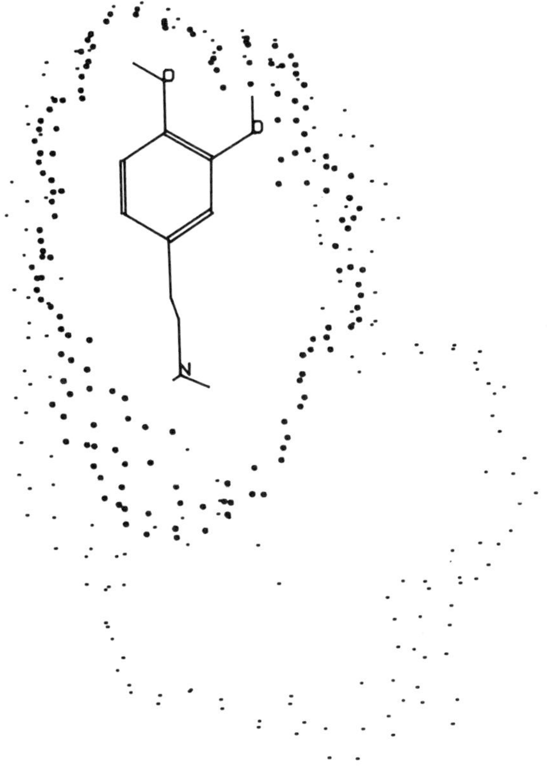

Figure 12 The boundary of the proposed receptor map for α_2 adrenergic agonists (heavy dots) compared with that for D_2 agonists (fine dots).

ity are considered in the ultimate superposition of molecules. Figure 13 shows an example of a questionable fit of a compound to a map.

Binding site mapping thus allows the chemist to, in a shorter time, prepare structurally novel active

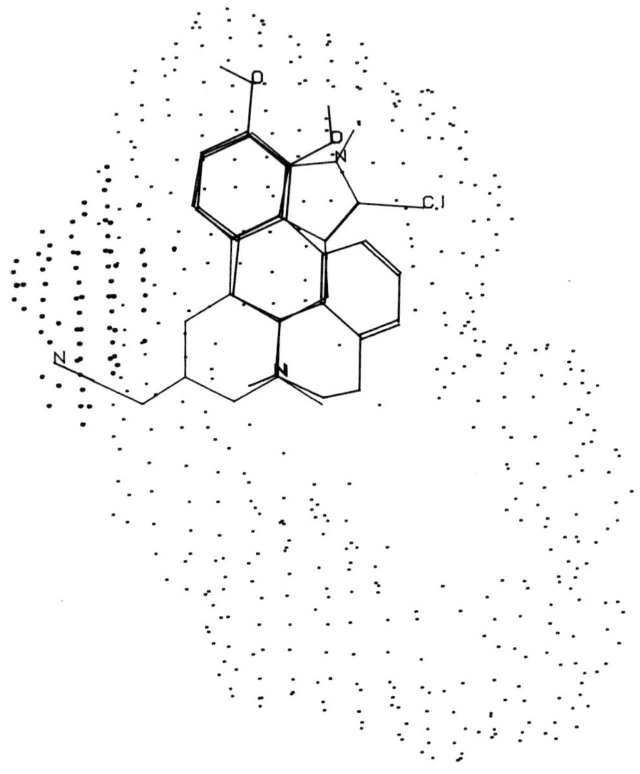

Figure 13 A clipped view showing how the receptor map can be used to evaluate proposed methods of superposition. In fine lines is shown lergotrile superimposed on apomorphine, heavy lines, so that the aromatic rings superimpose. Notice that with this superposition lergotrile penetrates into the space occupied by inactive compounds. Hence, with this superposition we would predict it to be inactive. Since it is active, the correct superposition must be that shown in Figure 11.

ligands. Better and safer drugs result because selectivity is built in and more chemical families are available for advanced testing.

V. STRUCTURE OF THE BIOLOGICAL MACROMOLECULE-DRUG COMPLEX

A. BACKGROUND

If the three-dimensional structure is available for the biomolecular target of therapeutic interest, then a very direct drug design could be used. What is known about the structures of the biological macromolecules that are the targets for drug action has been reviewed by two pioneers of the field (Beddell, 1984; Goodford, 1984).

DNA or RNA is the target of a few drugs, and the three-dimensional structure of the drug-DNA complex of several of them have been solved by X-ray crystallography. However, theoretical studies of nucleic acids are difficult because of the large percentage of charged groups and counter ions as well as their many possible conformations.

Proteins as enzymes or receptors are more commonly the target of drug action. Although proteins are made up of a linear sequence of amino acids, their three-dimensional structure is compact. Proteins are conformationally flexible; however, each exists in only a few conformations which establish an exact three-dimensional structure of the binding sites for small molecules and other proteins. The atoms of these binding sites may come from amino acids that are widely separated in the (one-dimensional) amino acid sequence of the protein.

All of the methods described below are heavily dependent on molecular graphics and computational power (Venkataraghavon and Feldmann, 1985).

B. SOURCES OF THE STRUCTURES (Cantor and Schimmel, 1980; Fesik, 1988).

1. Single-crystal X-ray diffraction. This is the most powerful technique available for the determination of the three-dimensional structure of a macromolecule. Although only a few hundred protein structures are known, new ones are being solved at a rapidly increasing pace due to advances in instrumentation and computer hardware and software.

Protein crystals are heavily solvated and show little protein-protein contact except between subunits of one structure. Thus the crystal structure of a protein is not much different from that in solution.

For crystallography one must have at least milligram supplies of the protein: cloning techniques may provide the sample but sometimes post-translational processing is required or the synthetic protein may not fold into the active conformation. Obtaining suitable crystals may involve an empirical investigation of such factors as added salt, co-solvent, and temperature. Finally, to solve the structure, one usually needs diffraction data from at least one heavy-atom derivative that crystallizes isomorphously with the original protein. In the 25 years since the first protein structure determination there have been major advances in the accuracy and speed of the instrumentation available. One gets better data faster. For example, to collect the data that identified the structure of a bound antiviral compound to an intact rhinovirus involved a 2.25 minute exposure to X-rays in a synchrotron. In addition, area detectors and powerful computers allow one to more rapidly collect and manipulate the experimental data to produce the electron density contours. In short, although it is a large experimental task to establish the three-dimensional structure of a protein the task becomes simpler every year.

From the original data a tentative three-dimensional structure is made using computer graphics. To do this, the contours of electron density are displayed on a computer screen with special molecular graphics programs designed for protein crystallography. The amino acid chain (built of standard bond lengths and angles) is then threaded through the regions of electron density. The rotatable bonds in the structure are manipulated using dials. Special attention is paid to well-defined secondary structures such as alpha-helices and beta-sheets in that the sequence of amino acids selected to be in these regions must be consistent with the requirements of these secondary structural elements.

This preliminary structure is refined with mathematical functions to maximize the correspondence between the observed electron density and that calculated from the model structure. The result is an accurate model of the three-dimensional structure of the protein. Neutron diffraction experiments on the same crystals provide the precise location of the hydrogen atoms.

Protein crystallography may also give a hint of the flexibility of the molecule. There might be two molecules in the unit cell, each of which has a different conformation. In any one structure, some atoms may appear to move more than others; this is expressed as a temperature factor, hotter atoms move more. In addition, during the solution of the protein structure one may observe regions of partial occupancy, that is certain atoms are in one relative position in certain molecules in the crystal and at another position in another molecule. Finally, in some proteins whole regions of the sequence are not seen in the crystal. Those regions are so flexible that no order is seen.

It is often a simple task to use the information on the structure of the native protein plus diffraction data of the complex to determine the structure of a bound ligand and the changes in protein structure that result from such binding.

The power of X-ray crystallography is that, as seen in Figures 2 and 5, one has at hand the atomic-level details of drug-macromolecule interaction. The shape and chemical properties of the atoms in the drug binding site do not have to be inferred, they are observed directly. The disadvantage is that the effort requires substantial investment of time and energy. The target must be carefully chosen and its biological function well characterized so that by the time the structure is solved it is still of interest for drug design.

2. Nuclear magnetic resonance spectroscopy, NMR. NMR can also be used to establish the three-dimensional structure of a protein. Various experimental protocols are used to produce measurements of the time-averaged distance between atoms in the structure; if enough distances are unambiguously assigned, then a unique structure is determined. The distances observed in NMR, plus the atomic connectivities established from the amino acid sequence, are then processed by computer programs. Either molecular dynamics or distance

geometry is used to produce structures that match the experimental observations. If several approximately equal energy structures meet the criteria, then one may need to do more experiments to distinguish between them. On the other hand, one may find that no one structure matches all the distances and that the solution conformation is the average of one or more conformations. Thus to establish the structure of a protein by NMR one also uses the tools of computational chemistry and molecular graphics.

The NMR determination of structures is currently limited to small proteins of molecular weight ca. 10,000. In contrast to crystallography, one need not find crystallation conditions. However, the protein must be fairly soluble in some suitable solvent. The solution of a protein structure by NMR is probably no less work than solving it by X-ray diffraction.

3. Comparative modeling. Enough protein structures have been solved by crystallography that we now recognize that there are families of proteins that are homologous in three-dimensional structure. Figure 14 shows the superposition of three serine proteases. Notice that the general tracing of the chains follow very closely. In particular, these proteins are essentially identical in the arrangement of atoms in the catalytic site; the subtle differences are in the regions nearby that establish the difference in specificity of these enzymes. If one had the amino acid sequence of yet another serine protease it seems logical to model the structure of it from the structures of its relatives.

To do comparative modeling one aligns the amino acid sequences so that the conserved amino acids line up. The three-dimensional structure of the conserved regions in the unknown protein is formed from the three-dimensional structure of those known. The variable loops of the unknown protein are then modeled from the protein that is closest to it in length or chemical character of that loop. By such a means one can arrive at a tentative structure that can be energy minimized and also form the basis of biochemical, molecular biology, or NMR experiments to further refine the model.

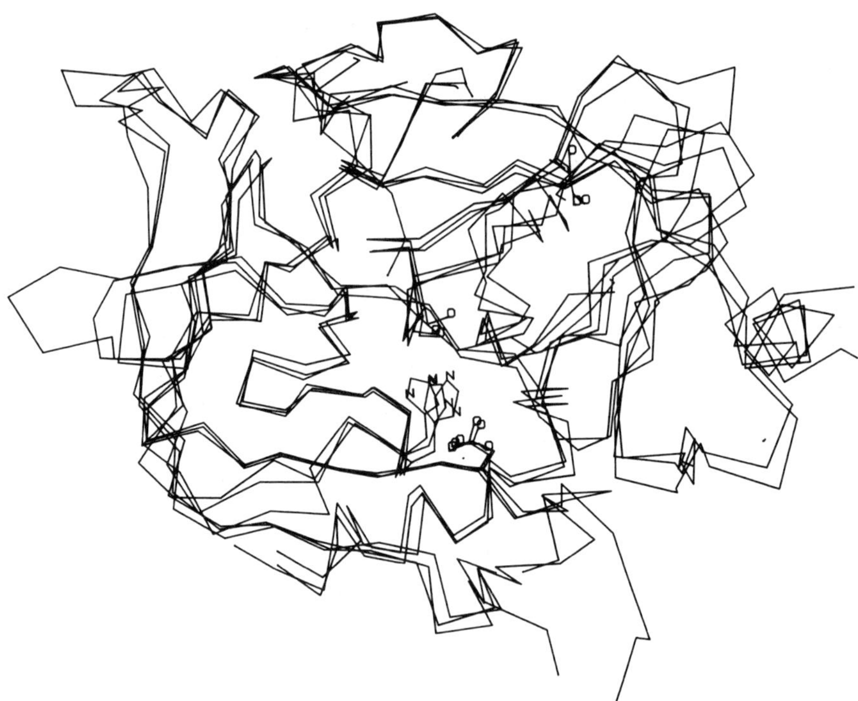

Figure 14 The serine proteases chymotrypsin, elastase, and kallikrein. The proteins are overlapped such that the active-site serine and histidines are superimposed.

C. TECHNIQUES FOR THE DESIGN OF THE LIGAND

1. Mechanical models and molecular graphics. As noted above, the solution of the structure of the protein or protein-small molecular complex involves either mechanical models or, these days, computer graphics. Such pictures of the molecules allow one to see how to modify the structure of a known ligand so that it will bind more tightly. One identifies sites for possible additional hydrogen-bonding or electrostatic interactions. If the program also calculates molecular surfaces, then optimum dispersion interactions may also be anticipated.

By this means a thyroxine analogue, 28, that binds more tightly to pre-albumin than the parent, 29 (Blaney, et al., 1982); and a trimethoprim, 30, analogue, 31, that binds more tightly to dihydrofolate reductase (Kuyper, et al., 1985) were discovered. In the former case additional dispersion interactions were added, and in the latter, an additional electrostatic interaction. A more dramatic design was of molecules, 32 and 33, that bind to the diphosphoglycerate, 34, site of hemoglobin but form primary interactions to different protein atoms than does the natural effector (Beddell, 1984; Goodford, 1984). This same group also designed compounds, 35 and 36, that prevent the sickling of sickle-cell hemoglobin by using a binding site for which there is no known natural effector.

Molecular graphics provides no measure of the relative energetics of the various modes of binding of the various possible compounds. Only chemical intuition is available to suggest which potential binding sites should be exploited.

Additionally, a graphics display of a crystal structure alone gives little indication of possible

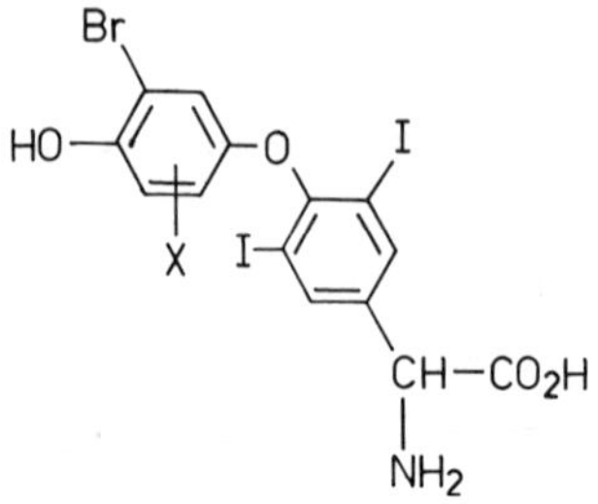

Br, HO, O, I, I, X, $CH-CO_2H$, NH_2

28 $X = 2,3-C_4H_4-$

29 $X = Br$

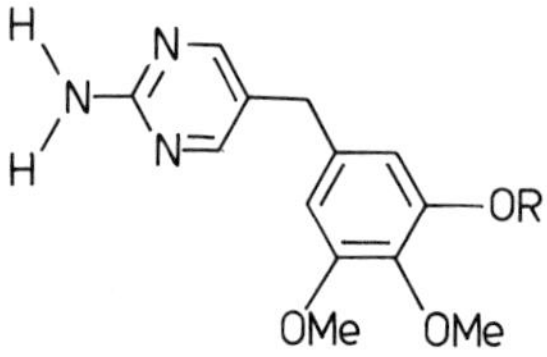

32 $R = CHO$

33 $R = CH(OH)-SO_3H$

$H_2O_3PO-CH_2-CH(CO_2H)-OPO_3H_2$

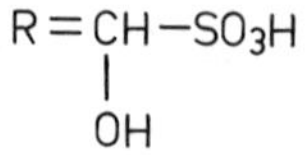

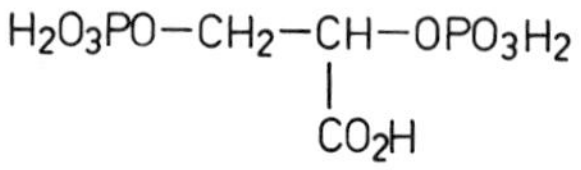

34

H, N, H, N, N, OR, OMe, OMe

30 $R = Me$

31 $R = (CH_2)_5-CO_2H$

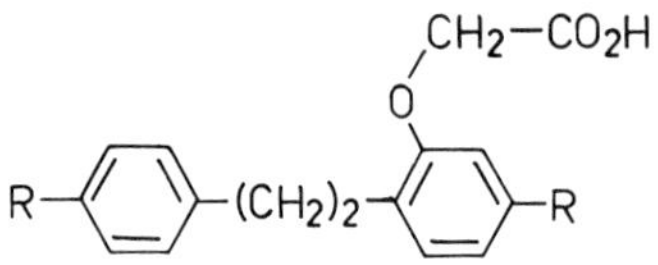

CH_2-CO_2H, CHO

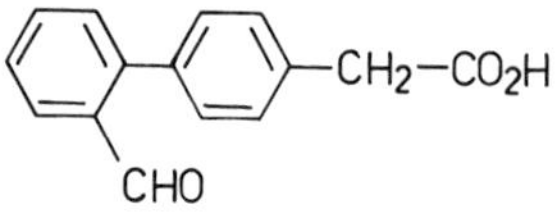

35

CH_2-CO_2H, O, R, $(CH_2)_2$, R

$O-(CH_2)_4-CO_2H$, HO, CHO

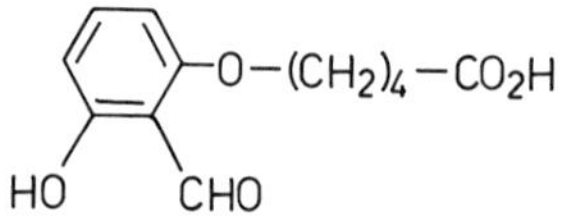

36

movement of atoms of either the protein or the proposed drug when binding occurs.

2. <u>QSAR plus molecular graphics</u>. If one has, in addition to the protein structures, a QSAR equation involving the separation of descriptors by position, then molecular graphics can show one the structural meaning of the descriptors. For example, Selassie et al. (1986) showed that the QSAR for inhibition of dihydro-

folate reductase by a series of trimethoprim analogues was interpretable by molecular graphics of the enzyme inhibitor complex.

3. Computer calculations. Favorable sites of interaction of proposed drug atoms with a protein can be established by exploring the energy of interaction of the protein with various typical atoms or groups such as a carboxylate O^-, OH, CH_3, NH as in tryptophan, and NH_3^+ (Goodford, 1985). Ligand atoms would be placed at several of these sites and molecules designed to hold these atoms in these positions. Others have devised computer programs that match the shape of a binding site with the shape of a possible ligand (Kuntz, et al., 1982).

Molecular mechanics minimization and dynamics calculations do allow atoms to move; minimization, to the nearest minimum energy structure and dynamics, with a motion simulated to the real short-time scale motions of a system. Although the framework is similar to small molecule molecular mechanics, the scale of the calculations is much larger. One minimization calculation can take as long as days and dynamics, months. Special neglect of less important atoms may sometimes be used. Special techniques are now being explored that promise to allow one to calculate the binding free energy, that is the difference between the relative energy in solution and when bound to the protein.

These energy programs for proteins are still under development, however. It is only recently that there have been computers large enough and structures accurate enough to evaluate them. Current research is evaluating the functional forms of some of the relationships, parameters for more types of ligand atoms, how to treat electrostatics, what to do about the water-protein interface and the bulk water surrounding the protein, and how to realistically explore the possibilities of gross change in conformation or relative orientation of the two molecules.

QSAR calculations may be useful to be sure the proposed molecules have appropriate lipophilicity for good distribution to tissues and absorption from the gut. QSAR analysis may also be used with graphics and energy calculations to suggest new substituents on an existing molecule.

VI. SUMMARY AND OUTLOOK

In this chapter we have seen that a macromolecule recognizes a small molecule by a subtle combination of electrostatic, hydrogen bonding, dispersion, steric repulsive, and hydrophobic interactions. Often the structure of the target biomolecule is unknown. However, structure-activity modification of the small molecule can be used to probe the basis for any particular molecular pair. Such analyses are much more powerful if the interacting partners are thought of in terms of physical properties. More recently it has become possible to map the shape and chemical properties of the target biomolecule from the corresponding properties of its ligands. Finally, the experimental observation at atomic resolution of the entire drug-macromolecule complex is possible in some cases today. Each of these methods uses concepts perfected by the others and each offers insights that can be used directly or indirectly by the medicinal chemist in the discovery of better and safer drugs.

REFERENCES

Beddell, C., Designing Drugs to Fit a Macromolecular Receptor, Chem. Soc. Rev. 13, 279-319 (1984).

Blaney, J.M., P.K. Weiner, A. Dearing, P.A. Kollman, E.C. Jorgensen, S. Oatley, J.M. Burridge, C.F. Blake, Molecular Mechanics Simulation of Protein-Ligand Interactions: Binding of Thyroid Hormone Analogues to Prealbumin, J. Am. Chem. Soc. 104, 6424-6434 (1982).

Burkert, U. and N.L. Allinger, Molecular Mechanics, Amer. Chem. Soc., Washington (1982).

Cantor, C.R. and P.R. Schimmel, Biophysical Chemistry, Freeman, San Francisco (1980).

Fesik, S.W., Approaches Towards Drug Design Using NMR Spectroscopy, in "Computer-Aided Drug Design", (T. Perun and C. Propst, eds.), Marcel Dekker, New York (1988), in press.

Franke R., Optimierungsmethoden in der Wirkstofforschung:, Akademie-Verlag, Berlin (1980).

Fujita, T., The Role of QSAR in Drug Design, in "Drug Design: Fact or Fantasy", (G. Jolles and R.H. Wolldridge, eds.), Academic Press, New York, 19-33 (1984).

Goodford, P., Drug Design by the Method of Receptor Fit, J. Med. Chem., 27, 557-564 (1984).

Goodford, P.J., A Computational Procedure for Determining Energetically Favored Binding Sites on Biologically Important Macromolecules, J. Med. Chem., 28, 849-857 (1985).

Gund, P., J.D. Andose, J.B. Rhodes, and G.M. Smith, Three-Dimensional Molecular Modeling and Drug Design, Science, 208, 1425-1431 (1980).

Hansch, C. and A. J. Leo, Substituent Constants for Correlation Analysis in Chemistry and Biology, Wiley, New York (1979).

Hansch, C., R.M. Miur, T. Fujita, P.P. Maloney, F. Geiger, and M. Steich, The Correlation of Biological Activity of Plant Growth Regulators and Chloromycetin Derivatives with Hammett Constants and Partition Coefficients, J. Am. Chem. Soc., 85, 2817-2824 (1963).

Hopfinger, A.J., Computer-Assisted Drug Design, J. Med. Chem., 28, 1133-1139 (1985).

Humber, L.G., A.H. Philipp, F.T. Bruderlein, M. Gotz, and K. Voith, Mapping the Dopamine Receptor: Some Primary and Accessory Binding Sites, in "Computer-Assisted Drug Design", (E.C. Olson and R.E. Christoffersen, eds.), Amer. Chem. Soc., Washington, 227-241 (1979).

Humblet, C., and G.R. Marshall, Three-Dimensional Computer Modeling as an Aid to Drug Design, Drug Dev. Res., 1, 409-434 (1981).

Jolles, G., and K. Woolridge, eds., Drug Design: Fact or Fantasy?, Academic Press, New York (1984).

Kuntz, I.D., J.M. Blaney, S.J. Oatley, R. Langridge, and T. Ferrin, A Geometric Approach to Macromolecule-Ligand Interactions, J. Mol. Biol., 161, 269-288 (1982).

Kuyper, L.F., B. Roth, D.P. Baccanari, R. Ferone, C. Beddell, J. Champness, D. Stammers, J. Dann, F. Norrington, D. Baker, and P. Goodford, Receptor-Based Design of Dihydrofolate Reductase Inhibitors: Comparison of Crystallographically Determined Enzyme Binding with Enzyme Affinity in a Series of Carboxy-Substituted Trimethoprim Analogues, J. Med. Chem., 28, 303-311 (1985).

Langridge, R., T.E. Ferrin, I.D. Kuntz, and M.L. Connolly, Real-Time Color Graphics in Studies of Molecular Interactions, Science, 211, 661-666 (1981).

Leo, A.J., and D. Weininger, CLOGP, Pomona College Medicinal Chemistry Project; Claremont, California (1987).

Martin, Y.C., W.B. Martin, and J.D. Taylor, Regression Analysis of the Relationship Between Physical Properties and the In Vitro Inhibition of Monoamine Oxidase by Propynylamines, J. Med. Chem., 18, 883-888 (1975).

Martin, Y.C., in "Quantitative Drug Design. A Critical Introduction", Marcel Dekker, New York (1978).

Martin, Y.C., The Quantitative Relationships Between pKa, Ionization, and Drug Potency: Utility of Model-based Equations, in "Physical Chemical Properties of Drugs", (S. Yalkowski, A.A. Sinkula, and S.C. Valvani, eds.), Marcel Dekker, New York, 49-110 (1980).

Martin, Y.C., and E.B. Danaher, Molecular Modeling of Receptor-Ligand Interactions, in "Receptor Pharmacology and Function", (M. Williams, R.A. Glennon, P. Timmermans, eds.), Marcel Dekker, New York, 137-171 (1988).

Martin, Y.C., A Practitioner's Perspective of the Role of Quantitative Structure Activity Analysis in Medicinal Chemistry, J. Med. Chem., 24, 229-237 (1981).

Martin, Y.C., J.B. Holland, C.H. Jarboe, N. Plotnikoff, Discriminant Analysis of the Relationship between Physical Properties and the Inhibition of Monoamine Oxidase by Aminotetralins and Aminoindans, J. Med. Chem., 17, 409-413 (1974).

Olson, E.C., and R.E. Christoffersen, eds., Computer-Assisted Drug Design. Amer. Chem. Soc., Washington (1979).

Olson, G., H-C. Cheung, K.D. Morgan, J.F. Blount, L. Todaro, L. Berger, A.B. Davidson, E. Boff, A Dopamine Receptor Model and Its Application in the Design of a New Class of Rigid Pryyolo[2,3-g]isoquinoline Antiphyschotics, J. Med. Chem., 24, 1026-1034 (1981).

Perutz, M.F., Electrostatic Effects in Proteins, Science 201, 1187-1191 (1978).

Petrillo, E.W. and M.A. Ondetti, Angiotensin-Converting Enzyme Inhibitors: Medicinal Chemistry and Biological Actions, Med. Res. Rev. 2, 1-41 (1982).

Selassie, C.D., Z-X. Fang, R. l. Li, C. Hansch, T. Klein, R. Langridge, B. Kaufman, Inhibition of Chicken Liver Dihydrofolate Reductase by 5-(Substituted benzyl)-2,4-diaminopyrimidines. A Quantitative Structure-Activity Relationship and Graphics Analysis, J. Med. Chem., 29, 621-626 (1986).

Seydel, J.K., and K.J. Schaper, Chemische Struktur und biologische Aktivitat von Wirkstoffen, Verlag Chemie, Weinheim (1979).

Tanford, C. The Hydrophobic Effect: Formation of Micelles and Biological Membranes, Wiley, New York (1980).

Topliss, J.G., ed., Quantitative Structure-Activity Relationships of Drugs, Academic Press, New York (1983).

Venkataraghavan, B., and R.J. Feldmann, *Proc. N.Y. Acad. Sci*., 439, (1985).

Wold, S., W.J. Dunn, III, and S. Hellberg, Pattern Recognition as a Tool for Drug Design, in "*Drug Design: Fact or Fantasy*", (G. Jolles, eds.), Academic Press, New York, 95-115 (1984).

Wolff, M.E., ed., *Burger's Medicinal Chemistry*, *4th ed*., Wiley, New York (1980).

5

The Importance of Biotechnology for the Discovery of Better and Safer Drugs

P. Swetly

Ernst-Boehringer-Institut für Arzneimittelforschung
Dr. Boehringer-Gasse
Vienna, Austria

I. INTRODUCTION

One of the revolutions in modern biology is the discovery of techniques for the insertion of specific segments of the DNA of one species into that of another, a process called DNA recombination or molecular cloning. Most commonly, the DNA that forms a mammalian gene is inserted into the DNA of a bacterium or yeast. Since every species uses the same code to translate the DNA into proteins, this results in the subsequent expression of the donor DNA to produce proteins characteristic of the donor. These methods are now so commonplace that laboratory manuals are available on the techniques (Maniatis, et al., 1982). A related set of techniques allows one to cultivate a monoclonal antibody, that is an antibody preparation in which every antibody molecule has the same molecular structure.

Many of the advances in modern biology depend greatly on the use of biotechnology. For example, one reason that protein crystallography is so exciting these days is that it is possible to prepare a sufficient sample quantity of any protein that one wishes to study. Alternatively, site-specific mutagenesis methods can be used to modify a protein structure at any site: this is an essential tool for the investigation of the relationship between structure and function of enzymes and other proteins. Even if biotechnology did not more than provide better insight into the functioning of biological systems, one could not overemphasize the importance of these techniques for the discovery of better and safer drugs.

However, biotechnology techniques are also used commercially. Although only five such products are used in human therapy in 1987, in principle such methods should allow the production of material for the replacement or increased reproduction of any protein in the human body that is not functioning properly. Since proteins provide the essential architecture and function of the cell, this means that therapy for basically any disease may be possible. The replacement proteins may be either those that occur naturally or mutants designed to have improved properties such as greater stability. In this chapter we will explore the opportunities and limitations of these techniques for the discovery of better and safer drugs.

II. PRODUCTION OF HUMAN PROTEINS, THE FIRST GENERATION OF NEW DRUGS FROM RECOMBINANT DNA

A. CLONING OF MAMMALIAN GENES IN MICROORGANISMS BY RECOMBINANT DNA TECHNOLOGY

Almost every gene is now clonable. This is demonstrated by such achievements as the cloning of hemophilic factor VIII which will soon be available for administration to hemophiliacs. Cloning a gene may require tremendous effort, but success is virtually guaranteed.

The genome of an organism can be thought of as a library that contains the complete construction plans for that organism. In the case of the human genome, this library represents some 3000 books of 1000 pages each. Each page represents one gene. Each gene is an ordered sequence of DNA bases that code for both the sequence of one protein and the signals necessary for the regulated production of this same protein. At least one copy of the entire genome is present in every somatic cell of the organism.

In 1953 Watson and Crick recognized that the mechanism by which DNA replicates is inherent to the structure of the DNA itself. More recently it has been recognized that DNA also directs gene expression, the process during which selected stretches of DNA are transcribed into proteins. The mechanism by which DNA directs gene expression is not understood but is an active area of molecular biology research. Scientists are exploring how it is that a cell responds to external or metabolic signals by the transcription of a series of distantly spaced genes or by the transcription of only one of a family of closely clustered genes.

The basic methodology of making recombinant DNA is shown in Figure 1: donor DNA is cut at specific sites by one of a class of enzymes known as restriction endonucleases. The result is the formation of DNA fragments, one of which contains the gene of interest. Alternatively, if the DNA sequence of the gene is known, one could chemically synthesize a mutant DNA that would code for a mutant protein. The DNA fragments are then attached (ligated) into a plasmid DNA that can replicate in the host microbial cell strain. The recombinant DNA plasmids are introduced into the host cells by a process called transformation. Inside the host cells the recombinant DNA plasmids replicate to produce multiple copies of the donor DNA. Bacteria or yeast are the most common hosts for recombinant DNA cloning.

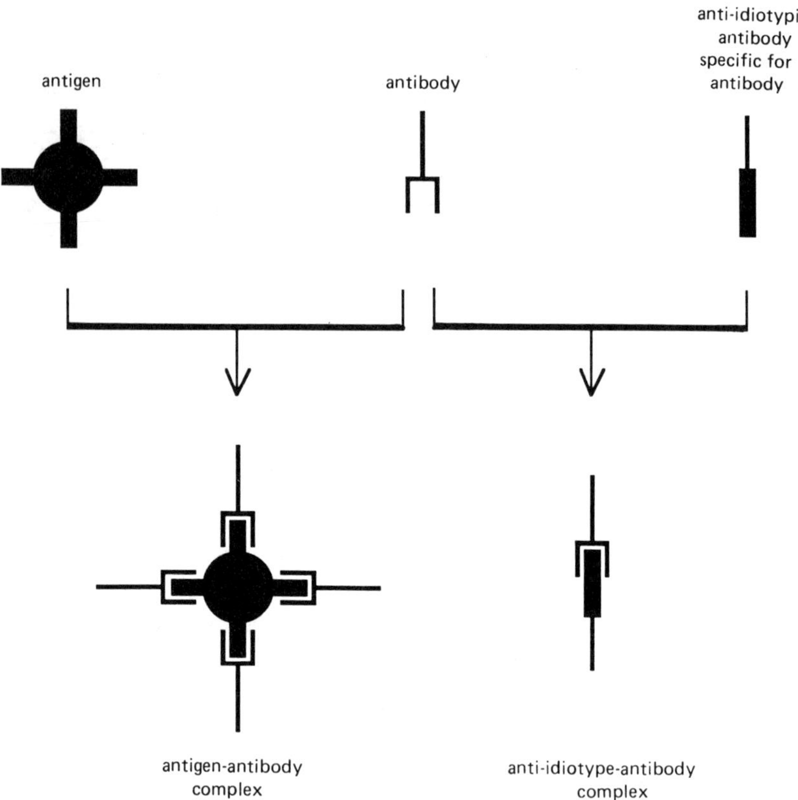

Figure 1 Anti-idiotypic antibodies mimic the configuration of the antibody binding site on the original antigen.

B. EXPRESSION OF PROTEINS FROM GENES CLONED IN MICROBIAL CELLS

Bacteria and yeasts are the most commonly used organisms for the large scale production of proteins. They are chosen because they grow to high densities with a short generation time and can therefore produce large quantities of proteins. For the production of a human protein in a microorganism, the cloned gene must be expressed in that microorganism. Each species has a specific signal that regulates gene expression, so usually one must insert a host regulatory sequence into the recombinant plasmid DNA near the inserted human DNA. By such means the expression of the human gene is under host control and can often be manipulated by the scientist by changes in temperature or composition of the medium.

Although several groups have reported that they have produced a system in which 10% of the protein produced by the microorganism is that derived from the recombinant DNA, finding the correct conditions to accomplish good expression is still empirical.

Once the desired protein is expressed it may be secreted by the recombinant organism into the extracellular medium. Although the yield may be lower than if the protein were not secreted, secretion is often desirable because it makes it easier to purify the product. Secretion usually requires a suitable "signal peptide" sequence on the N-terminal of the protein. The signal sequence labels the polypeptide for the secretion through the bacterial cell membrane and cell wall. During transport the signal sequence is cleaved and the desired protein produced.

C. FUSION OF ANTIBODY-PRODUCING CELLS WITH IMMORTAL CELLS TO PRODUCE MONOCLONAL ANTIBODIES

Antibodies are produced and secreted by B-type lymphocytes. Any given B-lymphocyte produces only one type of antibody protein, but the same antigen stimulates different B-lymphocytes to produce different antibodies. B-lymphocytes cannot divide and they do not live long enough in a cell culture to produce large amounts of antibody. For this reason it was impossible, until recently, to study a homogeneous population of antibody molecules.

Koehler and Milstein (1975) demonstrated that one can fuse an antibody-producing B-lymphocyte from the spleen of an immunized animal with an immortalized myeloma (cancer) cell that grows easily in tissue culture. The resulting hybridoma can be maintained permanently in cell culture and produces the one type of antibody that is characteristic of the B-lymphocyte from which it was derived. Thus hybridomas allow one to produce large quantities of highly specific monoclonal antibodies. These antibodies are homogeneous and their production is subject to much stricker environmental control than is the production of polyclonal antibodies produced from the serum of immunized animals. Thus it is possible that these techniques will result in better and safer vaccines for use in man.

Stable hybridoma cell lines have been derived from the fusion of mouse or rat myeloma cells with B-lymphocytes from spleen cells from immunized mice or rats. Although these antibodies have good reactivity to antigens, they are not suitable for human immunization because of the expected allergic reaction to the mouse or rat protein.

The major biomedical use of monoclonal antibodies is currently in diagnostic tests for disease and to measure the blood or urine level of therapeutic or abused drugs. Since it is possible to produce an antibody against virtually any chemical and to select the hybridoma that produces the antibody with exactly the required specificity and affinity, such tailored monoclonal antibodies are the key reagents in a large number of improved diagnostic tests.

D. THE EXPRESSION OF PROTEINS IN EUCARYOTIC TISSUE CULTURES

Certain proteins cannot be obtained from yeast or bacteria but must be obtained through cultivation of eukaryotic cells. This is because they require post-translational modifications, such as covalent attachment of carbohydrates to the protein, which can be performed only by eukaryotes. Examples include the tissue plasminogen activator and hemophilic factor VIII. Bacteria have no such capability, and yeast produces a different product than mammalian systems.

Plasmids which replicate in eucaryotic cells have been constructed. The genes ligated into such vectors can also be expressed in cultured mammalian cells. The resultant proteins, when glycosylated, are usually secreted from the cells into the medium. If one can select cells which produce glycosylated protein while growing in a serum-free medium, the final protein is relatively easy to purify.

An alternative strategy is use a tumor virus to immortalize the original human cell that produces the protein of interest. This strategy is currently suffering from poorer yields and controllability compared with the recombinant system.

Eukaryotic cell cultures have distinct disadvantages. Their division cycle is usually longer than 24 hours per cell division: the resulting long cultivation cycles provide more opportunity for contamination with unwanted microorganisms. Eukaryotic cells usually require media with expensive components which add significantly to the cost of production of the protein.

E. THE LARGE SCALE PRODUCTION OF GENETICALLY ENGINEERED PROTEINS

A number of factors must be examined in order to be successful in the transfer of technology from the growth of a modified organism in the laboratory to the commercial production of a protein (Arathoon, and Birch, 1986). Close collaboration between molecular biologists and fermentation scientists is essential.

1. Fermentation optimization. This subject includes such engineering manœuvres as the sterilization and aeration of the medium and the timing of nutrient feeds.
2. Fermentor design. The conventional stirred tank reactor is currently used for the production of many recombinant DNA proteins from bacteria or yeast. Its advantages are compatibility with containment guidelines in most countries and easy scale-up. Expression levels are usually such that a 10,000 l fermentor is large enough.
3. Host strain optimization. The choice of the host strain is a critical decision. One usually prefers to work with bacterial and yeast strains that have metabolic mutations that allow the operator to control the production of the desired protein by optimization of the composition of the growth medium.

Host strain optimization must also involve consideration of the final purification of the target protein. Usually it is advantageous for the protein to be secreted by the host cell, but one can also work with those cultures in which the recombinant protein is stored in intracellular inclusion bodies.

Since no gene product will be produced if the plasmid is rejected by the host strain, this factor must also be considered in host strain selection.

The recombinant protein itself can have a profound influence on the growth of the host strain. If the expression of the desired protein occurs too early in the growth cycle of the cells, this may result in poor cell growth and a lower product yield.

Cloned human genes are frequently expressed in bacteria as a protein with an unwanted N-terminal methionine. Careful fermentation optimization can downregulate this methionine content to undetectable levels.

It is more difficult to achieve controlled growth of mammalian cells. Cell growth from seed culture to production batch takes several weeks, so sterility can be a problem. Certain mammalian cells need a solid substrate on which to grow - this provides an engineering challenge. In addition, the cell membrane may break as a result of agitation or external pressure so that the fermenters must be carefully designed. Finally, the culture medium is very complex and it contains serum components which are heat-sensitive and hence cannot be autoclaved for sterility. Thus large scale fermentation of mammalian cells requires a careful step-by-step upscaling process.

F. ISOLATION OF THE PROTEIN

The procedures for the efficient isolation of the pure protein must also be optimized. Each protein has a different stability and solubility profile and so each presents a new research problem. Purification of the protein to almost homogeneity is usually required: large scale production of such pure proteins was unknown before recombinant techniques were available.

Purification generally involves the centrifugation of the bacterial suspension as the first step. The desired protein may be either in the supernatant or the pellet. Affinity chromatography is commonly used for the further purification of the specific protein. Especially useful is immuno-affinity chromatography in which the protein product binds an immobilized antibody that has been covalently attached to the chromatographic column. Computer-integrated process control is especially useful with such large-scale chromatographic purification of proteins.

G. QUALITY CONTROL OF THE PRODUCTS OF BIOTECHNOLOGY

The most expensive part of the manufacture of a product of biotechnology is not the production and isolation of the protein product, but rather the assurance that the protein is safe and efficacious in use.

The properties of biological products such as proteins cannot yet be completely characterized by chemical and physical tests of the final product. Product quality can best be ascertained by separating quality control into control of the production process and characterization of the final product.

In-process control involves characterization of starting materials and a control system for the reliability of the manufacturing process and prevention of microbial contamination. It is essential that production be based on a careful control of the host seed. This usually involves a master seed culture and working seed bank. The origin, form, storage, and use of every subculture must be well documented. During the design of the fermentation system one must establish the stability of the host-plasmid DNA in the seed stock under the storage and recovery conditions which will apply in the manufacturing process.

Quality assurance of the final protein product involves the assay of the potency and specificity in biological test models. Usually international standards are available for reference.

Table 1 summarizes some of the more important assays of purity of recombinant proteins. New analytical techniques had to be established for this new class of pharmaceutical products. For example, methods for the assay of nucleic acids at the picogram level and for the determination of subtly modified forms of the protein were designed because of the demands on the purity of recombinant proteins intended for human use. In addition, the analysis of the amino acid sequence of the protein is an essential component of purity determination. These assays generally demand considerable time of skilled scientists working on expensive instruments: this is why the cost of the purity and safety of a protein product is so high.

Table 1 Purity control of recombinant DNA - protein formulations for therapeutic use in humans

Microbiological contaminations
Enzyme activity harmful to product or recipient (proteases etc.)
Biological activity unrelated to product (endotoxins, enterotoxins, etc.)
Immunogenic substances activating antibody formation, complement or T-lymphocytes
Foreign proteins from production organism, growth medium or chromatographic columns
Modified forms of the protein product such as methionine N-terminus, glycosylation, phosphorylation of amino acids, wrong linkage (scrambling) of disulfide bridges, incorrect conformation, oligomerization or partial degradation of protein, wrong amino acids (norleucine instead of methionine)
Nucleic acids and nucleotides

H. PHARMACEUTIC FORMULATION OF PROTEINS

The biological activity of proteins is generally not preserved if they are administered to the patient by mouth since the digestive tract effectively hydrolyzes proteins into their constituent amino acids or peptides. The formulation of proteins for injection must consider the stability and solubility of the protein. Proteins are often unstable in solution at room temperature unless they are protected by stabilizers such as serum albumin. Some success has been reported in stabilizing proteins by coupling them to non-immunogeneic synthetic polymers. Proteins may also be incorporated into liposomes. This helps delivery to phagocytic blood cells. The development of delivery forms other than injections will considerably help promote acceptance of proteins as a form of therapy.

I. PATENTABILITY OF PRODUCTS AND PROCESSES FROM BIOTECHNOLOGY

Biotechnology leads to inventions of products and processes. Products include modified microorganisms, cell lines or hybridomas, parts of organisms such as expression plasmids and synthetic genes, and products of such organisms as proteins. Processes include ways of making such organisms or using them for the production of products.

Since patent protection may create an incentive for the marketing of a product or process, there is much interest in how patent law applies to biotechnology. Although most countries have adopted a broad interpretation of patentable subject matter, there is still uncertainty as to what kinds of biotechnological inventions can be patented. One major area of uncertainty is the disclosure standard for such inventions. This issue relates mainly to the reproducibility of the invention. It is still not clear whether the disclosure of an oligonucleotide sequence is sufficient for obtaining a patent or whether the entire microorganism which has been transformed by a vector containing this oligonucleotide sequence has to be present. In some cases the inventor is required to deposit a culture of new microorganism or cell line. However, such deposited cultures could become publicly available before any patent rights had been granted to the inventor.

J. FIRST GENERATION OF THERAPEUTIC AGENTS DERIVED FROM BIOTECHNOLOGY

Biotechnology developed from biologists, aim to understand the underlying principles of the structure and function of biological systems. The main contributions in the first decade of work in this field stemmed from studies of the biological function of interesting human proteins, made possible thanks to large quantities of pure proteins of many sorts now being available. These proteins had previously only been available in crude mixtures which could also have contained other synergistic or antagonistic substances.

Proteins belong to a range of sizes which is recognized by the immune system and classified as "self" or "nonself". Thus, non-human proteins trigger an antibody response in the human. This response leads to inactivation of the biological activity of the protein as well as further unwanted immunological reactions. It is therefore important to use human proteins for the treatment of human disease.

So far five products of recombinant DNA technology have been registered for therapeutic use: human insulin, human growth hormone, human alpha-2 interferon, tissue plasminogen activator and hepatitis B vaccine. Insulin and growth hormone are peptide hormones. Insulin derived from animals has long since been the largest volume peptide hormone used in medicine. The advantage of the product from humans is that one need not fear allergic reactions which frequently occurred with the animal products as a result of impurities. Human growth hormone cannot be substituted by animal products. It is currently used in hypopituitarism in children, but the availability of large quantities of the protein suggests it could be used to improve the healing of burns, wounds, and bone fractures.

Table 2 lists several proteins that are under development because of their activity in modulating the immunological system. These factors are important mediators between cells of the immunological system and adjust the host defense to cope with external infection and internal imbalance. They interact with cells of the immunological system either to stimulate their growth or to trigger differentiation and production of other proteins. Other factors interfere with immunoglobulin production, induce an antiviral and anti-proliferative state in target cells, or induce lysis of aberrant cells. These factors interact with specific receptor molecules on the surface of the target cells. Such binding triggers multiple changes in the metabolism of the target cell, a pleiotropic effect.

A list of blood proteins currently being investigated is shown in Table 3. The current source of these proteins is the fractionation of human blood. Since this is a sizable market at present, there is great financial incentive for biotechnological innovation to make the proteins even purer and less expensive for the patient. Among these blood proteins are albumin and Factor VIII, the largest proteins ever produced by recombinant DNA technology.

The plasminogen activators are an important group of fibrinolytic enzymes which act specifically on blood clots over a prolonged period of time. Applying these proteins entails less risk of hemorrhage during thrombolysis than does conventional therapy. Members of this class of enzyme initiate the dissolution of blood clots by converting plasminogen, a plasma protein, into plasmin. Plasmin then attacks fibrin, the major component of the clot. One such factor is tissue plasminogen activator (tPA). In view of its complex protein structure and the necessity of sugar

Table 2 Immune modifiers

Immune modifiers
B-cell factors
Colony-stimulating factors
Immunglobin Receptors
Interferons alpha
beta
gamma
Interleukins 1
2
3
Interleukin Receptors
Macrophage-activating factor
Tumor necrosis factors alpha
beta

Table 3 Blood proteins

Blood proteins
Albumin
Alpha-1-antitrypsin
Angiogenesis factors
Antihaemophilic factors VIII and IX
Apolipoproteins
Erythropoietin
Lipomodulin
Lung surfactant protein
Plasminogen activators
Superoxyddismutase
Urokinase

residues for stability, it is produced in genetically engineered eucaryotic tissue culture cells.

Biotechnology can contribute to the control of infectious diseases by supplying monoclonal antibodies or vaccines from recombinant DNA technology. Most conventional vaccines contain an attenuated or killed form of the entire infectious organism. Injection of the vaccine triggers the immunological system of the recipient into forming antibodies to the organism. The problem with this approach is that the entire genome of the pathogen is introduced into the recipient. Antibodies are not made to DNA. Therefore, if it replicates, there is no host defense mechanism to prevent damage associated with such DNA. Severe neurological side effects can occur.

On the other hand, vaccines made from recombinant technology would contain only a portion of the protein of the infectious agent. An additional advantage of recombinant methods is that one need not cultivate the pathogen in order to prepare the vaccine. Thus it is possible to prepare vaccines against pathogens that are dangerous to ferment or against viruses that cannot be grown in culture such as the AIDS or hepatitis type B virus. Table 4 lists some of the vaccines under development with these methods.

The structure and replication cycle of parasites is much more complex than that of viruses or bacteria. Again, biotechnology methods, including monoclonal antibodies, hold promise for the development of respective vaccines.

An alternative approach towards antiviral drugs may be the inhibition of "processing enzymes". Many viruses use a specific strategy for their maturation, which is similar to cellular mechanisms for the synthesis of active peptide hormones. Some viral proteins and peptide hormones are first produced as large inactive precursor proteins which can be proteolytically cleaved to generate the

Table 4 Vaccines

AIDS
Cytomegalovirus
Hepatitis A,B non-A/non-B
Herpes
Malaria
Pseudorabies
Retrovirus

active proteins and peptides. This cleavage is achieved by specific processing proteases which recognize specific cleavage sites on the precursor proteins.

The knowledge about structure and sequence of cleavage sites and processing enzymes allow multiple ways of interfering with this process and can lead to specific agents which inhibit either the maturation of a virus or the formation of a specific peptide hormone.

III. NATURAL PROTEINS AS LEADS FOR STRUCTURALLY MODIFIED PROTEINS: THE SECOND GENERATION OF PRODUCTS FROM BIOTECHNOLOGY

A. SITE-DIRECTED MUTAGENESIS: SPECIFIC MODIFICATION OF PROTEINS

With recombinant DNA technology scientists can engineer a variety of changes into the amino acid sequence of a protein. Site-directed mutagenesis allows one to change a specific portion of the DNA code for a protein in order to change a few specific amino acids of the protein (Fersht, et al., 1985). Hypotheses about the role of specific amino acids in the function of the protein can be tested by systematic variation of these amino acids. Site-specific mutagenesis is used to make new or improved proteins for clinical use by increasing thermal stability or catalytic efficiency or by changing the pH profile or substrate specificity of the protein.

Natural human beta interferon contains three cysteines. Two of these form an intramolecular disulfide bond necessary for the biological activity of the protein, whereas the third forms intra- or intermolecular disulfides that render the molecule biologically inactive. By site-directed mutagenesis, the free cysteine is replaced by serine. The altered protein has the desired biological properties but is more stable.

Replacement of a single amino acid residue can markedly alter protein stability by making the protein more resistant to proteolysis. For example, if the arginine 283 of tissue plasminogen activator is replaced by other amino acids, one isolates a single-chain protein from the fermentation rather than the hydrolyzed two-chain protein usually isolated.

Site-directed mutagenesis has been used to stabilize the elastase inhibitor alpha-1-antitrypsin (Tosenberg, et al., 1984). Methionine 358 is essential for the inhibitory properties of the protein, but it is also readily oxidized to the inactive sulfoxide. Replacement of this amino acid by valine leads to a potent and stable inhibitor of elastase. Such a protein may be useful in the treatment of emphysema.

B. CHIMERIC PROTEINS

Another technique for obtaining new proteins has been used with human interferon. There are more than 12 subtypes of human alpha interferon; each has a different form of antiviral activity. Hybrid interferons constructed by combining the N-terminal region of one subtype with the C-terminal region of the other give consensus interferons with antiviral qualities superior to the natural proteins.

It is also conceivable that there are chimeric proteins in which the antigen recognition region of a mouse antibody might be fused to the self-recognition region of a human antibody to produce a new human antibody with the antigenic specificity of that prepared in the mouse.

C. AUTOMATED CHEMICAL SYNTHESIS OF PROTEINS

An automated peptide synthesizer can be used to chemically synthesize small proteins. The synthesis of biologically active interleukin 3, a protein of 140 amino acids, was reported recently (Clark-Lewis, et al., 1986). The protected peptide was assembled using solid phase synthesis starting from the carboxyl terminal residue by adding amino acids in a stepwise fashion.

D. DESIGN OF USEFUL PROTEINS WITH NOVEL SEQUENCES: PROTEIN ENGINEERING

For a rational approach to protein engineering, the conformation of the molecule to be produced and the atomic details of its mode of interaction with its ligands and its environment must be predictable. The main tools for this design are protein crystallography of related proteins, interactive computer graphics, and the Brookhaven data base of three-dimensional protein structures. These have been discussed in Chapter 4. In addition, there are data banks of protein sequences.

The large quantities of protein now available due to recombinant DNA technology has been a big help to protein crystallographers. If a protein structure has been solved to a resolution of 0.3 nm or better, then the positions of the non-hydrogen atoms are rather well defined. If the resolution is 0.2 nm, then the positions of the hydrogen bonds are also clear. If ligand binding or changes in the amino acid sequence do not change the protein conformation or crystal packing, the structure of the modified complex can be solved rather quickly.

Interactive computer graphics allow one to examine the protein structure and propose changes in structure that might result in useful functional changes.

Protein data banks contain more than 3000 protein sequences. Comparative analysis of proteins is revealing many empirical rules that can be used as a guide for amino acid exchanges for protein engineering.

IV. EXAMINATION OF THE STRUCTURE AND FUNCTION OF PROTEINS AS TOOLS FOR THE EVALUATION OF PATHOLOGICAL PROCESSES: THE THIRD GENERATION OF PRODUCTS FROM BIOTECHNOLOGY

A. CELLULAR RECEPTORS: STRUCTURAL ASPECTS

Recombinant DNA technology and monoclonal antibodies have been essential for the isolation and subsequent molecular characterization of a number of receptor systems. Recall from Chapter 4 that the current view claims receptors are multi-domain structures that contain at least one binding site for their natural ligands and another binding site for a molecule of the effector system that converts the binding event into some physiological response. Studies aimed at probing the structure of receptors will benefit from the ability to make site-specific changes in a protein structure; those aimed at probing structure-function relationships will benefit from the ability to make chimeric proteins that contain the proposed ligand binding site from one receptor and the proposed transducing system from another.

The structure of the genes for growth factor receptors has revealed an interesting relationship between these receptors and oncogenes. Several oncogenes are related to growth factors or to their cellular receptors. However, the structures of the oncogenes are slightly altered with respect to their normal counterparts. They are either truncated at one end or contain specific mutations of one amino acid of the protein. These structural changes disrupt regulation of signal transduction and so lead to the abnormal growth patterns of transformed cells.

The primary structure of nicotinic and muscarinic acetyl choline and β_2 and α_2 adrenergic receptors have recently been elucidated (Noda, et al., 1983; Kobilka, et al., 1988) and were used to propose models for the three-dimensional structures of the receptors and of the mammalian synapse.

The cloning and sequencing of a gene of great importance in cholesterol research that codes for the low-density lipoprotein receptor has been reported (Russell, et al., 1983).

Cloned receptors can often be introduced into liposomes, membranes of cultured cells, or frog oocytes. Such preparations allow thorough investigation of receptor function. On the other hand, the expressed receptor protein sometimes retains ligand binding activity and so may be used to study the possible binding affinity of synthetic ligands.

B. THE USE OF BIOTECHNOLOGY TO PROBE THE FUNCTIONS OF RECEPTORS

Genetic engineering and monoclonal antibodies have also contributed to the study of the function of receptors. The binding of a natural ligand to a receptor is thought to trigger a change in the conformation of that receptor. The receptor and/or receptor-ligand complex often appears to change location from the cell membrane to some membrane inside the cell. Subsequently or concurrently the cellular effector system is triggered. This affects cytoskeletal elements, nuclear elements and intracellular signalling of the cell. As noted in Chapter 1, receptors appear to use at least four different cellular effector systems: tyrosine kinase, phospho-inositole and diacylglycerol, adenylate cyclase, and ion channels.

C. THE USE OF TRANSGENIC ANIMALS FOR THE STUDY OF THE FUNCTION AND TOXICITY OF HUMAN OR RECOMBINANT PROTEINS

The evaluation of the safety and efficacy of human proteins obtained by recombinant DNA technology presents problems. The use of animals for this purpose will be pointless if the animal receptors do not recognize the human protein. Such is the case for interferons and interleukins. One way of handling this problem is to transfer the genes for the human receptor into the genome of an experimental animal to make a transgenic animal. Mice are used for this purpose as a rule. The foreign gene is usually linked to a specific promoter or regulator region so that the expression of the foreign protein always remains within the control of the experimenter. Such transgenic animals allow us to investigate some aspects of the effect of ligands and inhibitors for human receptors.

D. ANTI-IDIOTYPIC ANTIBODIES AS MODELS FOR LIGANDS AND RECEPTORS

Each antibody can itself be the antigen for other antibodies. Some such anti-antibodies recognize the specific antigen recognition domain of the original antibody; they are termed anti-idiotype antibodies. They are described as antibodies that react only with antibodies to one antigen and not with antibodies from the same organism directed against other antigens. The primary antigen can often block the reaction between the antibody and the anti-idiotype antibody. This indicates that the very specific antigen-binding site of an antibody also recognizes the anti-idiotypic antibody. Additionally, a limited number of these anti-idiotype antibodies are in fact structurally so similar to the antigen that they can be seen as an image of the antigen (Figure 1).

If one makes monoclonal antibodies to receptor ligands and then makes anti-idiotypic antibodies to these antibodies, the anti-idiotypic antibodies are in some instances images of the receptor-bound conformation of the ligand. When they bind to the receptor they may mimic the hormone action of the ligand. Such anti-idiotypic antibodies may be therapeutically useful. Moreover, if one establishes the three-dimensional structure of such a protein, then the important aspects of the three-dimensional structure of the bound conformation of the ligand may be inferred. This is just the information needed for the design of peptidomimetics and antagonistic compounds with the aid of approaches described in Chapter 6.

Alternatively, one could raise antibodies and anti-idiotypic antibodies using the receptor as the antigen. Some of these anti-idiotypic antibodies will have a structure that is an image of the receptor. The three-dimensional structure of such proteins would allow one to design receptor mimics and ligands.

V. SUMMARY

Human insulin, growth hormone, interferon, tissue plasminogen activator and hepatitis B vaccine are products of biotechnology that currently offer the patient more effective therapy with fewer side effects. Products of biotechnology have allowed more accurate diagnosis of disease and control of drug dosage, thus providing the patients with more reliable therapy. We expect many other human proteins to be available for therapy soon. The second generation of products from biotechnology, specifically designed proteins, will also be available soon. The safety evaluation of such products is a challenge, but biotechnology techniques themselves may allow solutions to this problem. Finally, the development of biotechnology has started a revolution in biology. As scientists use these methods to gain a better understanding of the molecular and atomic bases for the structure and function of living organisms, the rational design of better and safer drugs becomes increasingly feasible.

REFERENCES

Arathoon, W.R., and J.R. Birch, Large-Scale Cell Culture in Biotechnology, Science, 232, 1390-1395 (1986)

Clark-Lewis, I., R. Aebersold, H. Ziltener, J.W. Schrader, L.E. Hood, and S.B.H. Kent, Automated Chemical Synthesis of a Protein Growth Factor for Hemopoietic Cells, Interleukin-3, Science, 231, 134-139 (1986)

Fersht, A.R., J.P. Shi, J. Knill-Jones, D.M. Lowe, A.J. Wilkinson, D.M. Blow, P. Brick, P. Carter, M.M.Y Waye, and G. Winter, Hydrogen Bonding and Biological Specificity Analysed by Protein Engineering, Nature, 314, 235-238 (1985)

Kobilka, B.K., H. Matsui, T.S. Kobilka, T.L. Yang-Feng, U.Ranodke, M.G. Coron, R.J. Leskowitz, and J.W. Regan, Cloning, Sequencing and Expression of the Gene Coding for the Human Platelet α_2 adrenergic Receptor, Science, 238, 650-656 (1988)

Koehler, G., and C. Milstein, Continuous Cultures of Fused Cells Secreting Antibody of Predefined Specificity, Nature, 256, 495-497 (1975)

Maniatis, T., E.F. Fritsch, and J. Sambrook, Molecular Cloning - A laboratory Manual, Cold Spring Harbor Laboratory, 1982

Noda, M., H. Takahashi, T. Tanabe, M. Toyosato, S. Kikyotani, Y. Furutani, T. Hirose, H. Takashima, S. Inayama, T. Miyata, and S. Numa, Structural Homology of Torpedo Californica Acetylcholine Receptor Subunits, Nature, 302, 528-532 (1983)

Russell, D.W., T. Yamamoto, W.J. Schneider, C.J. Slaughter, M.S. Brown, and J.L. Goldstein, cDNA cloning of the Bovine Low Density Lipoprotein Receptor: Feedback regulation of a receptor mRNA, Proc.Natl.Acad.Sci., 80, 7501-7505 (1983)

Tosenberg, S., P.J. Barr, R.C. Najarian, and R.A. Hallewell, Synthesis in Yeast of a Functional Oxidation-Resistant Mutant of Human a1-Antitrypsin, Nature, 312, 77-80 (1984)

6
The Medicinal Chemist's Approach

V. Austel

Department of Chemistry
Dr. Karl Thomae GmbH
Biberach, Federal Republic of Germany

I. GENERAL ASPECTS

A. CHEMICAL STRUCTURE AND BIOLOGICAL PROPERTIES

The biological properties of a chemical compound are determined by its chemical structure. Therefore by varying that structure one ought to be able to change the biological properties too. As far as these variations are skillfully designed they will lead to therapeutically more valuable compounds. But what are the characteristics of an appropriate variation? To what extent must a structure be changed? Are there any rules that classify structures into beneficial and adverse types?

The organic chemist conceptually dissects chemical structures into some basic skeleton and substituents that are attached to it. The basic skeleton is the main characteristic of a structural field or a set of chemical compounds respectively.

Are particular therapeutic effects always associated with particular structural fields and vice versa? This is not generally the case. Consider for example the field of benzene sulfonamides. It contains compounds that have antibacterial (sulfadiazine, 1), diuretic (furosemide, 2) or antidiabetic (tolbutamide, 3) properties.

Conversely, a certain biological property, for example, inhibition of phosphodiesterase, can be present in compounds from different structural fields, for example, isoquinolines (papaverine, 4), purines (caffeine, 5), and pyridines (amrinone, 6).

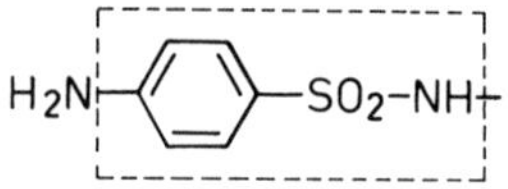

H_2N–C_6H_4–SO_2–NH– 1

(furyl)–H_2C–HN–C_6H_2(Cl)(COOH)–SO_2–NH_2 2

H_3C–C_6H_4–SO_2–NH–CO–NHC_4H_9 3

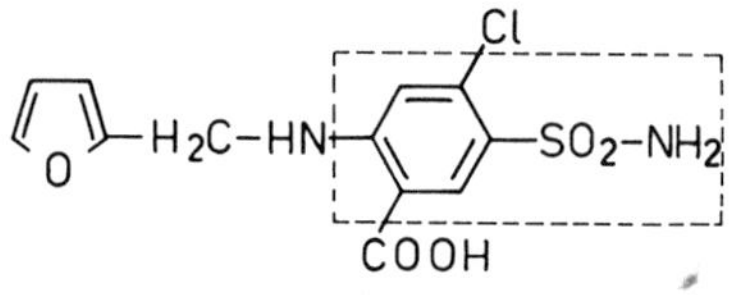

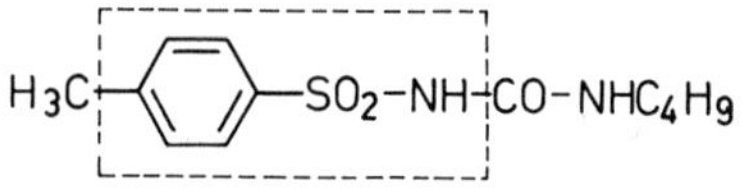

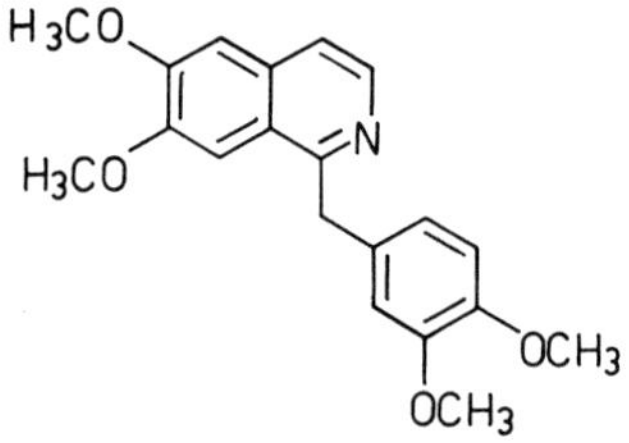

4

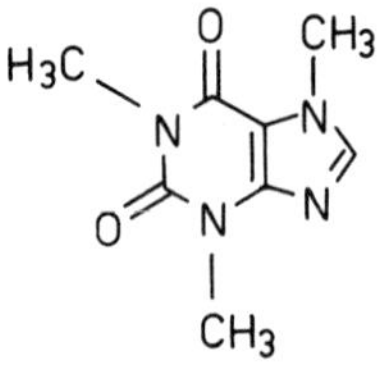

5

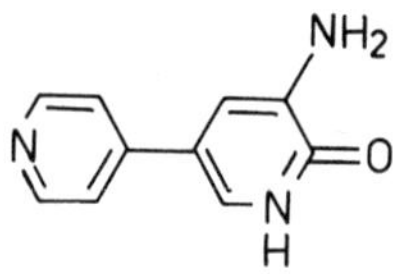

6

These considerations suggest that as far as pharmacological activity is concerned, every conceivable type of chemical structures is in principle suitable as a basis in searching for better and safer drugs. Does the same apply to toxic effects: is it also not possible to associate such effects with certain structural moieties? Experience shows that in fact certain structural features must be considered intrinsically toxic and therefore ought to be avoided. As was discussed in Chapter 3, this is particularly true for highly electrophilic structures such as activated alkyl halides, unless such compounds are used precisely for their toxic effects especially in cancer chemotherapy.

With the exception of such obviously toxic groups, it is generally not possible to predict which structural field or which structural variations within such a field will yield better and safer drugs. Does that mean that progress in drug therapy is only a matter of luck? In order to obtain an unbiased answer to that question it is necessary to have a closer look at the ways by which better and safer drugs can be found.

B. POSSIBLE WAYS TO BETTER DRUGS

The most obvious route to better drugs would be direct design based on the biomolecular processes underlying diseased states, such as the ones described in Chapter 2.

Such an approach is, however, not often feasible at present. Why is that so? First, for most of the diseases that would require improved drug therapy, these biomolecular processes are not well understood. On the contrary, biomolecular mechanisms have frequently been discovered via the biological properties of previously synthesized chemical compounds. (For example adrenergic receptors and their differentiation into α_1, α_2, β_1

and β_2 subtypes were found only when specific synthetic compounds became available). Secondly, even if one could design the structure of a compound so as to interfere in a desirable manner with a specific abnormal biomolecular process, one would still have to consider that similar but normal processes may also be affected. This might lead to intolerable side effects. An example is provided by certain types of anti-cancer agents that not only bind to DNA of rapidly proliferating cancerous cells but also interfere with DNA synthesis in normal cells thereby causing serious side effects including even carcinogenesis. Finally we learned from Chapter 3 that even the most selectively acting compounds will be of limited therapeutic value if they are unreliably absorbed, rapidly metabolized or unfavorably distributed in the organism. Such pharmacokinetic and metabolic problems are, for example, one of the reasons why natural peptides are not normally useful as drugs.

In summary, completely targeted approaches to the discovery of better and safer drugs are illusionary due to the complexity of the organism and its interaction with drugs.

Under these circumstances one wonders whether chance might indeed not be the only determinant of progress in drug therapy. In fact, many of the presently important types of drugs have been discovered accidentally. The most remarkable example of serendipitous drug discovery is certainly that of the antipyretic and analgesic properties of acetanilide (7), the parent compound of phenacetin (8) and paracetamol (9). These properties were observed when acetanilide had mistakingly been given to patients instead of naphthaline in an attempt to find a cure against worms.

Doubtlessly, the luck of the draw is still a major factor in present day drug research. Need it be the only

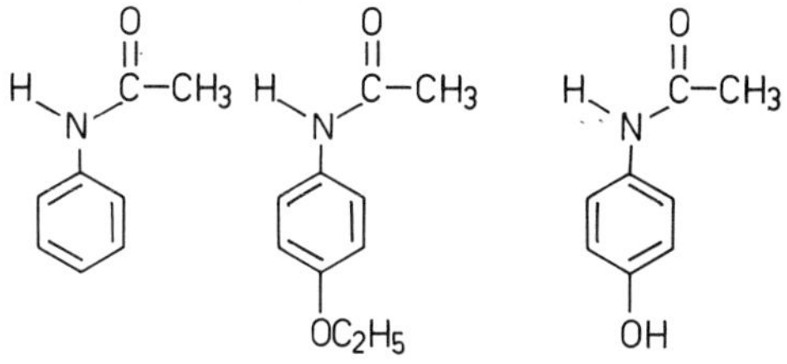

7 8 9

determinant? If that were so, searching for new drugs would in the long run not be worthwhile any more. Presently an average of 8,000 - 10,000 compounds have to be synthesized and tested before a new drug can be introduced into the market. For statistical reasons the probability of finding a better drug by chance decreases rapidly with the extent of the expected improvement. Considering the high costs of synthesis and biological testing, one must find ways to reduce the influence of chance as much as possible.

Critical examination of the literature reveals, however, that the unfavorable ratio between investigated compounds and marketed drugs is partly due to procedures that are not chance but bias controlled. Bias control prevails for instance if series of test compounds are conceived so that the overall structural variation is small in relation to the number of synthesized and tested compounds. If the search for better drugs is to become more efficient, then primarily one ought to give serendipity a real chance.

That, of course, still leaves us with the problem of ultimately reducing the importance of chance in an appropriate manner. This can be achieved by systematic empirical procedures which incorporate methods that allow detection and objective expression of relationships between chemical structures and their biological properties (structure-activity relationships, SAR, see Chapter 4). Even though such procedures are not dependent on the availability of computers, computer assistance considerably facilitates their use.

A general frame for a systematic procedure in drug discovery is outlined in Scheme 1. One starts by identifying a lead compound or lead structure. A lead compound is one that shows some but not all biological properties that are expected of a better drug. Lead compounds can be derived from experience (known active compounds), may be found serendipitously, or, as was discussed in Chapter 4, may arise as outliers from QSAR or from molecular modeling. This initial phase of drug discovery is referred to as lead finding. Once a lead compound has been obtained it is subjected to structural variation with the aim of optimizing its biological properties in the desired direction. This phase is called lead optimization. In fortunate cases one indeed succeeds in optimizing the biological properties to such an extent that the corresponding compounds are better and safer than the currently available therapies.

Lead optimization profits greatly from systematic approaches. Such approaches proceed in an iterative manner: A set of test compounds is conceived, synthesized, and biologically tested. From the results, structure-activity relationships (SAR) are derived which in turn serve as a basis for a new set of test compounds. This loop is repeated until an optimum of all relevant biological properties has been reached. Unfortunately, however, this optimum does often not suffice for therapeutic purposes. A detailed description of the individual steps in Scheme 1 as well as of some of the applied techniques will be given in Section II.

There is one additional general aspect of modern drug research that becomes quite obvious from Scheme 1.

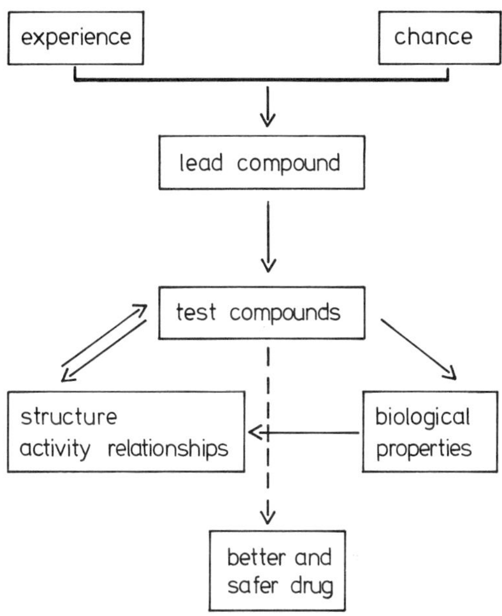

Scheme 1 Empirical search for better and safer drug, general procedure.

This is the strong interdependence of the participating scientific disciplines. The medicinal chemist, in particular, needs a close cooperation with biochemistry and pharmacology in both the lead finding and the lead optimization phases. Obviously biological criteria must be given a decisive role in selecting lead compounds. Meaningful SAR can only be derived on the basis of properly interpreted biological data (see Chapters 1, 2, 3, and 7 for more discussion of biological data).

In this context it needs to be emphasized that the therapeutic relevance of SAR cannot be better than that of the biological data.

However also the biologists benefit from a close cooperation with the medicinal chemist. Thus properly designed test compounds in conjunction with structure-activity analyses can provide deeper insight into the biological mechanisms that underlie the test models. Such insight has often led to improved test procedures or new therapeutic approaches. For example, the knowledge about subclasses of various classes of receptors, such as muscarinic cholinergic, α- and ß-adrenergic, or dopaminergic, has been acquired via compounds that have been designed as agonists or antagonists of such receptors. This differentiation has considerable practical consequences for the development of better drugs. For example, selective blockers of H_2 receptors and M_1-selective antimuscarinic agents have improved the therapy of stomach and duodenal ulcers.

II. PROCEDURES THAT LEAD TO BETTER AND SAFER DRUGS

A. CHOICE OF THE LEAD STRUCTURE

1. Lead structures from biological concepts. Nowadays most projects that are aimed at developing better and safer drugs start from some sort of biological information. Such information can have direct or indirect character. In the latter case, which is the most common, one only knows that certain compounds show biological effects of the type one is interested in. Sometimes also the structure-activity relationships or even the physiological mechanism of action are known. Yet the biomolecular basis of the effect, that is the structure of the binding site and its mode of interaction with the drug, may still need to be unveiled. For example, we know that isoprenaline (10) has antiasthmatic properties because of its bronchodilator effects. We also know that these effects are due to $ß_2$-receptor stimulation by the drug. In addition there are structure-activity relationships available that suggest that the side chain nitrogen, the

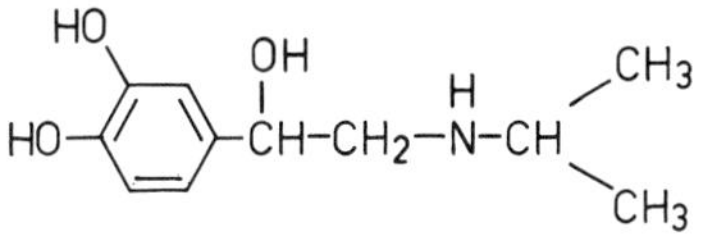

HO
HO–(ring)–CH(OH)–CH_2–N(H)–CH(CH_3)(CH_3) 10

aromatic ring, and the hydroxyl groups play a role in the binding to the β_2-receptor. Although we know the amino acid sequence of this receptor, the three-dimensional structure of the binding site and the exact mode of interaction between the compound and its receptor are not known.

So far the lack of such knowledge does not seem to have been crucial for progress in drug research. In fact the great majority of the drugs that are presently on the market have been derived from indirect biological information, i.e., from lead compounds which exerted the required pharmacodynamic properties. What compounds can serve as sources for leads in this context? In principle one can distinguish three such sources:

1) endogenous compounds that govern the physiological functions of the body,
2) compounds that have proven active in the clinic, and
3) compounds that are active in pharmacological or biochemical models but have not been tested clinically.

Several lead compounds of the first category had great impact on the development of modern drug therapy. The best known example is probably noradrenaline (11). This compound is naturally involved in the regulation of circulatory functions including blood pressure, heart contractility, heart rate. Consequently, noradrenaline was used as a lead compound for molecular modification. The most valuable result of these efforts was the devel-

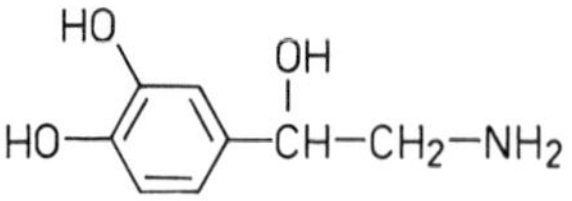

HO, HO—[benzene ring]—CH(OH)—CH_2—NH_2 11

[naphthalene]—O—CH_2—CH(OH)—CH_2—NH—CH(CH_3)(CH_3) 12

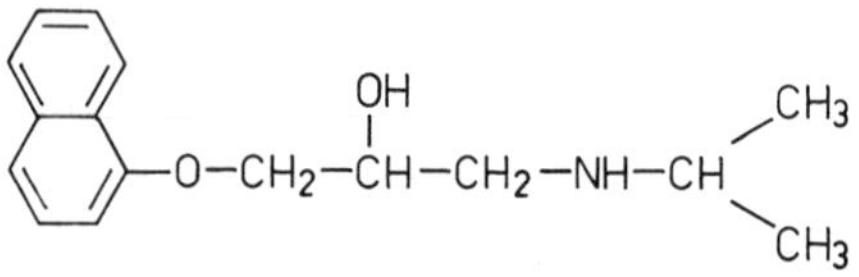

opment of the ß-adrenergic blockers (for example, propranolol, 12). They are presently indispensable in the therapy of coronary heart disease and the control of elevated blood pressure.

In recent years a class of compounds that are extensively used by the body as regulators and modulators have gained considerable interest: the peptides. The use of naturally occurring peptides as drugs often causes problems. This is principally because peptides are normally not absorbed from the gastrointestinal tract and cannot pass the blood-brain barrier. Secondly, the organism is extremely efficient in metabolizing them so that effective blood levels cannnot build up. Thirdly, peptides are frequently not selective; they exert a variety of biological effects which are not all wanted simultaneously. For example, somatostatin (13) not only inhibits the release of growth hormone but also that of insulin. The former effect makes somatostatin a valuable drug for treating acromegaly, whereas the latter effect is responsible for the diabetic states that can occur during such treatment. However, by using somatostatin as a lead compound more selective and metabolically more stable compounds (14, 15) have been derived. They are expected to be more valuable as drugs.

Therapeutically used drugs are suitable as leads if either they have shortcomings that diminish their thera-

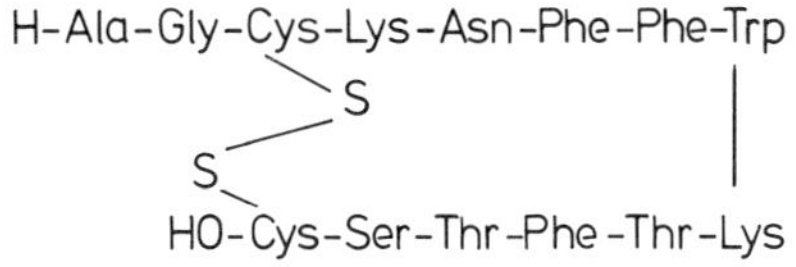

H-Ala-Gly-Cys-Lys-Asn-Phe-Phe-Trp (S–S bridge) HO-Cys-Ser-Thr-Phe-Thr-Lys <u>13</u>

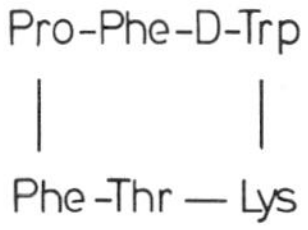

Pro-Phe-D-Trp / Phe-Thr — Lys <u>14</u>

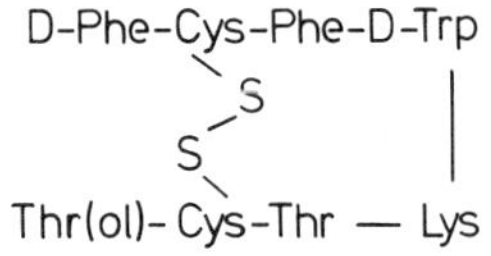

D-Phe-Cys-Phe-D-Trp (S–S bridge) Thr(ol)-Cys-Thr — Lys <u>15</u>

peutic value or if they exert side effects which are of interest for other indications.

The first situation applies to most compounds which are on the market. Common shortcomings are insufficient potency and duration of effect, unreliable oral activity, development of tolerance and undesirable side effects. As almost none of the marketed drugs is free of deficiencies, medicinal chemists all over the world have put a lot of effort into designing potentially better analogues of already marketed compounds. The approach has been very successful as becomes apparent from the frequently encountered terms "second" or "third generation" type of drugs.

The development in the field of antidiabetic sulfonylureas provides a pertinent practical example. The primary lead compound was carbutamide (<u>16</u>) whose additional antibacterial effects were clearly undesirable in the treatment of diabetes. This side effect could however be abolished by exchanging the

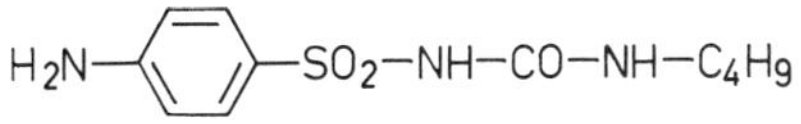

$H_2N-C_6H_4-SO_2-NH-CO-NH-C_4H_9$ <u>16</u>

aromatic amino group for a methyl group (tolbutamide, 17). Later, a second generation of considerably more potent sulfonylureas such as glibenclamide (18) was developed.

$H_3C-C_6H_4-SO_2-NH-CO-NH-C_4H_9$ 17

Cl, OCH_3 — $C_6H_3-CO-NH-CH_2-CH_2-C_6H_4-SO_2-NH-CO-NH-C_6H_{11}$ (H)

18

The sulfonylureas also illustrate the second manner in which clinically used drugs can become interesting lead compounds. Thus carbutamide was introduced into therapy as an antibacterial compound. During its clinical use its antidiabetic properties were discovered.

Many of the biological properties of chemical structures are published before they themselves or analogues are subjected to clinical trials. Especially the practice of a number of patent offices (for example, the European patent office) to publish patent applications already within eighteen months after filing opens a rich source of potential lead compounds whose biological effects have been demonstrated in pharmacological or biochemical models only. Work involving lead structures from this category is very common even though such a source is often not given credit in the literature.

As successful as drug design based on lead compounds has been in the past, a cautionary note is appropriate. If such a lead compound has already been subjected to broad structural variation, it is unlikely that further modifications will yield significant additional improvement in therapeutic value. This is especially true the more structurally related the new analogues are to the known ones. Thus cardiac glycosides have been investigated for decades but their narrow therapeutic index has not been significantly improved. Therefore cardiac glycosides cannot be considered promising lead structures for new cardiotonic compounds.

Also from a commercial point of view, marginally improved relatives of marketed drugs are unlikely to yield enough financial return to cover the cost of their development.

2. Lead structures from biomolecular structures or processes. In recent years an alternative to the search for better drugs based on known lead compounds has become possible. This approach starts from considerations of possible molecular mechanisms of action of the sought for new drug. Such an approach is feasible if at least one of the following pieces of information is available:

1) the atomic resolution three-dimensional molecular structure of the target binding site, preferably in a state where a ligand is bound to it,
2) a well founded hypothesis about the binding mode of the new drug, or
3) the mechanism of the biochemical process with which the drug is meant to interfere.

a. From the structure of binding sites. As discussed in Chapter 4, information of the first type is usually obtained via X-ray crystallography. Thus three-dimensional structures of a number of enzymes have been elucidated and the nature of their active site determined (recent review: Stezowski, 1986). X-ray analyses of enzyme ligand complexes have revealed the binding modes especially of inhibitors. One of these enzymes that attracted much attention is dihydrofolate reductase (Beddell, 1984). This enzyme is one of the targets for the development of new chemotherapeutic agents that are useful in the therapy of infectious diseases and cancer.

Information about the three-dimensional structure of enzymes or of binding modes can also be derived from known structures of related enzymes or enzyme-substrate complexes (see also Chapter 4). Thus three-dimensional structures of renins and binding modes of its substrates have been proposed from the X-ray structure of related aspartyl proteases such as endothiapepsin or penicillopepsin (Tickle et al., 1984). Renin is an aspartyl protease that catalyzes the liberation of angiotensin I (19) from its precursor angiotensinogen. This is the first step in the biosynthesis of angiotensin II (20), one of the most potent natural pressor substances. Inhibitors of renin are therefore of considerable interest as new drugs in the therapy of hypertension.

The renin-angiotensin system also provides an example for constructing a hypothetical binding model for a potential new drug from the known binding mode of similar substrates to a functionally similar enzyme (Cushman et al., 1980). The target enzyme in this case is angiotensin converting enzyme (ACE) which cleaves a dipeptide (His-Leu) off the carboxylic terminal of angiotensin I (19) to produce angiotensin II (20). ACE is therefore a carboxypeptidase and it was hypothesized that its catalytic site might be similar to that of carboxypeptidase A, an enzyme that removes one amino acid from the carboxylic end of its peptide substrates.

H-Asp-Arg-Val-Tyr-Ile,-His-Pro-Phe-Phe-His-Leu-OH 19

↓ ACE

H-Asp-Arg-Val-Tyr-Ile -His-Pro-Phe-OH 20

If the two catalytic or binding sites are indeed equivalent then also the binding modes of the substrates should be similar.

For carboxypeptidase A the three-dimensional structure of the enzyme-substrate complex was known. A schematic drawing of this binding site is shown in Figure 1. The most important features of this site are a positive charge that interacts via coulombic forces with the C-terminal carboxylic group of the angiotensin I and a Zn^{++}-ion that activates the carbonyl group of the scissile peptide bond. These two features must be present in the catalytic site of ACE also; but they should be further apart than in carboxypeptidase A in order to allow a dipeptide rather than only one amino acid to be removed. A corresponding model is also shown in Figure 1.

On the basis of this model potential inhibitors were designed to carry complementary groups at an appropriate mutual distance. Candidates for such groups are negatively charged moieties (e.g., a carboxylate) as counterions of the positive charge of the active site and a group that can function as a ligand for Zn^{++}-ions such as a carbonyl or a mercapto group. The binding site model became subsequently more refined and finally led the way to the well established antihypertensive drug, captopril (21).

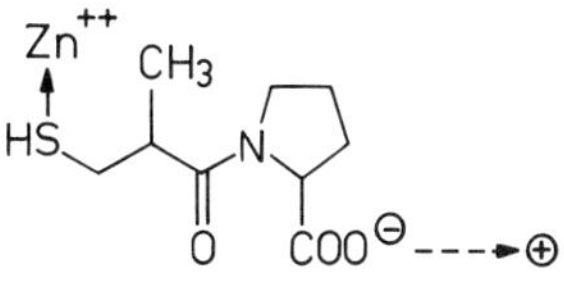

21

As discussed in Chapter 4, molecular graphics and modern methods of theoretical chemistry have proven particularly valuable tools for visualizing binding sites and for fitting potential ligands to them.

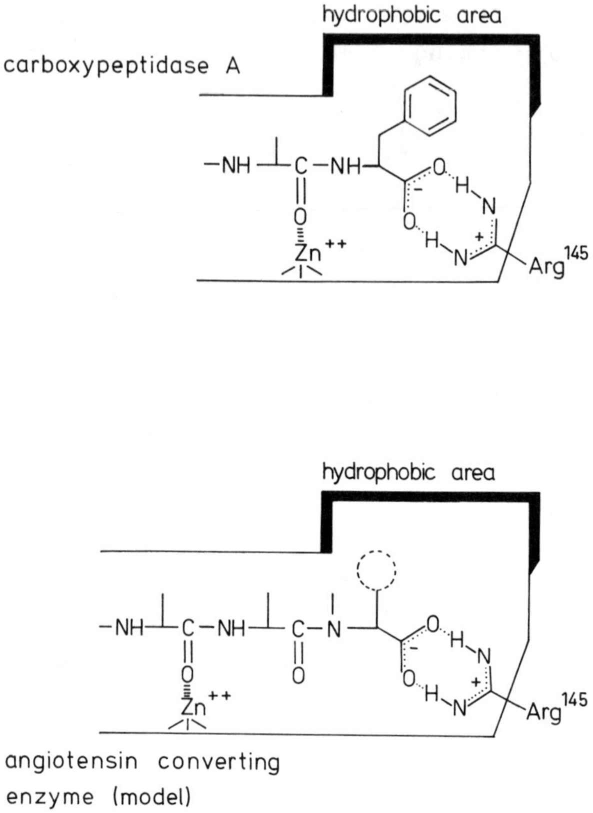

hydrophobic area

carboxypeptidase A

hydrophobic area

angiotensin converting enzyme (model)

Figure 1 Visualization of the postulated relationship betwee angiotensin converting enzyme and carboxpeptidase A.

All the examples outlined above refer to the design of lead structures from direct information about the molecular structure of binding sites of enzymes. In principle the same approach should also be applicable to potential receptor ligands. However the elucidation of the molecular structure of receptors has only recently become feasible. Currently, the most we know is the amino acid sequence, not the three-dimensional structure, of a receptor. Since receptors are frequently integrated into membranes, the crystallography is difficult and the structure of an isolated receptor-ligand complex might not always be the same as that of the same complex in its natural environment. This latter structure must however be known if ligands for receptors are to be designed in the manner outlined above. Nevertheless, the direct elucidation of the three-dimensional structure of receptors and of their complexes with specific ligands will become one of the major sources of information for the medicinal chemist in the future. These possibilities have been opened by the development of recombinant techniques discussed in Chapter 5 which allow receptor proteins to be produced in amounts sufficient for X-ray crystallography.

b. From binding models derived from QSAR. The field of enzyme inhibitors also provided evidence for the feasibility of an approach to lead discovery by which the binding mode of a class of drugs can be elucidated via the methods of quantitative structure-activity relationships discussed in Chapter 4. In a particularly intriguing example, Hansch had derived such a binding mode for compounds of general structure 22 to papain, an en-

$(X_3, X_4, X_5)C_6H_2{-}O{-}CO{-}CH_2{-}NH{-}CO{-}C_6H_3(Y_3, Y_4)$

22

zyme that cleaves peptides: binding correlated only with the lipophilicity of the most lipophilic one of the two substituents in positions 3 and 5 (X_3, X_5) of the phenoxy group, but not with the sum of lipophilicities of all the substituents on the ring. This led to the conclusion that the phenyl ring partly entered a hydrophobic pocket and partly remained exposed to the surrounding aqueous medium. This conclusion was later on confirmed by X-ray crystallography (Smith et al., 1982). The SAR approach to elucidating binding modes is also applicable to receptors and will become increasingly important in the near future.

c. From biochemical mechanisms. For enzymes, a second type of direct biological information, the mechanism of action, can become a valuable source for lead structures. Such structures can be divided into two categories

1) enzyme-activated irreversible inhibitors (for review see Penning, 1983)
2) transition state analogues (for recent review see Douglas, 1983).

The simplest route to irreversible enzyme inhibition utilizes highly electrophilic species such as α-halogenated carbonyl compounds or other strongly alkylating agents. As discussed in Chapter 3, compounds of this type are rather unselective as they can react with nucleophilic groups of any constituent of the organism. Consequently one has to consider the possibility of side effects. Therefore the therapeutic use of such compounds is limited to very serious diseases such as cancer.

There is, however, a possibility of making electrophilic compounds more selective, namely by letting the target enzyme produce them from unreactive precursors. Such precursors are frequently referred to as k_{cat} inhibitors, suicide enzyme inactivators or suicide substrates. The structures of such drugs are designed with reference to the mechanism of action of the target enzyme. A good illustration for the principle is the inactivation of pyridoxal phosphate (23) dependent enzymes such as amino acid decarboxylases or transaminases. The catalytic process begins with the formation of a Schiff base.

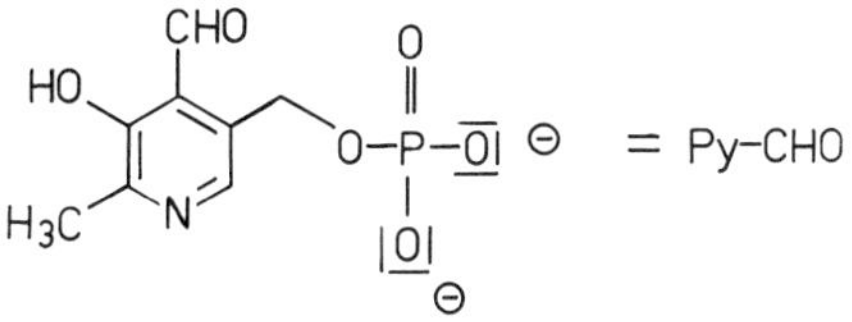

23 = Py-CHO

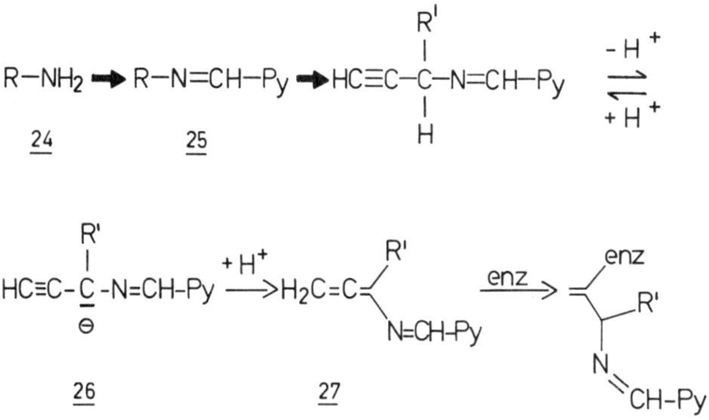

$R-NH_2 \rightarrow R-N=CH-Py \rightarrow HC\equiv C-C(R')(H)-N=CH-Py \underset{+H^+}{\overset{-H^+}{\rightleftharpoons}}$

24 25

$HC\equiv C-\bar{C}(R')-N=CH-Py \xrightarrow{+H^+} H_2C=C=C(R')-N=CH-Py \xrightarrow{enz}$

26 27

Consequently, potential inhibitors must also be able to form a Schiff base (25) with pyridoxal phosphate, that is a sufficiently reactive amino group (24) must be present in the molecule. The Schiff base must subsequently be able to rearrange so that a strongly electrophilic species is formed. A possible mechanism is proton migration facilitated by the formation of the Schiff base. This can be achieved by constructing R so that a proton adjacent to the nitrogen is activated and that the intermediate anion (26) on reprotonation yields the active species (27). Therapeutically, the principle of suicide inhibition has been used for the development of tranylcypromine (28) a suicide substrate of monoamine oxidase. In this case the reactive intermediate, produced via oxidation by the enzyme, is a cyclopropanone-imine or related species (29) which acts as an electrophile presumably towards a sulfhydryl group of the enzyme (30).

The transformation of a substrate into products by an enzyme passes through energetically unfavorable inter-

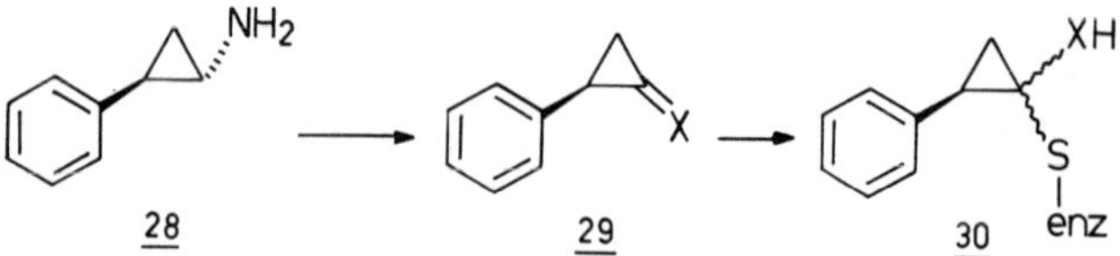

28 29 30

mediates called transition states. The energy that is required to reach this state must be supplied by the binding energy of the intermediate to the enzyme, in other words, the intermediate must bind tightly to the enzyme. A stable compound that resembles the intermediate therefore also ought to bind strongly to the enzyme and act as an inhibitor. Such compounds are called transition state analogs. The structure of the intermediates can again be derived from the proposed mechanism of the catalytic process. For example, deamination of adenosine (31) occurs via an intermediate of type 32. Compound 33, which resembles the intermediate, is indeed a good inhibitor of adenosine deaminase.

If an enzyme catalyzes the reaction between two substrates it initially has to bind both. In this case the binding energy must make up for the loss in entropy associated with the simultaneous binding of the two

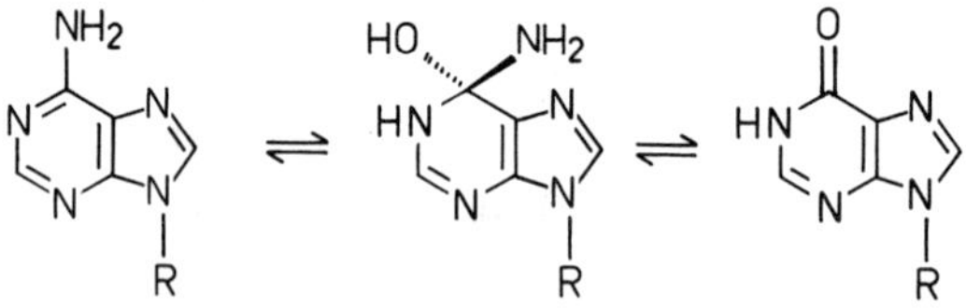

31 32

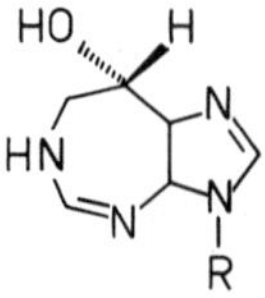

33

substrates. If one combines the structural features of the two substrates into one molecule this should be bound much more tightly since less of the binding energy needs to be spent on entropy losses. An example is the anti-cancer drug N-phosphonoacetyl-L-aspartate (PALA, 34). This compound inhibits aspartate transcarbamylase, an enzyme that is involved in the synthesis of pyrimidine nucleobases and binds aspartate (35) and carbamyl phosphate (36) simultaneously. Inhibitors that structurally combine several substrates are called multi-substrate inhibitors.

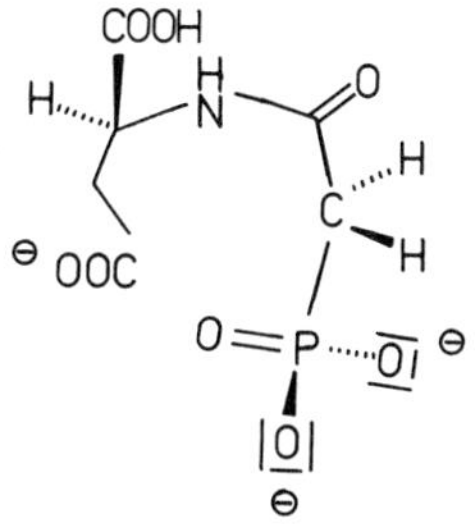

34

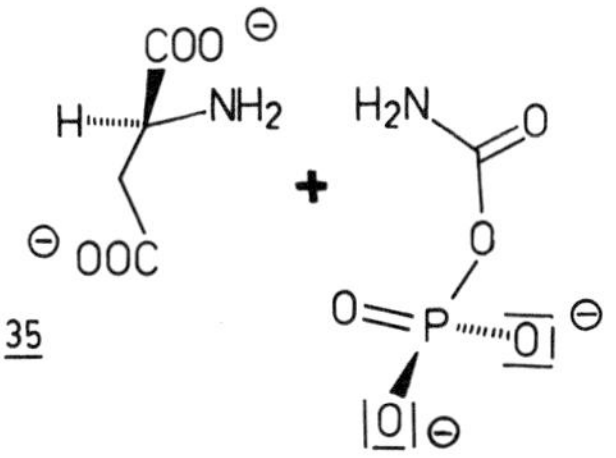

35

36

3. Lead structures from chemical concepts. As mentioned before, available lead structures may not always be a good starting point in the search for better and safer drugs. Also molecular mechanisms that underly the target disease might not be known or sufficiently understood in order to derive potentially active structures from them. Under such conditions it is worthwile to search for lead compounds among types of chemical structures that have been newly discovered or have at least not been widely pharmacologically investigated. Such an exploration of chemical concepts has been the source for important advances in drug therapy. One of the best known examples is the discovery of the benzodiazepines, the most important class of tranquilizers (Sternbach, 1978).

Sternbach's chemical concept was the investigation of benzoxadiazepines (37), which had hardly been examined biologically before but which he thought would be readily synthetically accessible. Only later did these compounds turn out to constitute an interesting variation of an otherwise well known type of compounds, the quinazolines (38). However, on treating these (X = Cl) with primary amines an unexpected rearrangement took place leading to a new class of compounds, i. e., benzodiazepines (39).

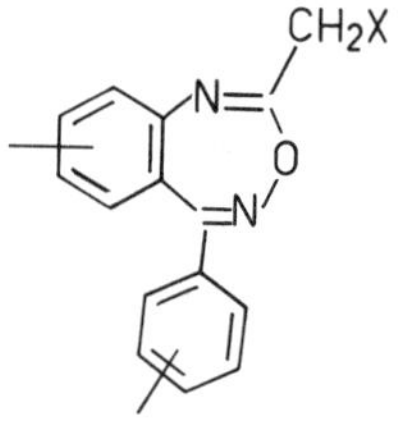

CH_2X 37 CH_2X RNH_2 H NR

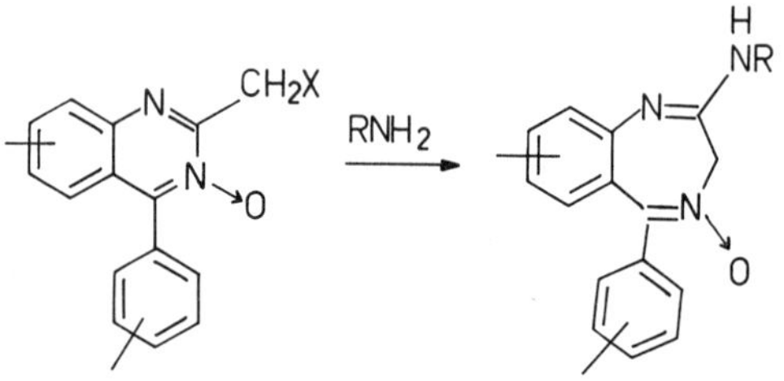

38 39

Properly selected chemical concepts have a high potential of leading to original new biologically active structures. However, the probability of encountering such structures is not very high. Therefore, in order to keep the rate of success at a reasonable level, the number of compounds that are synthesized and tested in a certain time must not be too small. This means that efficient synthetic routes as well as convenient biological tests must be available. Additionally, the rate of success can be enhanced if the test compounds are conceived so as to differ appreciably in their chemical structures.

4. Lead structures from natural sources. Drugs derived from natural sources, especially plants, have been used for curative purposes since prehistoric times. Also in addition to antibiotics, peptides and steroid hormones, a number of contemporary drugs have been developed from

naturally occurring leads. Examples are provided by the digitalis derived drugs for cardiac failure, analgesics based on morphine, anticholinergic derivatives of atropine, and cocaine-related local anesthetics.

Much effort still goes into isolation and biological testing of compounds from medicinal and other plants, animals, fungi, and bacteria. This approach still offers considerable promise as can be judged by the progress in the field of antibiotics. Search for new leads among naturally occuring compounds has also been successful in other fields. A more recent example is forskolin (40) which has novel structure for a cardiotonic drug and which has a new mechanism of action.

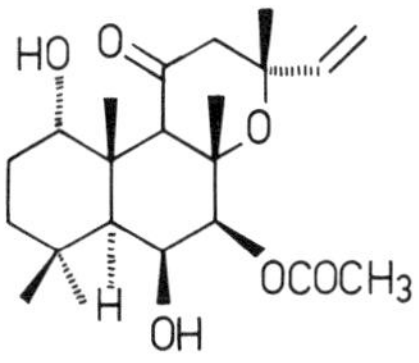

40

It is frequently assumed that natural products are a much more promising source of new leads and better drugs than are synthetic compounds. Practical experience does, however, not suppert this view. In relation to the screening efforts with natural materials the yield of therapeutically interesting compounds has been disappointingly low (Burger, 1978). Nor are natural compounds intrinsically less hazardous than synthetic ones: atropine, digitalis glycosides, cocaine, morphine are examples of toxic natural substances.

5. Targeted screening. In recent years molecular pharmacology has developed assays in which the interaction of a potential drug with its target site can be measured directly. Well known examples are various receptor and enzyme assays discussed in Chapter 7. As these models

have a high capacity, a large number of compounds can be screened in a short time, up to several hundred per week. The models are chosen so that they relate to the mechanistic aspects of diseases or are at least assumed to do so. This mechanism based screening has been called targeted screening. The materials that are subjected to this high capacity screening stem from plant extracts, fermentation broths, purchased chemicals, chemical synthetics, a company's own stock of previously prepared test compounds and intermediates. Despite a very low incidence of discovery, this procedure may still be the fastest way to new leads provided the number of test models as well as their capacity are sufficiently high. The rapid progress in biotechnology discussed in Chapter 5, especially in terms of cloning and expressing drug receptors, will make targeted screening the most important screening procedure in the search for new drugs.

B. PROCEDURES FOR LEAD OPTIMIZATION

1. General aspects. A lead compound exerts the pharmacodynamic property that is deemed beneficial for treating a diseased state. A new drug, however, must be able to improve current therapy. This drug, therefore, has to fulfill many more and stricter requirements than does the lead compound. Generally such requirements include

- a sufficiently high potency in order to keep dosage within reasonable limits
- a reasonable duration of the effect (not too short and not too long in order to allow a dosage regimen with which patients can easily comply and which does not lead to drug accumulation),
- effectivity on oral administration if multiple dosing is required,

- a reasonable selectivity in order to avoid unacceptable and undesirable side effects, and
- a reasonably low toxicity.

It is the long-standing experience of medicinal chemists that all these properties can be improved via structural modifications of the lead compound.

Are there any rules that specify which structural modification is relevant for which improvement? Generally this is not the case. Consequently, the appropriate modifications will have to be developed empirically. Sometimes, however, lead optimization can aim toward specific moleuclar properties previously shown to be optimum for a particular biological property. To optimize such biological properties targeted lead optimization is used.

2. Targeted lead optimization. Targeted lead optimization is especially useful if pharmacokinetic properties are to be optimized. Thus in Chapter 3 is was noted that absorption, distribution, and excretion of drugs are mainly determined by lipophilicity and corresponding rules of thumb can be formulated (for review see Austel and Kutter, 1983): for example, the optimal lipophilicity for absorption from the gastrointestinal tract lies between apparent log $P_{o/w}$ values of 0.5 and 2.0 (Log $P_{o/w}$ refers to the partition of the compound between n-octanol and buffer at neutral pH). Corresponding adjustments in log $P_{o/w}$ using the concepts discussed in Chapter 4 can conveniently be achieved by varying the substitution in those parts of the molecules that are not crucially involved in receptor binding, the bulk tolerance areas.

Another possibility is bioisosteric replacement (for reviews see Thornber, 1979; Lipinski, 1986). Bioisosteric groups are chemical moieties within a molecule that significantly affect a biological property and can replace each other without loss of that property. The concept of bioisosterism is not an absolute one. It refers rather to a specific biological property, usually a pharmacodynamic one. If bioisosteric groups differ with respect to their lipophilicity they lend themselves to optimizing pharmacokinetic properties. A well established example is the bioisosterism between 3,4-dihydroxy-phenyl (41) and 4-amino-3,5-dichlorophenyl (42). The latter moiety is considerably more lipophilic and should therefore have a positive effect on gastrointestinal absorption especially if introduced into a hydrophilic molecule. In fact, the orally inactive ß-adrenergic agonist isoprenaline (43) can be converted into an agonist that is absorbed from the gut (clenbuterol, 44) via replacing moiety 41 by 42.

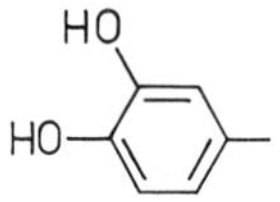

41

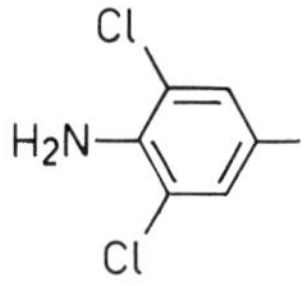

42

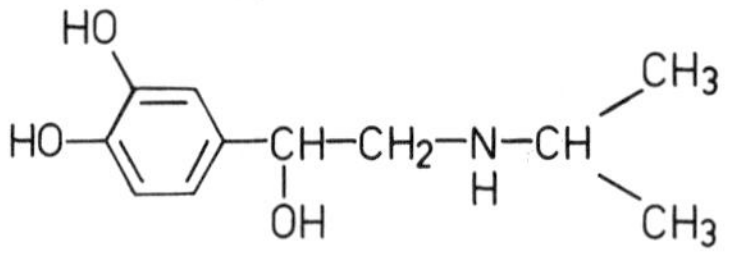

43

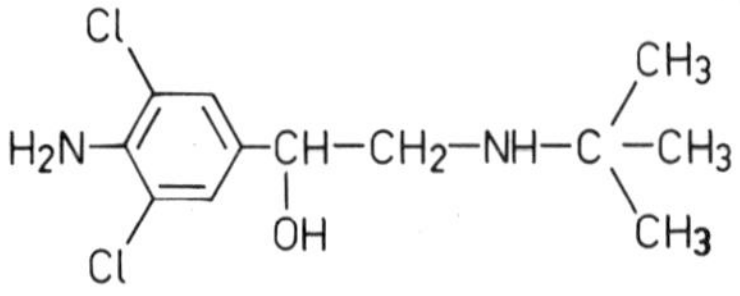

44

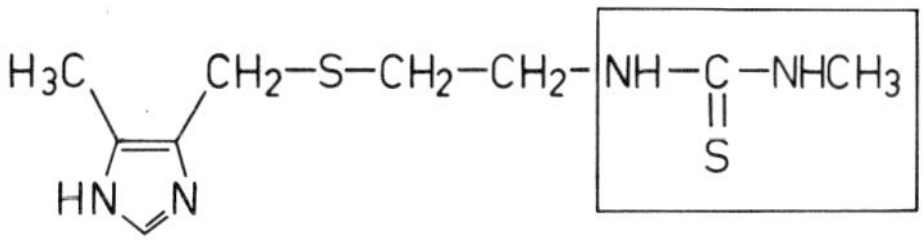

H_3C (imidazole, HN N) $CH_2-S-CH_2-CH_2-$ NH–C(=S)–$NHCH_3$ 45

H_3C (imidazole, HN N) $CH_2-S-CH_2-CH_2-$ NH–C(=N–CN)–$NHCH_3$ 46

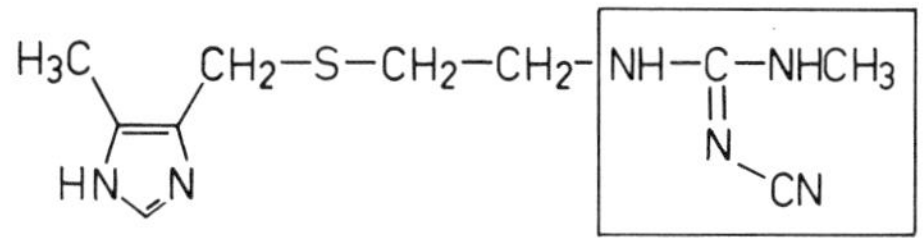

Bioisosteric replacement is, however, useful not only in improving pharmacokinetic properties. It can also help in changing selectivities and toxic effects. Thus the toxic effects of methiamin (45) could be overcome through bioisosteric replacement of the thiourea moiety by a cyanoguanidino group (cimetidine, 46).

Finally, bioisosteric replacement can also serve to prevent metabolic degradation and thereby to prolong the duration of the therapeutic effect. The catechol moiety of compound 43, for example, is a target for metabolizing enzymes and is therefore responsible for the short-lived activity. The bioisosteric amino-dichloro-phenyl group cannot be attacked by these enzymes and thus compound 44 has long-lasting effects.

Frequently, however, appropriate bioisosteric replacements cannot be found. Nevertheless one can still improve pharmacokinetics in a targeted manner by using a the so-called pro-drug approach (for review see Sinkula and Yalkowsky, 1975; Pitman, 1981). A pro-drug is a chemical compound that after administration to the organism becomes metabolically converted into a chemically different compound that in turn shows the expected biological activity. Note that the pro-drug is frequently but need not necessarily be inactive.

Pro-drugs can be designed in a targeted manner if the structure of the corresponding active compound is

Table 1 Chemical groups and their derivatives which have been used for prodrugs.

group	derivative
-COOH	-COOAlk
	-COOAr
	-COO-CH(R)-O-C(=O)-Alk (R = H, Alk, Ar)
	-COO-CH(R)-OOC- (R = H, Alk, Ar)
	-C(=O)-NH-CH(R)-COOR'
	-C(=O)-N
-OH	$-OCH_3$
	$-O-CH_2-OCH_3$
	AlkO, -O
	-O
	-O-C(=O)-R
	-O-C(=O)-O-Alk
	$-O-SO_3^{\ominus}$
	$-O-PO_3^{2\ominus}$
OH OH	O O

Table 1 continued

group	derivative
NH (non basic)	$>N-CH_2-OH$
	$>N-CH_2-O-C(=O)-R$
	$>N-CH_2-N(R)(R')$
	$>N-C(=O)-R$
NH (basic)	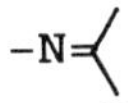 $>N-C(=C(CH_3)-COCH_3)$ (enaminone)
	$-N=C<$
	$>N-C(=O)-R$
$-N(H)\frown N(H)-$ cyclic structure	$-N\frown N-$ bridged by $>C<$ open chain precursor

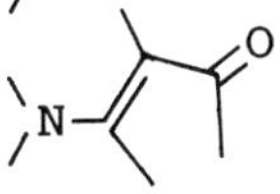

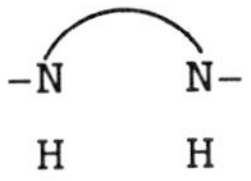

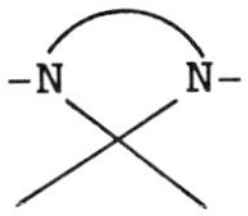

known. Suitable groups within this structure are modified so that improvement of the pharmacokinetic behavior is to be expected. The modifications must, however, be chosen so that the active compound can be liberated. One of the most common modifications leading to a pro-drug is the esterification of an alcoholic or phenolic group. The esters are usually more lipophilic than the hydroxy compound and may therefore be orally absorbed whereas the parent compounds are not. The latter compounds are then liberated from the pro-drug by the action of esterases and can now exert their therapeutic effects.

There are many other ways to design pro-drugs. Table 1 is a collection of possibilities that have been reported in the literature.

The pro-drug concept is, however, not free from pitfalls. Since the liberation of the parent compound is usually dependent on the action of specific enzymes, species differences in the activity of the enzymes have to be taken into consideration. Thus a compound may act as a pro-drug in one animal species but may not do so in another one or in man. For example, compound 47, which is an inactive pro-drug of the antimalarial dapsone (48), is activated via deacylation in mice but not in rats. Similarly the butyric acid derivative 49 has anti-

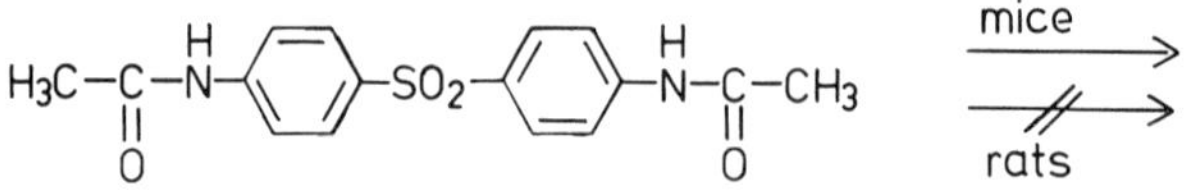

$H_3C-C(=O)-NH-C_6H_4-SO_2-C_6H_4-NH-C(=O)-CH_3$ → (mice) ; ↛ (rats)

47

$H_2N-C_6H_4-SO_2-C_6H_4-NH_2$

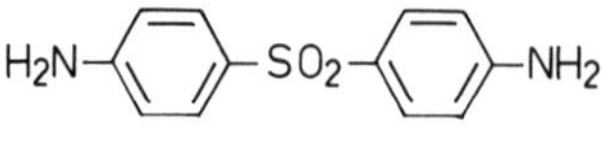

48

inflammatory activity in rats since it is metabolized via α-oxidation to the active propionic acid derivative 50. In man, however, 49 proves to be virtually inactive

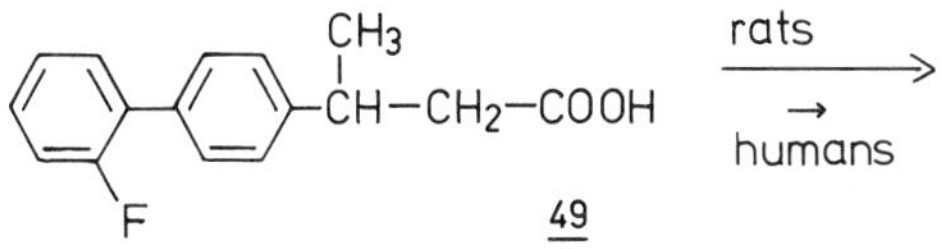

CH_3
CH–CH_2–COOH
F
49

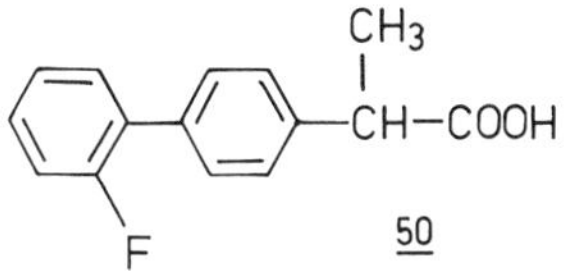

rats
humans

CH_3
CH–COOH
F
50

because metabolic transformation to 50 occurs to a very minor extent only.

Chapter 3 provided examples for which metabolism studies revealed that an apparently active compound is in reality a pro-drug. Substituting this compound by the actually active substance can lead to improved therapeutic quality.

Finally, targeted lead optimization can be based on the knowledge of metabolic pathways of a lead that result in rapid inactivation or in the formation of toxic metabolites such as the electrophilic species discussed in Chapter 3.

3. Optimization of peptidic lead structures. In recent years a large number of peptides have been discovered in the organism. These peptides function as neurotransmitters, hormones, enzymes, enzyme inhibitors or modulators of other transmitters. Such biological properties make peptides valuable lead compounds. The peptides themselves are, however, not ideal as drugs, except perhaps for acute medication. Common drawbacks are problems with absorption and distribution, metabolic instability, antigenic potential and lack of selectivity.

Much effort has therefore gone into modifications of peptides and a number of principles have been developed. In general, three types of peptide modifications can be envisaged:

1) variation of side chains
2) variation of the backbone
3) replacement of the peptidic by more or less non-peptidic structures (peptide mimetics).

These types of modifications have been used either alone or in combination.

The main objective of side chain variation is to increase metabolic stability. This approach is promising in so far as proteases can recognize their target peptide bond via the surrounding side chains. Commonly used side chain variations are replacement of the methyl-thio group of methionine by an ethyl group (norleucine) or the conversion of cysteine (51) to penicillamine (52).

Backbone variations have been summarized by Spatola (1983). They also serve metabolic stabilization. Backbone variations comprise among others: exchange of L- for D-amino acids; isosteric replacement of the peptide bond for example by E-CH=CH, CHOH-CH_2, CH_2CH_2, O-CH_2, S-CH_2 or reversion of the peptide bond (53); and cyclization

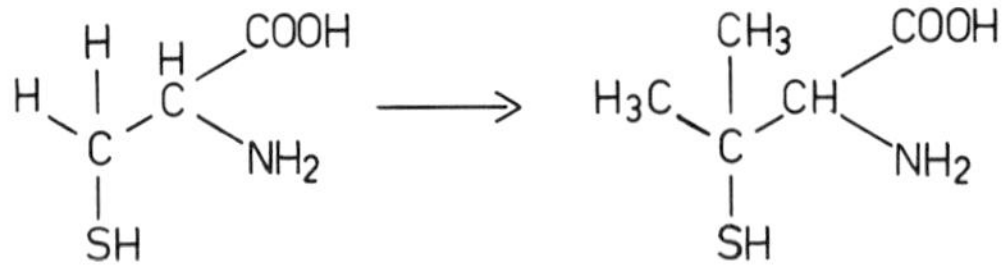

51

52

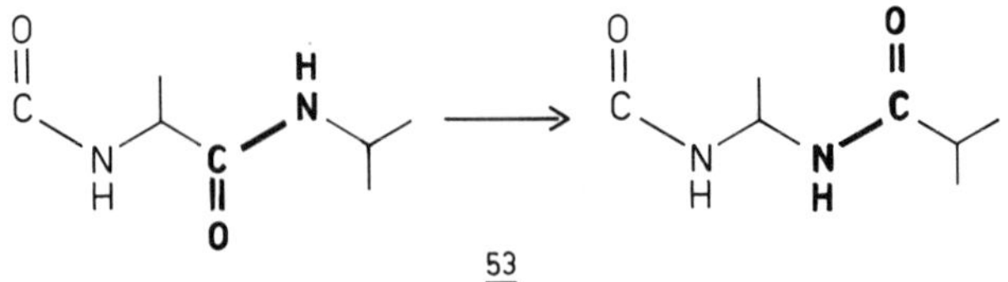

53

either of the backbone itself or via side chains, for example, by forming a disulfide bond between two cysteines). Both the latter variations have been successfully applied to somatostatin, with the effect of contracting the ring size. The resulting analogues (14, 15) show considerably longer lasting effects than the lead peptide.

That completely non-peptidic structures can exert the same biological properties as peptides has been demonstrated by nature itself, which mimicked the endogenous enkephalines by morphine. Synthetic work in the field of peptide mimetics does not seem to have developed very much even though principles have been reported a number of years ago (Farmer, 1980).

Optimization of peptidic lead structures is very much dependent on information about those amino acids in the sequence that have decisive influence on the activity. One obtains such information via specific deletions (especially of N- and C-terminal sequences) or exchanges of putative key amino acids for different ones. With small peptides conventional chemical peptide synthesis can be applied to this end. Deletion of sequences can be achieved for peptides of all sizes by enzymatic methods. Alternatively, recombinant techniques discussed in Chapter 5 have opened possibilities for synthesizing the remaining sequence directly. This approach becomes especially powerful if one wants to exchange amino acids at specific positions, point mutations, in larger peptides and proteins.

4. Empirical lead optimization. Targeted optimization of a lead compound requires the structural basis of its shortcomings to be known. Normally such knowledge is not available, at least not for all inadequacies of the lead compound. Consequently such information needs to be generated empirically. As outlined in Section I.B, this can

be achieved with the aid of series of test compounds. How can such series be designed? Two different approaches can be envisaged, i.e., the chemical concept approach and the SAR approach.

a. Optimization via chemical concepts. In this approach lead compounds are subjected to structural variations that are based exclusively on synthetic accessibility, for example, via modification of reactive groups. From the compounds that have been conceived in this way one or more test series is selected. Consider as a very simplified example methyl-phenylacetate (54) to be a lead compound. By making use of the chemical reactivity of the ester group one can synthesize test compounds in which this group is modified to other esters and various amides of type (55). Similarly the nucleophilic potential of the methylene group would make corresponding alkyl derivatives (56) candidates for test series. Finally, one can apply knowledge of the synthesis of aryl-substituted phenyl acetic acids to add derivatives of type 57 to the test series. As far as it is possible to combine

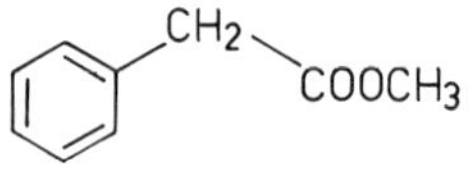

CH_2 $COOCH_3$ 54

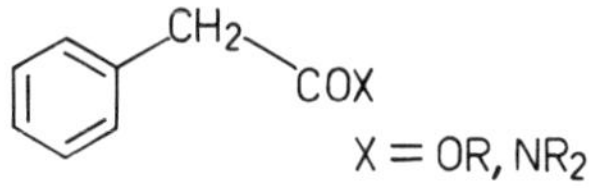

CH_2 COX 55
X = OR, NR_2

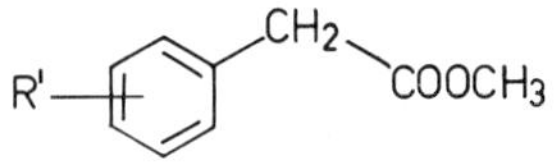

R' CH_2 $COOCH_3$ 57

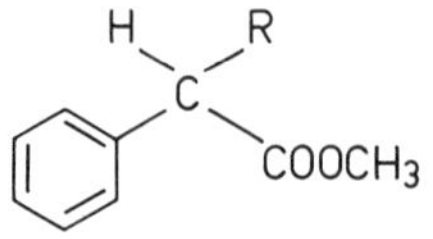

H R C $COOCH_3$ 56

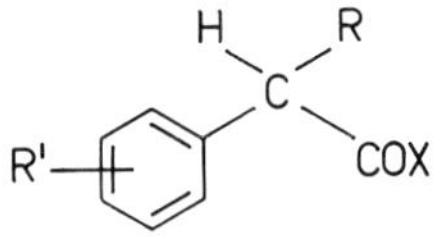

H R C R' COX 58

all three types of modifications, the possible members of the test series can be described collectively by general formula 58. This formula also characterizes a structural field. All compounds that belong to this field form a corresponding set. This set is characterized by the same general structure as the structural field. As long as no additional specifications for R, R' and X are introduced the set is of course infinitely large. Every test series can be viewed as a finite subset of such a set.

In the chemical concept approach, the members of the test series are selected according to synthetic accessibility. Synthetic accessibility is an important factor in the search for new drugs. As has been discussed before, chance still contributes significantly to success. This implies that the probability of finding a new better and safer drug increases with the number of test compounds, which in turn depends on the ease of synthesis.

b. Optimization via SAR. The SAR approach starts with a structure-activity hypothesis. A test series is ideally designed so that some of its members verify and others challenge the hypothesis. By subjecting the test series to a structure-activity analysis as outlined in Chapter 4, one obtains SAR which serve to modify the original structure-activity hypothesis. As described in Section I.B, the modified hypothesis forms the basis for the next test series.

What do structure-activity hypotheses refer to? Usually they specify those factors that are assumed to relate the chemical structure of the test compounds to one or more of their biological properties. In Chapter 4, it has been pointed out that the biological properties arise from physico-chemical interactions between the test compounds and their target structures in the body. Therefore the factors must ultimately also be of physico-chemical nature. As outlined in Chapter 4, there are principally two ways of expressing physico-chemical properties of chemical compounds.

1) implicitly as more or less fixed combinations of several properties. Examples are indicator variables that refer to the presence or absence of particular chemical moieties in a molecule and parameters derived from graph theory which relate to the connectivity of a molecule.
2) explicitly by using parameters that describe specific physico-chemical properties such as Hansch π-values for lipophilicity or Hammett σ-values or quantum mechanical quantities for electronic properties.

A more detailed discussion of the parameters which can serve to characterize the physico-chemical properties of chemical compounds is given in Chapter 4.

If one chooses only indicator variables to characterize the compounds, the test series must be designed so as to meet the requirements for the Free-Wilson-Analysis (Free and Wilson, 1964) described in Chapter 4. If continuous variables such as explicit physico-chemical parameters or connectivity indices are used, the test series ought to be reasonably orthogonal with respect to those parameters. This means that within the series none of the parameters show any appreciable correlation with any other one nor any combination of the others. Methods which can help design appropriate test series will be discussed briefly in section II.B.5.

There is one more aspect of the QSAR-approach that needs to be mentioned: such relationships are not valid

Table 2 Some specifications for the variable substituents in formula 58.

R	π (R)	R'	σ (R')	X	MR(X)
H	0.00	H	0.00	OCH_3	7.87
CH_3	0.56	m-CH_3	-0.07	O-i-C_3H_7	17.06
i-C_3H_7	1.53	p-CH_3	-0.17	NH_2	5.42
C_6H_5	1.96	p-OCH_3	-0.27	$NHCH_3$	10.33
		m-NO_2	0.71	$N(CH_3)_2$	15.55
		p-NO_2	0.78		

The finite structural field (set) has 4 x 6 x 5 = 120 compounds (elements)

in a general sense. They rather pertain only to those structures that are similar to the compounds of the test series, i.e. to a certain structural field. This structural field can be characterized by a general formula just as in the chemical concept approach. In the QSAR-approach, however, the field must be specified more exactly, namely such that the corresponding set becomes finite. This can be achieved in two ways:

1) by specifying all the atoms and groups which are represented by the variable substituents in a general formula (e.g. R, R', X in formula 58).
2) by specifying ranges for the parameters pertaining to physico-chemical properties of the compounds within the structural field (for the

structural field defined by formula 58, e.g., π (R) -1,20 to +2.50, σ(R') -0,30 to +0,50 MR(X) 10.00 to 30.00).

For most purposes it appears practical to start with the first type of specification and to derive from it the second type. An example is given in Table 2 in which some specifications for the variable substituents in formula 58, are compiled. By forming all possible combinations of R, R' and X one obtains a finite structural field or the corresponding finite set of 120 members or elements. A specification in terms of physical properties can now be achieved by simply looking for the highest and the lowest values that the parameters can adopt in the finite set. In the example the ranges are: π (R) 0,00 (H) to 1.96 (C_6H_5), σ(R') -0.27 (p-OCH_3) to 0.78 (p-NO_2), and MR(X) 5.42 (NH_2) to 17.06 (O-i-C_3H_7).

The necessity of specifying a finite structural field (or set) to which structure-activity hypotheses and SAR pertain imposes an additional requirement on those test series that are conceived on the basis of continuous variables. In addition to being orthogonal with respect to these parameters they also have to be representative for the finite structural field or set. This will be discussed in more detail Section II.B.5.

What is the origin of the structure-activity hypotheses? Are there any standard hypotheses that can be applied to all structural fields? Considering the complexity of drug action and the vast range of chemical structures that medicinal chemists can conceive, it is not surprising that such standards do not exist. However, such hypotheses can be formulated on an individual basis by simply asking oneself the question: "How would I interpret the biological data of my test compounds in terms of SAR? Which structural and physico-chemical properties would I take into consideration?" The answer to that question is the desired structure-activity hypothesis.

Take as an example general structure 58. One might wonder whether R may be different from hydrogen and if so whether it should be aliphatic or aromatic. With respect to R' one might be interested in how the position and the electronic properties affect the biological response. In terms of X one might suspect that there is a difference between alkoxy and alkyl- or dialkyl-amino with respect to activity. These considerations are expressed as a structure-activity hypothesis that specifies the following factors as possible determinants of the biological response:

- nature of R (hydrogen, aliphatic or aromatic),
- electronic properties of R' (electron donating or electron withdrawing),
- position of R' (meta or para), and
- nature of X (alkoxy or alkylated amino)

How are these factors used to design a test series? The simplest solution is to express the factors in terms of indicator variables. This can be easily achieved by representing every level of the factor by one structural feature (e.g., a substituent) whose absence or presence determines the value of a corresponding indicator variable. Thus the factor "nature of R" can be represented by the substituents H, CH_3, C_6H_5. Their presence or absence is expressed with the three indicator variables I_H, I_{CH_3}, $I_{C_6H_5}$. Similarly the electronic properties of R'can be accounted for by electron donating (e.g. p-OCH_3, m-CH_3) and electron withdrawing substituents (e.g. m- and p-NO_2). This choice of substituents also covers the third factor "position of R'". Consequently, the two factors can be expressed simultaneously in terms of four indicator variables, i.e., $I_{p\text{-}OCH_3}$, $I_{m\text{-}CH_3}$, $I_{p\text{-}NO_2}$, $I_{m\text{-}NO_2}$. Finally by choosing X=OCH_3, $N(CH_3)_2$ we can account for the fourth factor (nature of X) by two additional indicator variables I_{OCH_3} and $I_{N(CH_3)_2}$. Altogether we have thus constructed a structure-activity hypothesis in terms of nine parameters. Since all of them are indicator variables, a corresponding test series must be designed so as to meet the Free-Wilson conditions.

Alternatively one can express the factors by the explicit physico-chemical parameters discussed in Chapter 4 and apply those in selecting a test set. For instance the factor "nature of R" can be expressed by any physico-chemical parameter or any combination of such parameters that distinguishes between hydrogen, aliphatic, and aromatic systems (e.g., electrostatic potential, energy of highest occupied molecular orbital, or size). In advanced stages of an optimization process this type of representation can become more revealing than the simpler approach via indicator variables which in turn is normally more practical in the initial stages.

In lead optimization it is not important to maximize any one particular biological property but rather to find an optimal compromise between all those properties listed in Section II.B.1 that contribute to the therapeutic value of the new drug. All these properties should, as far as possible, be accounted for by structure-activity hypotheses.

c. Practical example of the SAR approach. A simplified example drawn from the field of potentially cardiotonic benzimidazole derivatives may illustrate the SAR approach to lead optimization. The lead compound was the 2-phenyl benzimidazole 59. This compound is only weakly

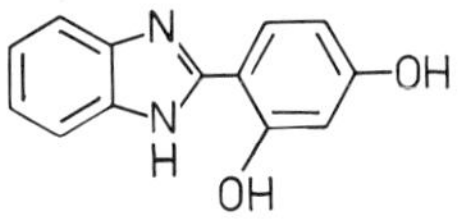

59

active in vitro in the guinea pig atrium and in vivo in the anesthetized cat. It is also only active on i.v. administration and the effect is very short-lived. Therefore optimization must primarily improve potency, duration of effect, and effectivity on oral administration in the pharmacological test models. Note that only those properties can be improved via the SAR-approach which can be reliably assessed in biological models that also allow a not too small number of compounds to be tested. A simple initial structure-activity hypothesis would normally suggest those moieties in the molecule that are capable of significant intramolecular or intermolecular interactions with the receptor. In the case of compound 59 these are primarily the mobile hydrogen atoms bound to nitrogen and oxygen. Two of these hydrogen atoms can participate in intramolecular hydrogen bonds (60, 61). These impart to the molecule a planar shape that might be important for receptor binding. Resonance effects between the phenyl and the benzimidazole rings also

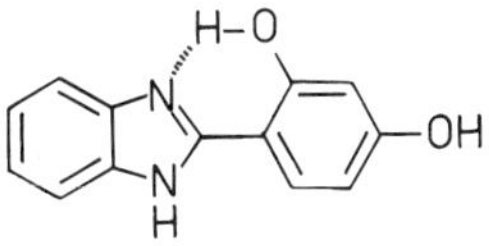

60

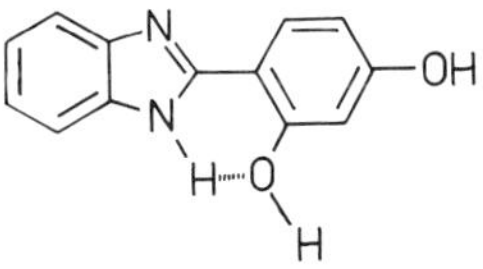

61

promote coplanarity. The resonance effects increase with increasing electronegativity of the benzimidazole system. These physico-chemical considerations lead to the structure-activity hypothesis that the two phenolic hydrogens, the imidazole hydrogen, and the electronegativity of the benzimidazole system determine whether a compound is significantly active.

The first two factors can be accounted for by indicator variables for the presence or absence of the mobile hydrogen, I_{2-OH}, I_{4-OH}, I_{NH}). A simple way of varying the third property is by introducing an additional nitrogen atom into the benzimidazole ring (62). This

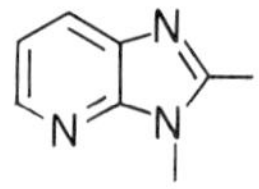

62

makes it possible to describe electronegativity in terms of an indicator variable (I_N) also. With respect to our hypothesis it is immaterial by which group one replaces the mobile hydrogens. Therefore one can choose groups accorording to synthetic accessibility, in the present case, for example, CH_3. A test series that fulfills the Free-Wilson conditions is shown in Table 3. In the guinea pig atrium model all compounds in which the imidazole hydrogen is replaced by methyl proved inactive whereas all the others show various degrees of activity. These results support the hypothesis as far as the imidazole hydrogen is concerned, whereas the other assumptions are apparently not valid. In-vivo experiments showed however that two other biological properties of interest, oral effectivity and duration of effect, are enhanced by the presence of the additional nitrogen. Therefore, in order to simplify the present discussion, further steps of the optimization procedure will be

Table 3 Test series derived from compound 59 via a structure-activity hypothesis (see text).

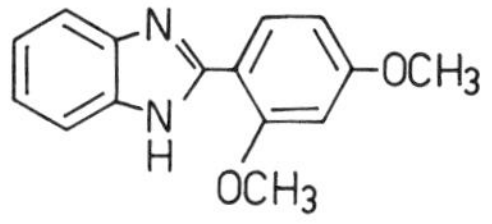

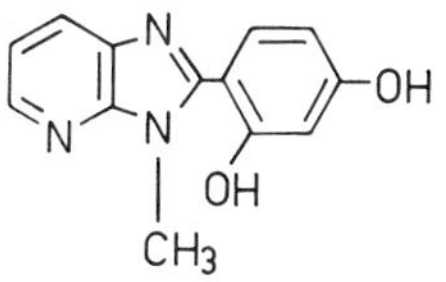

confined to 2-Phenyl-imidazo[4.5-b]pyridines (63).

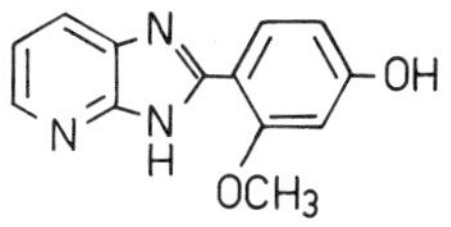

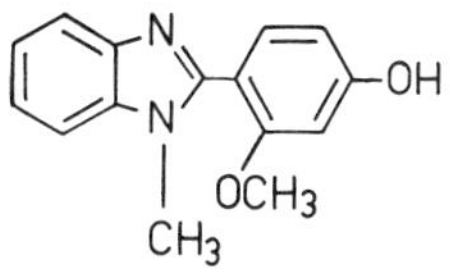

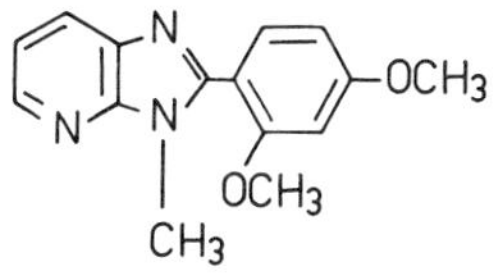

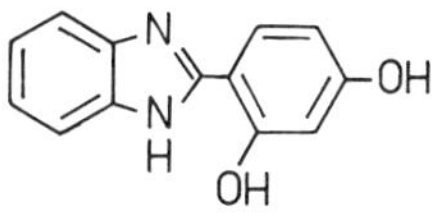

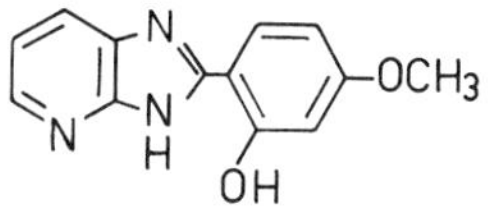

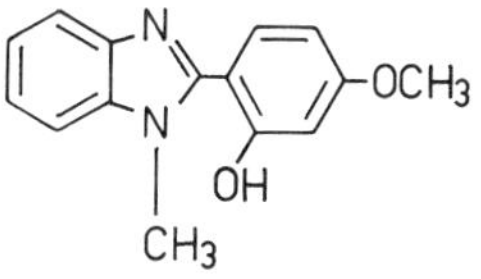

The previous test series (Table 3) demonstrated that both OH and OCH_3 are specifications for R^1 and R^2 that are compatible with inotropic activity. Any conclusion as to whether these substituents really need to be different from hydrogen or which of their physico-chemical properties are responsible for the activity cannot be drawn. An answer to these problems can again be found via structure-activity hypotheses which one investigates by appropriate test series.

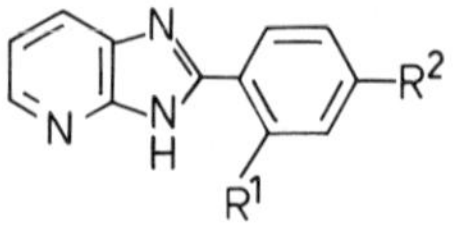

63

Let us start by considering the significance of the nature of the substituents R^1 and R^2. A possible hypothesis is that the activity is dependent on whether R^1 and R^2 are hydrogen, an oxy-substituent or some other group. If we respresent the oxy-subsituents by methoxy and the other groups by chlorine, the hypothesis can be expressed in terms of six indicator variables each of which denotes the presence of one of the possible substituents in one of the two positions. For example if $R^1 = OCH_3$ and $R^2 = Cl$ then the indicator variables for OCH_3 in 2-position and chlorine in 4-position adopt the values 1 whereas the other four indicator variables that refer to H in 2-position, Cl in 2-position, H in 4-position and OCH_3 in 4-position are zero. Any particular test series again needs to be designed so as to meet Free-Wilson conditions. One possible series is shown in Table 4. Only the two compounds in which $R^1 = OCH_3$ show appreciable biological activity. This finding largely supports our hypothesis that the factors "nature of R^1" and "nature of R^2" play indeed a decisive role. In addition the results suggest that an oxy-substituent in 2-position is favorable whereas in 4-position some variations seem to be possible without loss of activity.

All hypotheses and test series discussed so far have not used explicit parameters for physico-chemical properties. Yet, in order to find out what groups may be tolerated in 4-position a physico-chemical basis would be most desirable. We can introduce the physico-chemical parameters described in Chapter 4 via a third structure-activity hypothesis which assumes the lipophilic (π),

Table 4 Test series for checking the significance of oxygen substituents on the phenyl ring of structure 63.

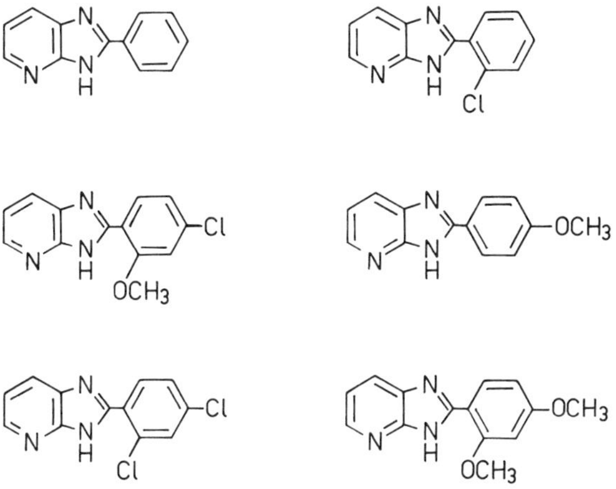

---

electronic (σ_p) and bulk (MR) properties of R^2 to be determinants of the activity. According to the general procedure outlined above we shall first specify ranges for these parameters and shall do so by assembling a finite set of substituents R, Table 5, for general formula 64. Note that this formula describes the structural field for which the hypothesis is tested.

From this finite set a test series is selected for which the three parameters are orthogonal and the values of which span the ranges for the parameters well. The choices are the underlined substituents in Table 5. The test results indicate that the size of R is crucial for in vitro and in vivo activity, that MR values of greater than about 20 are not tolerated, and that σ_p and π play a role mainly for in vivo activity, especially for duration of the effect. Parenthetically it might be added that the compound with R = $SOCH_3$, sulmazole, fulfilled

Table 5 Finite set of substituents R for formula 64.

Substituent	π	σ_p	MR
Br	0.86	0.23	8.88
Cl*	0.71	0.23	6.03
OH	-0.67	-0.37	2.85
NH_2	-1.23	-0.66	5.42
CH_3	0.56	-0.17	5.65
$SOCH_3$	-1.58	0.49	13.70
SO_2CH_3	-1.63	0.72	13.49
C_2H_5	1.02	-0.15	10.30
$N(CH_3)_2$	0.18	-0.83	15.55
SOC_3H_7	-0.63	0.50	22.90
$O(CH_2)_2SC_2H_5$	1.65	-0.11	28.73
$O(CH_2)_3SOCH_3$	-0.53	-0.11	28.73
$SO_2C_6H_5$	0.27	0.68	33.20
$OCH_2C_6H_5$	1.66	-0.23	32.19

ranges:			
	π	-1.63 to 1.66	(mean 0.02)
	σ_p	-0.83 to 0.72	(mean -0.05)
	MR	2.85 to 33.20	(mean 18.03)

*the underlined substituents have been selected for the test set

the requirements for a new drug in the pharmacological models to such an extent that it was submitted to clinical trial.

d. Strategic aspects of the SAR approach. In the preceding paragraphs it was demonstrated that lead optimization can be achieved in a systematic manner via experimental examination of a sequence of structure-activity hypotheses. From the point of view of strategy it is advisable to begin with hypotheses that pertain to major structural features of the lead compound. Such features comprise properties and substitution patterns of the basic skeleton of the lead compound. In the example above this was the 2-phenyl-benzimidazole skeleton, its electronegativity, and the presence of non-hydrogen substituents in positions 3 of the benzimidazole and 2 and 4 of the phenyl ring. Also groups that have the possibility of interacting strongly with the putative receptor should be given consideration in the early hypotheses. Examples of such groups are strongly basic, positively charged, and strongly acidic, negatively charged functions. Also good hydrogen bond acceptors and donors, such as the NH and OH groups in our example, as well as highly polarizable moieties such as aromatic π-electron systems should be subjects of the early hypotheses. All the early hypotheses are most conveniently treated in terms of indicator variables.

After the significance for activity of the main structural features has been investigated in this way, one can turn towards more subtle factors such as explicite physico-chemical properties. As will be discussed in Section II.B.5, for most purposes it is sufficient to account for these properties with variables that adopt two to four levels. Every combination of levels of the variables is represented by one structural feature, that is, a particular substituent. The test set for investigating the structural field characterized by formula <u>64</u> has been designed in this fashion (e.g., $SOCH_3$ represents the combination: low level of π, high level of σ_p and low level of MR). Only in very advanced stages and for the purpose of deriving quantitative structure-activity relationships need the physico-chemical properties be specified more precisely.

What is the advantage of the sequential procedure described above over an approach via a composite hypothesis that covers all potential structural factors simultaneously irrespective of their actual significance for the activity? The answer is that there are always structural features whose presence reduce the activity so much that moderate influences from other features become masked. In our example a non-hydrogen substituent attached to the imidazole nitrogen or the absence of an oxygen function in 2-positon of the aromatic ring are such activity-destroying features. It is the purpose of the early hypotheses to discover such features and to exclude them from further compounds.

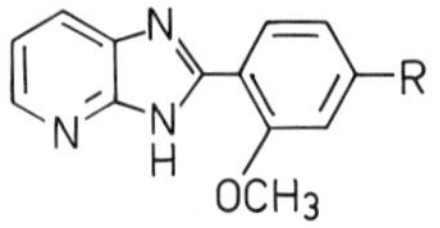

As straightforward and practical as this sequential approach may be, it is not free from pitfalls. Thus it cannot be excluded that an activity-destroying feature might be counterbalanced by one of the structural variations introduced in later stages of the optimization. Therefore it is advisable to back-check whenever structural modifications are applied that add a new quality to the molecule.

Even though the SAR approach is much more systematic than the chemical concept approach it is still empirical in nature. This implies that the rate of success depends on the rate at which structure-activity information is gained. This latter rate is determined by the quality of the test series but, of course, also by the time it takes to prepare these compounds. However, in the SAR approach test compounds cannot be chosen according to absolute but rather to relative synthetic accessibility. The synthesis problems that can arise in this context may be recognized if one considers that chirality is a main determinant of potency, selectivity and safety of drugs. Thus our structure-activity hypotheses must consider chirality too. Obviously one needs chiral test compounds in order to investigate such hypotheses. Enantiomerically pure chiral compounds are however not always easily accessible.

The developments in modern synthetic chemistry offer, however, many new possibilities for preparing complicated test compounds. For example some pure enantiomers may be made by stereoselective synthesis. The biotechnology methods discussed in Chapter 5 provide additional support. For example, one can use yeast, bacteria, or isolated enzymes to introduce chirality into a molecule in a completely specific manner. Moreover, one can use recombinant techniques to change the synthetic capabilities and specificities of such biological tools.

Within the frame of this chapter it is not possible to discuss these modern aspects of the synthesis of test series. It is, however, clear that progress in this field makes the SAR-approach increasingly feasible and attractive.

5. Techniques for the design of test series. As outlined in the preceding paragraph, systematic lead optimization relies on structure-activity hypotheses that are tested with the aid of series of test compounds. Ideally, such test series ought to be composed so that all the parameters specified by the structure-activity hypotheses are represented independently of one another and with equal weight. Normally medicinal chemists tend to design test series intuitively. As long as the structural field to be investigated is very limited and the number of parameters small, intuition might be an acceptable guide. However, the test series reported in the literature are

very frequently far from ideal. Consequently, structure-activity relationships derived from such series reflect more the composition of the series than the actual relationships. In addition it often happens that time and money are wasted by testing too many similar molecules. Therefore, techniques that help compose appropriate test series would be highly desirable. Medicinal chemists have realized this problem and have suggested a variety of solutions. Most of these have recently been reviewed (Austel, 1984).

Two different approaches to series design have been adopted:

1) Compilation of standard sets of substituents.
2) Selection of structural moieties or test compounds from some finite set such as from a larger list of substituents or a structural field.

In the present context one method for each approach will briefly be described. For references and descriptions of the other methods the reader is referred to the review mentioned above.

A method that belongs to the first category has been reported by Topliss (1977) and applies to cases where the dependence of a biological property on the substitution patterns of aromatic rings is to be investigated. One starts with a standard set of analogues which is composed of the unsubstituted (H) compound and its 4-chloro, 3,4-dichloro, 4-methyl, and 4-methoxy derivatives. This set pertains to a hypothesis that specifies lipophilicity and electronic properties of the substitution patterns as well as steric effects from the substituent in position 4 as relevant for activity. The test compounds are subsequently ranked according to their potency. From this ranking a new hypothesis is derived and tested via a second standard set of substitution patterns. These second sets are mainly designed so as to verify the hypothesis. On the one hand this can lead to quick optimization but on the other hand one might also increase the bias for a specific hypothesis which in reality may not apply fully. Such bias entails the risk of missing better alternative compounds.

In order to illustrate the method one of the examples given by Topliss will be used. The structural field under consideration is characterized by general formula 65, X referring to the whole substitution pattern (i.e.,

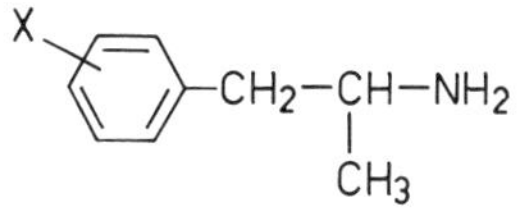

X-C_6H_4-CH_2-CH(CH_3)-NH_2 65

one or more substituents). By using for X the standard specifications mentioned above one finds the following ranking according to potency in the biological test (inhibition of phenethanolamine N-methyl-transferase): 3,4-Cl_2 > 4-Cl > 4-CH_3 > H > 4-OCH_3. Within the frame of the method this sequence suggests that the biological effect is equally dependent on lipophilicity as expressed by Hansch π-values and on electronic properties expressed by Hammett σ-values. Steric effects from the 4-position appear to be irrelevant. The new hypothesis, dependence of the activity on $\pi + \sigma$, leads to a second standard set of test compounds which carry the following subsitution patterns: 3-CF_3, 4-Cl; 3-CF_3, 4-NO_2; 4-CF_3; 2,4-Cl_2. This set contains the most potent compound of the series (4-CF_3), thus demonstrating the efficacy of the method.

An example of the second category of techniques is factorial design. This method can be applied to every conceivable structural field and to any number and type of variables that may appear in structure-activity hypotheses, i.e. indicator as well as continuous variables. For instance, the method can be used for problems in which substituents in several positions of the molecule are varied. The properties of these substituents may be accounted for by one or several indicator or continuous variables or by both types simultaneously. Also variations in the basic skeleton of a lead compound (e.g., benzimidazole-imidazopyridine, pyridine-quinoline, cyclic-acyclic, aromatic-aliphatic, planar-puckered, R- or S-configuration, etc.) with or without concomitant changes of subsituents can be handled by this technique.

The most useful and for most drug design purposes sufficiently accurate version is the manual one which will therefore be described. In fact the test series mentioned in Section II.B.4 have been designed by this method.

In factorial design one works with tables (schemes) that are composed of plus and minus signs which are arranged in a systematic and specific manner. Table 6 shows such schemes for two and three variables respectively. The columns (termed A, B, C) refer to the factors or parameters specified by a corresponding structure-activity hypothesis. The rows (numbered 1,...,8) signify the test compounds. A minus sign denotes a low level and the plus sign a high level of the parameter. Thus the first test compound of Scheme a in Table 6 needs to be constructed so that both paramters A and B are at low levels whereas the structure of compound 6 of Scheme b must be designed so as to set parameters A and C to the high and B to the low level.

In order to be able to apply these schemes we still need to define high and low levels of parameters. This task is especially easy if our hypothesis can be ex-

Table 6 Factorial schemes for two (a) and three variables (b) respectively.

a)

	A	B
1	−	−
2	+	−
3	−	+
4	+	+

b)

	A	B	C
1	−	−	−
2	+	−	−
3	−	+	−
4	+	+	−
5	−	−	+
6	+	−	+
7	−	+	+
8	+	+	+

A, B, C factors or corresponding variables
1, ..., 8 test compounds

pressed in terms of binary indicator variables. In this case the value zero, absence of the feature, and the value one, presence of the feature, determine the low and the high level respectively.

Take as a simple example a hypothesis that pertains to structure 66 and assumes the factors

- nature of X (CH or N) and
- nature of R (H or CH_3)

to be relevant for the biological response. These two factors can be expressed by the binary indicator varia-

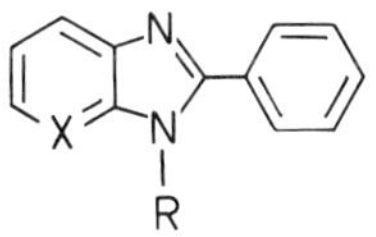

66

bles I_X and I_R. I_X denotes the presence or absence of the ring nitrogen and assumes the values zero for X = CH (low level) and one for X = N (high level). Similarly, I_R may be set to zero if R = H (low level) and adopt the value one if R = CH_3 (high level). We can now design a test series according to Scheme a of Table 6. This test series is shown in Table 7 with the levels the two parameters assume in each compound.

How can one specifiy levels for continuous variables such as π or σ ? In this case one can proceed in the same way as has been described in Section II.B.4 for reducing a structural field to finite size. One can either arbitrarily define a range for such parameters or derive the range from a predefined set of structures or substituents. For every parameter the range is divided in two halves. The lower one corresponds to the low level setting (minus level) and the upper half constitutes the high level setting (plus level).

Consider, for example, the set of substituents in Table 5. This set spans a range of π-values from -1.63 to 1.66. The mean that divides the range into an upper an a lower half is 0.02. Every substituent whose π-value is lower than 0.02 can represent the minus-level of π and every substituent whose π-value is greater than 0.02 can be considered if the scheme demands a plus level in π. The other continuous parameters in Table 5 can be treated accordingly. If we were to select a set of compounds from this table (as has been done for investigation of the structural field characterized by structure <u>64</u>) we would apply Scheme b of Table 6 since there are 3 variables to be covered. Thus we can identify π with factor A, σ_p with B, and MR with C. According to the scheme our first test compound carries a p-substituent that is hydrophilic (low level of π), electron donating

Table 7 Test series designed for structure <u>66</u> via scheme a) of Table 6.

test compounds	scheme	
	$I_X(A)$	$I_R(B)$
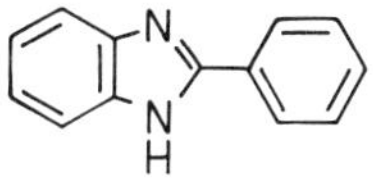 N, N, H	−	−
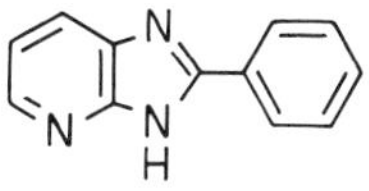 N, N, N, H	+	−
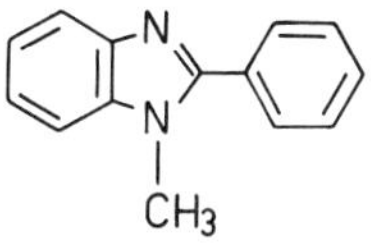 N, N, CH_3	−	+
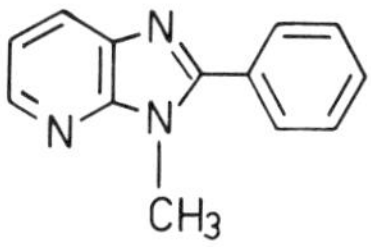 N, N, N, CH_3	+	+

(low level of σ_p), and small (low level of MR). The p-hydroxy derivative fulfills these requirements. So does NH_2. Under those conditions where there is a choice one would normally pick the compound that is most conveniently synthesized. Similarly, $SOCH_3$ is a representative for Row 3 of the scheme, as it is hydrophilic (minus level), electron attracting (plus level) and small (minus level in MR). In this way every one of the underlined subsituents of Table 5 represents one row of Scheme b of Table 6.

Factorial design schemes of the type shown in Table 6 call for a total of 2^n compounds to be made if the

Table 8 Complete factorial scheme for three parameters.

	A	B	C	AB	AC	BC	ABC
1	−	−	−	+	+	+	−
2	+	−	−	−	−	+	+
3	−	+	−	−	+	−	+
4	+	+	−	+	−	−	−
5	−	−	+	+	−	−	+
6	+	−	+	−	+	−	−
7	−	+	+	−	−	+	−
8	+	+	+	+	+	+	+

corresponding structure-activity hypothesis specifies n parameters. It is in fact not uncommon that a hypothesis demands four, five or six factors to be investigated simultaneously. This would require test series of 16, 32, and 64 compounds respectively. For practical purposes, especially in the initial stages of an optimization procedure, it is rather uneconomical to work with such large test series.

The size of the test series can be reduced by using fractional factorial schemes. A complete factorial scheme consists not only of columns for the individual factors or parameters but also for cross terms. Such a complete factorial scheme for three parameters is shown in Table 8. In order to reduce the number of test compounds if there are more than three parameters to be considered, we can identify the surplus parameters with cross terms. Thus four parameters can be adapted if the triple cross term ABC is identified with the fourth parameter. A fifth one could be represented by the cross term AC etc. Altogether a factorial scheme for three factors can be used for up to seven parameters (generally a 2^n-factorial scheme for 2^n-1 parameters).

A fractional factorial scheme derived from the full scheme of Table 8 has been used to design the test series shown in Table 3. In this case four parameters were considered (structure 67):

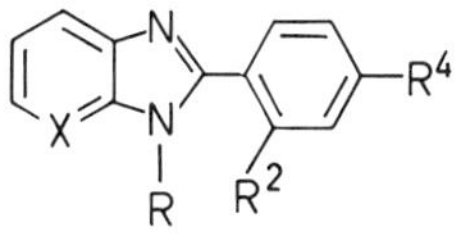

67

1) nature of X (I_X, minus if X = CH, plus if X = N)
2) nature of R (I_R, minus if R = H, plus if R = CH_3)
3) nature of R^2 (I_2, minus if R = OCH_3, plus if R = OH)
4) nature of R^4 (I_4, minus if R = OCH_3, plus if R = OH)

The first three parameters are represented by columns A, B, and C whereas the fourth one was chosen to adopt the levels specified by column ABC. Following the scheme the first compound (no. 1 in Table 3) must be a benzimidazole (I_X at minus level), have an unsubstituted imidazole nitrogen (I_R at minus level), carry a methoxy group in 2- (I_2 at minus level) as well as in 4-position (I_4 at minus level). Again every compound of Table 3 represents one row in a fractional factorial scheme composed of columns A, B, C and ABC of Table 8.

What happens if there are more than two alternatives for one position of variation? Such a problem arises in the design of a test series for structure 63. In that case both R^1 and R^2 can be either H

[I_H denoting presence (+) or absence (-) of H] or OCH_3 (I_{OCH_3}) or Cl (I_{Cl}). The usual factorial schemes cannot be applied here as setting one of the indicator variables to plus level automatically fixes the others at zero since there is no way of placing two different substituents simultaneously in one position of a benzene. Consequently, all rows that contain more than one plus sign cannot be represented by a compound. A method that can handle such problems has been suggested (Austel, 1985).The test series shown in Table 4 has been designed with this method.

During all the previous discussions of series design techniques one important molecular property, namely conformation, has never been addressed. With flexible molecules, however, this property is one of the major determinants of activity. Therefore in optimizing a lead compound one must also apply hypotheses that pertain to the conformation of the molecule when bound to the receptor. Chapter 4 detailed a number of methods that allow conformation-activity relationships to be determined. As with other methods of structure-activity analysis, the results depend heavily on the quality of the test set. Thus conformation-activity relationships can be relied upon only if the respective test series is orthogonal with respect to the conformational properties of interest.

In principle one could express these properties in terms of torsional angles and use these as parameters in, e.g., factorial series design. This approach, however, would be highly impractical not only because of the number of parameters but more so due to the fact that all the test compounds would have to be completely rigid. Generally such molecules require immense synthetic efforts.

In order to solve this problem we have developed a technique that allows test series to be designed so that they are approximately orthogonal with respect to conformational properties. These techniques do not require completely rigid molecules. The basis is again factorial design but it is applied here in a manner different to the one outlined above. For space reasons the present description will be confined to a "recipe" of application as illustrated by a simple hypothetical example. More complex problems can, however, easily be handled in a strictly analogous manner.

Consider compound <u>68</u> as being a lead compound whose

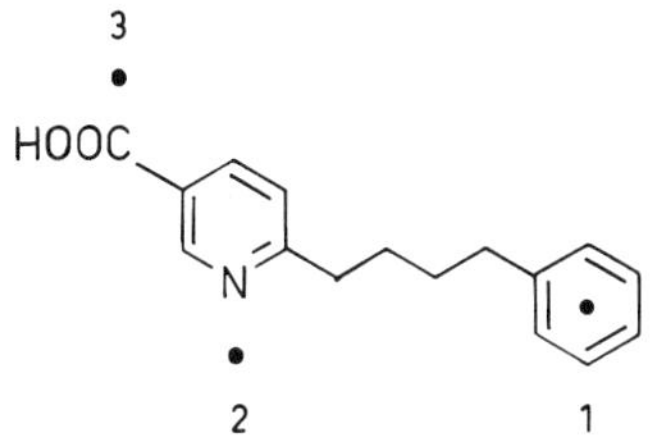

3
HOOC
N
2
1
<u>68</u>

conformation in the bound state is to be investigated. One starts the design by spreading a number of marker points to span the whole structure. They need not coincide with those parts of the molecule that interact with the receptor but it is helpful if they do. Therefore it is advisable to place the marker points on groups that are possible candidates for such interactions. For example, in <u>68</u> we may choose the phenyl ring because of possible hydrophobic interactions, the ring nitrogen as a potential hydrogen bond acceptor and the carboxyl group because of the negative charge it carries under physiological conditions. For <u>68</u> the marker points could therefore occupy the center of the phenyl ring, the position of the nitrogen atom and that of the carboxylic carbon.

Conformational states of the molecule can now be distinguished via the mutual distances of the marker points. In 68 the distance between marker points 2 and 3 is virtually constant. Handling the conformational properties of the molecule therefore reduces to a two parameter problem to which the 2^2-factorial scheme a of Table 6 can be applied.

As in the usual application of the schemes the range of the parameters need to be determined. In the present case these ranges can be obtained from a mechanical model, from computer graphics, or computer conformational generations. With a mechanical model we find for our example the maximum and minimum of distance 1-2 to be about 1 and 0.3 nm (van der Waals contact) respectively. The corresponding data for distance 1-3 are 1.6 and 0.3 nm. The parameter levels are defined as before (e.g., distances 1-2 of smaller than the mean, 0.65 nm, are at minus level; the others at plus level). We may now design compounds to examine the conformational properties by using the factorial schemes. Every row of this scheme represents a group of similar conformations which are distinct from those represented by another row. Thus examination of structure 68 reveals that all combinations of distance levels can be represented by conformations of the molecule. In order to find out which one of these combinations is the bioactive conformation, we have to design further test molecules. These can be flexible but must differ from one another with respect to the combinations of levels that can be represented by accessible conformations. In addition one has to make sure that the members of any pair of combinations do not occur exclusively together within the test series. Table 9 shows a test set designed via this method using Scheme a of Table 6. The columns A and B refer to distances 1-2 and 1-3 respectively. The test series fulfills the criteria mentioned above. Had we however omitted the second structure of Table 9 combinations (- +) and (+ +) would have only occurred together, i.e., the remaining compounds would not form a proper test series.

With the present design scheme it is also possible to get a sufficiently accurate estimate of the active conformation. If, for instance, apart from the lead compound 68 only the third compound in Table 9 would turn out to be active, one would have to conclude that the active conformation is closely related to the extended form (+ +).

III. SUMMARY AND OUTLOOK

Better and safer drugs are still needed for many diseases and in most cases the search for these drugs involves a lot of empiricism. Contrary to former times, however, we now have a much better understanding of how drugs can exert their effects. Especially the knowledge that the physico-chemical properties of drugs are responsible for their biological activity enables us to link this activity in a more general way to chemical structures. This ability makes it possible to differentiate between candidate structures for test compounds in terms of new structure-activity insights that they are likely to give. Such an approach has been made feasible and is strongly supported by the development of methods in theoretical chemistry such as quantum mechanics and molecular mechanics and with the advent of high performance computer technology discussed in Chapter 4. Thus we are now able to calculate electronic and conformational properties even of very large molecules. We can analyze large data sets and thereby extract information that has been hidden before. We can simulate and visualize enzyme action and receptor binding and thus open exciting perspectives for the formulation of structure-activity hypotheses and for

Table 9 Test series designed for structure 68 on the basis of conformational properties.

compound	occupied combinations of levels	
	1-2 (A)	1-3 (B)
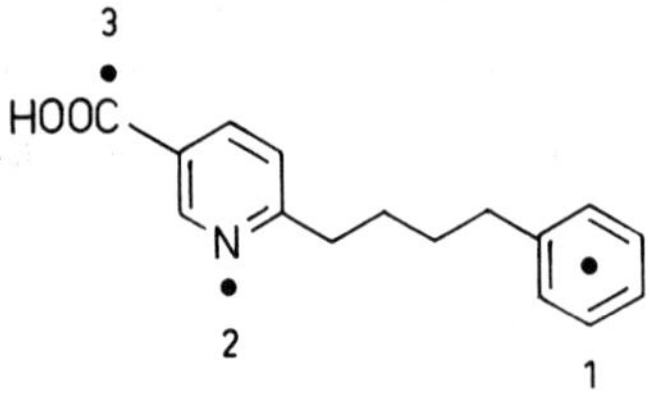 3 HOOC N 2 1	−	−
	+	−
	−	+
	+	+
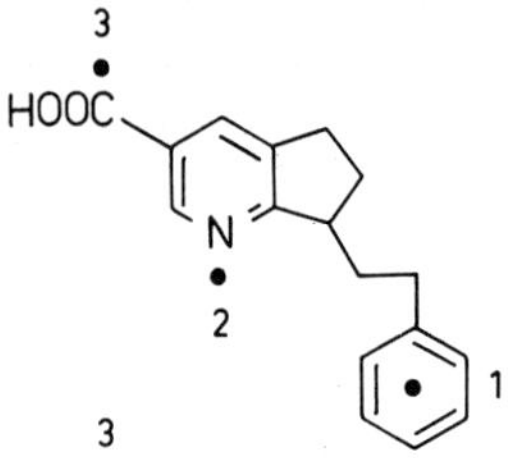 3 HOOC N 2 1	−	−
	−	+
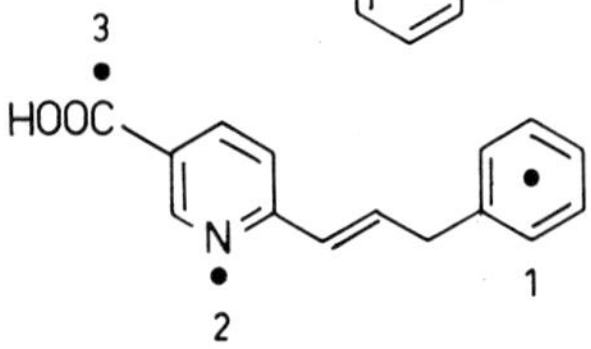 3 HOOC N 2 1	−	+
	+	+
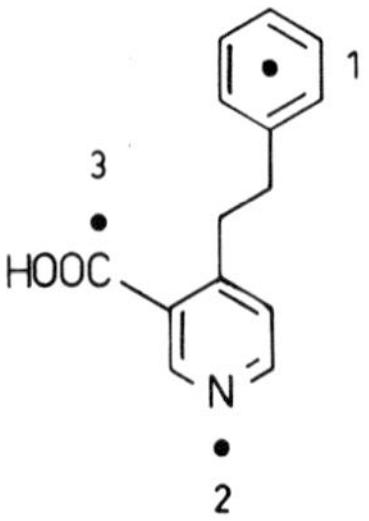 1 3 HOOC N 2	−	−
	+	+

the design of new original structures as candidates for better and safer drugs.

Considering all the means that have been made available for modern drug design, one wonders why medicinal chemists in former times were so admirably successful and why progress in drug therapy appears so slow nowadays. Certainly, the demands that are put on new drugs are much higher nowadays than they were in former times. Also the problems that are still challenges (cancer, autoimmune diseases, dementia etc., see Chapter 2) are much harder to tackle; otherwise at least some of them would have been solved already. Yet, this alone does not explain why such remarkable progress was possible in the early stages of medicinal chemistry. Probably it was the creative potential and an openmindedness that made the old chemists so successful without our modern tools. These virtues must be encouraged because still today it is the medicinal chemist's creative potential and openness for new principles and concepts that will ultimately determine progress in drug therapy. Therefore, we should not let ourselves be misled by the computer into worrying about insignificant details and into seeking perfectionist solutions. After all we have to concede that modern medicinal chemists still need luck to a certain extent and luck does not favor the computer, but the prepared (open) mind.

Acknowledgement

I wish to thank Miss Diana Schlichthärle for her expert typing of the manuscript.

REFERENCES

Austel, V., Design of Test Series in Medicinal Chemistry, <u>Drugs of the Future</u>, <u>9</u>, 349-365 (1984).

Austel, V., Manual Design of Test Series for Free-Wilson-Analysis, in <u>"QSAR and Strategies in the Design of Bioactive Compounds",</u> (J. K. Seydel, ed.), Verlag Chemie, Weinheim, 1985, pp. 247-250.

Austel, V. and E. Kutter, Absorption, Distribution, and Metabolism of Drugs, in <u>"Quantitative Strucutre-Activity Relationships of Drugs",</u> (J. G. Topliss, ed.), Medicinal Chemistry, Vol. 19, Academic Press, New York, 1983, pp. 437-496.

Beddell, C. R., Dihydrofolate Reductase: its Structure, Function and Binding Properties, in <u>"X-Ray Crystallography and Drug Action",</u> (A. S. Horn and C. J. De Ranter, eds.), Clarendon Press, Oxford, 1984, pp. 169-193.

Burger, A., Drug Design and Development. A Realistic Appraisal, <u>J. Med. Chem.</u>, <u>21</u>, 1-4 (1978).

Cushman, D. W., M. A. Ondetti, H. S. Cheung, E. F. Sabo, M. J. Antonaccio and B. Rubin, Angiotensin-Converting Enzyme Inhibitors, in <u>"Enzyme Inhibitors as Drugs",</u> (M. Sandler, ed.), Macmillan Press, London, 1980, pp. 231-247.

Douglas, K. T., Transition-State Analogues in Drug Design, <u>Chemistry and Industry</u>, <u>1983</u>, 311-315.

Farmer, P. S., Bridging the Gap Between Bioactive Peptides and Nonpeptides: Some Perspectives in Design, in <u>"Drug Design"</u> (E. J. Ariens, ed.), Vol. X, Academic Press, New York, 1980, pp. 119-143.

Free, S. M. and J. W. Wilson, A Mathematical Contribution to Structure-Activity Studies, <u>J. Med. Chem.,</u> <u>7</u>, 395-399 (1964).

Lipinski, A.C., Bioisosterism in Drug Design, <u>Annual Reports in Medicinal Chemistry</u>, <u>21</u>, 383-391 (1986).

Penning, T. M., Design of Suicide Substrates: An Approach to the Development of Highly Selective Enzyme Inhibitors as Drugs, <u>Trends Pharmacol. Sci.,</u> <u>4</u>, 212-217 (1983).

Pitman, I. H., Pro-Drugs of Amides, Imides and Amines, Medicinal Res. Rev., 1, 189-214 (1981).

Sinkula, A. A. and S. H. Yalkowsky, Rationale for Design of Biologically Reversible Drug Derivatives: Prodrugs, J. Pharm. Sci., 64, 181-210 (1975)

Smith, R. N., C. Hansch, K. H. Kim, B. Omiya, G. Fukumura, C. D. Selassie, P. Y. C. Jow, J. M. Blaney and R. Langridge, The Use of Crystallography, Graphics and Quantitative Strucutre-Activity Relationships in the Analysis of Papain Hydrolysis of X-Phenyl Hippurates, Arch. Biochem. Biophys., 215, 319-328 (1982).

Spatola, A. F., Peptide Backbone Modifications: A Structure-Activity Analysis of Peptides Containing Amide Bond Surrogates, in "Chemistry and Biochemistry of Amino Acids, Peptides and Proteins" (B. Weinstein, ed.), Vol. 7, Marcel Dekker, New York, 1983, pp. 267-357.

Sternbach, L. H., The Benzodiazepine Story, Progress in Drug Research, 22, 229-266 (1978).

Stezowski, J.J., X-Ray Crystallography of Drug Molecule-Macro-Molecule Interactions as an Aid to Drug Design, Annual Reports in Medicinal Chemistry, 21, 293-302 (1986).

Thornber, C. W., Isosterism and Molecular Modification in Drug Design, Chem. Soc. Rev., 8, 563-580 (1979).

Tickle, I. J., B. L. Sibanda, C. H. Pearl, A. M. Hemmings and T. L. Blundell, Protein Crystallography, Interactive Computer Graphics and Drug Design in "X-Ray Crystallography and Drug Action" (A. S. Horn and C. J. De Ranter, eds.) Clarendon Press, Oxford, 1984, pp. 427-440.

Topliss, J.G., a Manual Method for Applying the Hansch Approach to Drug Design, J. Med. Chem., 20, 463-469 (1977).

7

Evaluation of the Therapeutic Potential of New Compounds

C. Lillie and W. Kobinger

Department of Pharmacology
Ernst-Boehringer-Institute
für Arzneimittelforschung
Bender & Co GesmbH
Dr. Boehringer-Gasse
Vienna, Austria

I. INTRODUCTION

Before a new compound, expected to be better and safer than available drugs, can be used in patients or even in healthy volunteers, a great number of preclinical investigations are performed. Such investigations include experiments in animals, in isolated organs from animals, in tissue cultures and in biochemical systems in vitro. The ability of a scientific team to find new therapeutic agents is limited by the biological methods that are available to it. For this reason drugs with excellent or at least sufficient efficacy are available predominently in those therapeutic fields for which clinically relevant biological tests exist. This holds true for antibiotics, local anesthetics, muscle relaxants, spasmolytics, antihypertensives and antianginal drugs. It appears that in fields where no relevant biological test systems are available, no sufficient drug therapy has been developed, for example atherosclerosis or degenerative diseases of the central nervous system.

One must be aware that a direct approach to the discovery of a novel therapy is restricted by our limited knowledge of the underlying pathophysiology of the particular disease and the consequent limited ability of our animal models to imitate the human pathological situation. It follows that continued effort will have to be spent to investigate the nature of pathological events on a molecular basis.

II. PHARMACOLOGY AND MICROBIOLOGY AS LINKS BETWEEN BIOCHEMISTRY AND THERAPY

As discussed in detail in Chapter 6, traditionally the starting point for a chemical approach to drug discovery has been the chemical structure of known therapeutically active substances. These original leads are gained either from biological material or by chemical synthesis. The next step is often the synthetic variation of the structure of the leads and the testing of the analogues in animal or bacterial models that are considered to give some information on possible therapeutic effects. Thus by trial and error gradual improvements in therapeutic usefulness are achieved. Recent progress in synthetic and analytical chemistry has led to the elucidation of the biosynthetic routes, uptake mechanisms and metabolic routes of physiologically occuring compounds in viruses, bacteria, animals, and man. Many highly active modern drugs were developed on the basis of a continual feedback between advances in physiology, pharmacology and biochemistry: for example antidepressants, neuroleptics, antihypertensives and antiinflammatory drugs. The new drugs are often themselves tools to probe physiological or biochemical systems.

During recent years receptor pharmacology has gained increasing attention. In pharmacology a receptor is a macro-molecule that recognizes certain ligands, usually small molecules, with high specificity. If the ligand is an agonist, the interaction between the ligand and receptor is followed by a direct biochemical and physiological response of the target cell. The target cell may be an effector organ such as muscle or a secretory gland; or it may be a neuron that transmits information. Originally the existance of receptors was proposed from observations of pharmacological responses that followed the interaction of well defined drugs with different target systems. Then highly selective compounds were radioactively labeled and used to characterize the receptors involved in the biological effects of the compounds. This method has been extended to detect new compounds that also bind to the same receptor. The development of the ß-adrenoceptor blocking antihypertensives (Shanks, 1984), of the central α-adrenoceptor stimulating antihypertensives (Kobinger, 1978), of histamine H_2-receptor blocking anti-ulcer drugs, and of the muscarinic M_2-receptor blocking anti-ulcer drugs (Hammer and Giachetti, 1984) are examples of the use of receptor pharmacology or biochemistry in the recent discovery of novel drugs.

The ultimate goal of modern drug research is to find the molecular basis of a disease and to design chemical substances that correct the disturbance. This goal has been at least partly reached with the current treatment of Parkinson's disease. The critical observation was the discovery by Ehringer and Hornykiewicz of a deficiency of striatal dopamine in the basal ganglia of patients suffering from this neurological disease. It was logical to treat these patients with exogenous dopamine, using the precursor amino acid dopa, which easily penetrates the blood-brain barrier. The successful treatment of Parkinson patients with dopa was not only another success in therapy, but "it provided the first outstanding example of the successful therapeutic application of biochemically derived information to a chronic degenerative disorder" (Bianchine, 1980). Current research in Parkinsonism is directed to finding molecules that bind to the dopamine receptor in the appropriate brain regions with more specificity and that have better pharmacokinetic properties following systemic administration than the natural transmitter dopamine. Such substances can be found using animal models that imitate the pathological situation by the selective destruction of dopaminergic neurons (Hinzen et al., 1986).

Today the search for new anti-infective compounds often exploits the biochemical differences between microorganisms and man. For example, the antibiotic erythromycin effectively prevents bacterial growth with no human toxicity because it inhibits protein synthesis in bacteria but not man. The reason for this is that bacteria and humans differ markedly in the structure of their ribosomes, the cellular structures on which protein synthesis takes place. The explosive growth in the number of quinolone antibacterials in advanced human testing, some of which are expected to be major improvements in antibacterial therapy, was at least partly stimulated by the biochemical observation that the original members of this class of compounds inhibit an enzyme that is unique to bacteria and is required for their synthesis of DNA. Thus, new analogues can be tested as enzyme inhibitors and as inhibitors of bacterial growth in culture. Scientists are always on the alert for new biochemical differences to exploit in the development of antibacterials since it is usual that there is little cross-resistance between antibiotics with different modes of action.

III. ETHICAL CONSIDERATIONS ON ANIMAL EXPERIMENTS

In recent years there has been increasing concern as to the ethics of animal experiments. As will be seen in the following sections, the evaluation of potential new drugs is based on animal experiments. As will also be seen, many alternative methods are used, for example, effects on enzyme activities and on membrane preparations in vitro. Although these biochemical methods often use animal tissue, many more tests per animal are possible. It is to be expected that materials from biotechnology may also help to further reduce the number of test animals needed. Such methods moreover provide a rational and exact approach to the identification of compounds with the desired profile. However, it would be irresponsible towards society to test a new compound in humans on the basis of such alternative methods only and not to consider the complexity of an intact organism; the latter can be done only in experiments using intact animals.

The current practice of the use of animals in drug research includes the following points:

1) The majority of such experiments are performed in full anesthesia.
2) Generally, discomfort caused by manipulations to nonanesthetized animals do not exceed those that are demanded from human patients during routine diagnostic procedures such as intravenous, or intrarterial cannulas, stomach tubes or electrocardiographic measurements.
3) In particular experiments it may be necessary to obtain data with more complicated interventions without general anesthesia. In most countries such experiments are covered by special legal review procedures; they are performed under local anes-

thesia; and, in the course of drug development, such experiments are reduced to a minimum and planned only for the latest stage of development.

The rational and humane use of animals is an important criterion to be considered in each stage of preclinical drug development.

Animal experiments are controlled in most countries by government regulations; the qualifications of the scientists who undertake such experiments and the conditions for breeding and keeping laboratory animals are well defined.

Ethical drug development is a continuous process of decision making, see also Chapter 10. The following decision must be continually evaluated: Is the underlying medical need so great that better and safer drugs should be aimed at in spite of the inevitable use of animal experiments?

IV. STAGES OF BIOLOGICAL TESTING

Neither the individual patient nor society would be satisfied if the search for new drugs waited for a full understanding of the disease process. Rather, the search for new drugs has to use the tools and information available at the moment. No single biochemical or animal model is sufficient to decide on the potential clinical usefulness of a new substance. A combination of test systems from different species has to be used to gain an activity pattern of the substance. Either this pattern by itself may suggest a novel therapeutic usefulness of the substance, or the comparison with the pattern of drugs with established therapeutic value indicates the position of the new compound.

As can be appreciated from Chapters 4 and 6, drug design rarely meets the therapeutic goal with the first compound synthetized. Thus many new compounds have to be tested in several biological models in order to optimize the therapeutic potential of the set of compounds. In order to adapt the biological testing effort to a fixed research capacity and to minimize use of animals, a strategy has to be developed for each research project. Usually the strategy is stepwise, whereby expenses, complexity of methods, and complexity of the animal test system are increased from step to step. At each step those compounds that do not fulfil the requirements are discarded. This prevents further waste of capacity and animals on those compounds that are certainly not suited to the given therapeutic target.

The main steps of drug development may be characterized as follows. In Step 1 a great number of new compounds is put on a wide-meshed screen of rather simple models. The width of the mesh is adjusted to allow false positive rather than false negative decisions. In Step 2, the selected compounds are put through a greater number of different models to describe the activity pattern, as well as the pharmacodynamic profile, and the pharmacokinetics as determined indirectly by biological effects. In Step 3, the pharmacological or microbiological profile of the selected compounds is investigated in more detail, and the results are considered together with those from direct pharmcokinetic measurements and from toxicological investigations. This provides more information on the possible therapeutic usefulness of the compounds. Substances that fulfill these biological requirements have to meet further criteria, such as chemical stability, appropriateness for pharmaceutical formulation (see also Chapter 9) and economic feasibility in order to be selected for first human trials (Step 4). Further biological investigations usually continue even after the drug is in clinical trials and therapeutic use to reveal details of the molecular mode of action, interference with other drugs etc. Such studies may provide new insights that lead to a new testing strategy.

Thus far we have described the testing of compounds designed for a known mode of action or for a well defined action pattern of standard drugs ("targeted screening"). Another strategy for drug discovery is the "random screening" of compounds or isolates from natural systems such as soils or plants not designed for any particular drug effect. In this procedure more than one screening model is used in Step 1 and substances with positive actions in one direction are further investigated in Steps 2, 3, and 4 as described above.

In spite of all the efforts for a rational strategy for molecular drug design, it must be emphasized that a great number of novel drugs have emerged from observations entirely made by chance. Examples are the mercury and sulfonamide diuretics, oral antidiabetics, many psychoactive drugs and the antihypertensive clonidine. The testing of potential drugs, therefore, cannot be regarded as an inflexible procedure. Sometimes new and unexpected observations make it necessary to consider mechanisms not included in the original design. The ultimate success of a program is highly dependent on the flexibility, creativity and knowledge of the scientists involved as well as the environment in which they work.

V. BASIC CONSIDERATIONS FOR TEST PROCEDURES

The types of biological models available are summarized in Table 1. This table shows in the upper rows those tests that seem to allow an easy and fast selection of

Table 1 Thc Information Provided by Biological Test Models

Organizational Complexity of Biological Test System	Types of Information Gained
Isolated Enzymes	Inhibition or stimulation of enzyme activity, selective toxicity in bacteria or viruses vs man
Membrane preparations or intact cells	Radioligand binding to a receptor, displacement or release of natural substances, ionic exchanges across cell membranes, whole cell antibacterial growth, biochemical response to ligand binding
Isolated organs	Complex physiological functions such as muscular contractility or nerve excitability
Intact normal animals	Physiological functions in feedback with reflexes and including pharmacokinetics
Intact animals with induced pathological disfunction	Correction of malfunction

active compounds out of a larger number of substances. It seems natural, therefore, that a testing strategy always proceeds stepwise from the top to the bottom of the table. However different approaches are sometimes more rational and efficient.

VI. EXPLORATIVE PHARMACOLOGICAL OR MICROBIOLOGICAL INVESTIGATIONS

A. DRUG ACTION ON ENZYMES: INHIBITION OR FACILITATION

Simple models, as indicated in the upper part of Table 1, can be used whenever the therapeutic target is orientated towards a known mode of action. When this mode of action can be verified by an enzyme activity, which might be either inhibited or facilitated by a drug, a precise and quantitative evaluation of a great number of compounds may be possible. As can be appreciated from Chapters 4 and 6, this is a great advantage for the medicinal chemist because it allows the unambiguous assignment of certain chemical groups as essential or not essential for the targeted activity. As an example, potential natriuretic drugs of the acetazolamide-type (1) were tested in great numbers as inhibitors of carbonic anhydrase. This test is easily performed in vitro. As might be expected, the correlation between these in vitro results and diuretic potency in vivo is complicated by variations in drug pharmacokinetics (Mudge, 1980). Thus action on enzymes is most appropriately used to discard compounds that are obviously inactive.

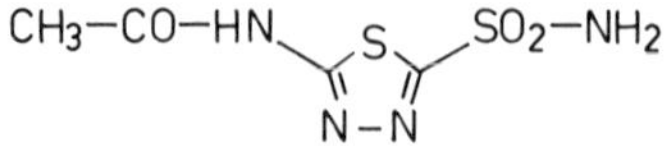

CH_3—CO—HN—(1,3,4-thiadiazole: S; N—N)—SO_2—NH_2

1

A special advantage of enzyme screening is that human enzyme may be used. Human enzyme may be easily available from genetic engineering or autopsy material. Being able to use human enzyme may be of decisive advantage. For example, in the search for new antihypertensives from renin inhibitors, it was observed that mouse and human renin differ to such an extent that it is inappropriate to use mouse enzyme for testing potential inhibitors.

In the search for a new anti-infective agent sometimes the target enzyme is found in both man and the infectious agent. Then testing against both enzymes gives a direct measure of possible selective toxicity. Compounds that selectively inhibit the enzyme of the invader would be pursued further.

Screening new compounds on enzymes has the disadvantage that substances will be discarded that might achieve the therapeutic target not by affecting the particular enzyme, but by another, perhaps hitherto unknown, mode of action. For example, the benzothiadiazine diuretics such as hydrochlothiazide (2), have much lower carbonic anhydrase inhibitory potency, but a much more favorable therapeutic profile than compounds of the acetazolamide-type (1; Lund and Kobinger, 1960). Thus drug discoveries may be missed by relying on highly specialized enzyme models only.

A further disadvantage of enzyme screening is that one will not observe the potential utility of metabolic products of the administered compound.

In addition, sometimes enzyme assays are very time-consuming to perform. It may be difficult to isolate the enzyme or the substrate; it might require elaborate equipment or experimental design to follow the kinetics of the reaction of the enzyme with substrate and inhibitor; one may have to follow the reaction under many ionic and pH conditions; and the potential inhibitors may not be soluble enough to test.

On the other hand, it can be expected that, in the

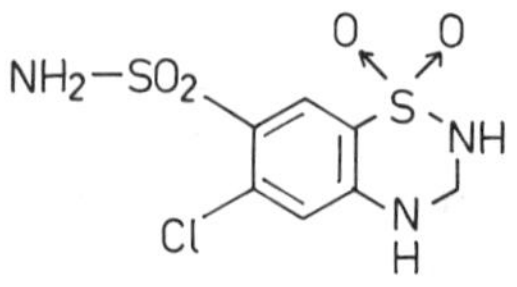

2

future, more information will be gained on the biochemical bases of pathophysiology and further improvements in genetic engineering and biochemical techniques will be made. The knowledge of the role of particular enzymes in pathophysiology will enable us to discover new drugs to treat diseases more specifically. Such targeted therapy could be achieved by administering native or synthetic enzymes, or by using small molecules to stimulate or inhibit the enzymes that are related to the pathology.

Additionally, new knowledge on the biochemical differences between infectious agents and man will allow specific targeting of agents with selective toxicity. Finally, knowledge of the biochemical differences between cancer and normal cells may allow the design of compounds that are selectively toxic to cancer cells.

B. RADIOLIGAND BINDING ASSAYS

As noted in Chapter 1, a fascinating approach to drug screening at a molecular level was realized with the introduction of radioligand binding assays. These methods measure the affinity of compounds for cell surface recognition sites. Such measurements are attractive for the medicinal chemist because they provide structure-activity information in a direct, quantitative, fast assay that requires a minimal amount of test substance and is not complicated by drug metabolism or penetration to the site of action.

Radioligand binding assays are also attractive because they use minimal amounts of tissue. The recent cloning of several human receptors suggests that soon it will be possible to screen compounds for binding to the human receptors.

Williams and U'Prichard (1984) reviewed the binding studies that have been performed with dopamine-, serotonin- and adreno-receptors. The authors pointed out that binding studies do not provide information about events after binding. A correlation of binding data with biological dose-response curves in identical tissue preparations is necessary to validate a binding assay. This is especially true of agonists. Testing of antagonists by means of radioligand binding tests might have a better chance to directly detect potential useful new drugs. As an example we may cite the good correlation of [^{3}H]-prazosin (3) binding in rat brain membranes with the α_1-adrenoceptor blocking activity in vitro in rabbit isolated pulmonary artery as well as the in vivo vasopressor effect in pithed rats (Figure 1, Timmermans et al., 1981).

On the other hand, all the displacement of radioligand is not specific and attributable to competitive binding to receptors (Laduron, 1983; Kenakin, 1984). In many cases a radioligand is best displaced by the unlabeled compound. Compounds from other chemical classes that have the same receptor- or pharmacological profile may show less displacement. This suggests that the radioligand binds to nonspecific as well as specific receptor sites. A number of such examples are reviewed by Laduron (1983). Consequently, screening of compounds by radioligand displacement will be most meaningful within a group of similar chemical structures. For example, by simple binding tests, neither gallopamil (4), verapamil (5), nor diltiazem (6) showed affinity for [^{3}H]-nitrendipine (7) binding sites (Belleman et al., 1981). All four compounds are Ca^{++}-antagonists and exert essentially similar biological effects. They belong, however, to three different chemical classes. The point is that neither gallopamil, verapamil nor diltiazem would have been detected using a radioligand displacement test with [^{3}H]-nitrendipine as the ligand.

Radioligand binding studies have assisted the dis-

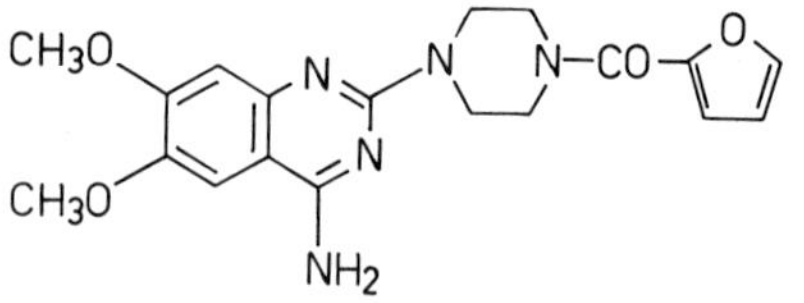

3

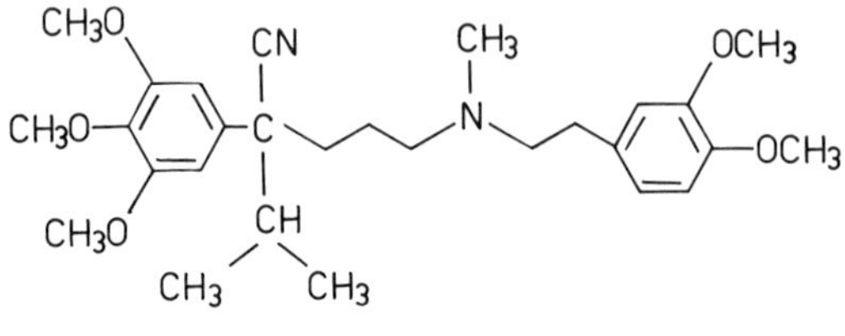

4

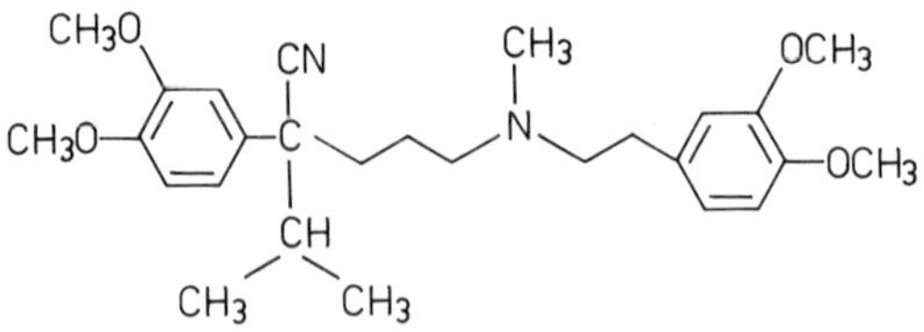

5

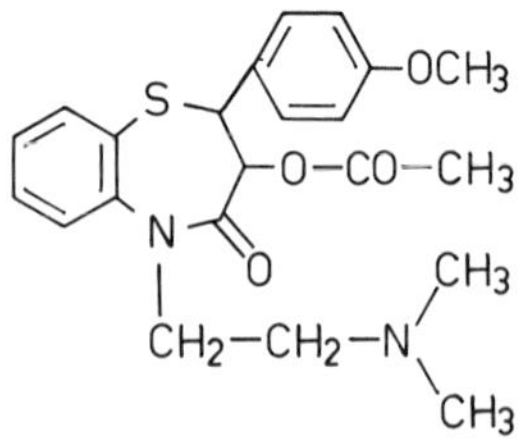

6

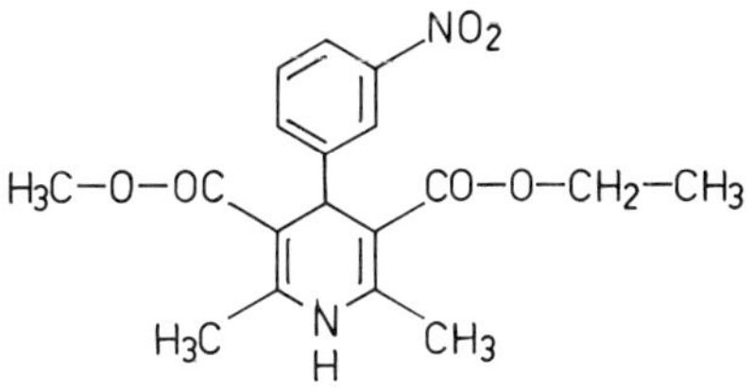

NO_2

$H_3C-O-OC$ $CO-O-CH_2-CH_3$

H_3C N H CH_3

7

covery of receptor subtypes. For example, different subclasses of muscarinic receptors were required to explain the selectivity of the antimuscarinic drug pirenzepine (8) towards inhibition of gastric secretion compared to inhibition of salivation or changes of heart rate (Hammer et al., 1980).

The interpretation of radioligand binding studies needs support from pharmacological tests of physiological function. For example, pharmacologically St 587 (9) appeared to be a highly selective α_1-adrenoceptor agonist, yet in binding studies the drug displaced both α_1- and α_2-ligands (De Jonge et al., 1983). Further pharmacological examination revealed that the drug is also a competitive α_2-antagonist (Pichler and Kobinger, 1985).

An example for a still unexplained lack of correlation between radioligand binding and pharmacological results is presented by the fact that the selective α_2-agonist B-HT 920 does not displace the α_1-adrenoceptor ligand $[^3H]$-prazosin (3) (10; Van Meel et al., 1981) in spite of B-HT 920's clear-cut antagonism of the functional response of the α_1-agonist methoxamine (11; Kobinger and Pichler, 1981). This observation suggests that there may be differences between radioligand binding sites and functionally intact receptors and that these

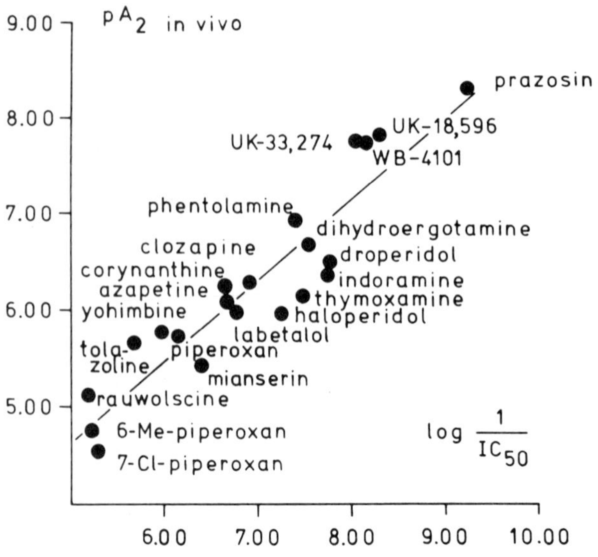

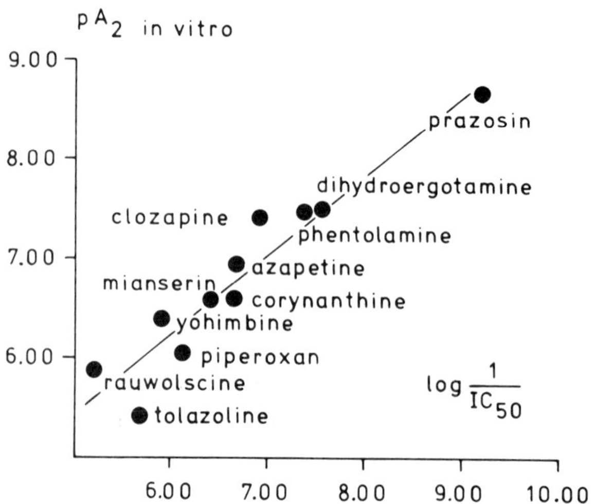

Figure 1 The relationship between the antagonistic potency of various alpha-adrenoceptor blocking drugs toward $alpha_1$-adrenoceptor-mediated vasopressor effects in vivo (pA_2 in vivo, top) or constrictor responses in vitro (pA_2 in vitro, bottom) and inhibition of $[^3H]$ prazosin binding in rat brain membranes (log $1/IC_{50}$). Values for pA_2 in vivo refer to the antagonism of the drugs with respect to (-)-phenylephrine-induced increases in diastolic pressure of pithed, normotensive rats. The pA_2 in vitro values were obtained with the isolated rabbit pulmonary artery. Log $1/IC_{50}$ values (abscissae) were calculated from displacement data (Timmermans et al., 1981).

differences are revealed by the interactions of intact receptors with drugs of different chemical classes.

In conclusion, radioligand experiments have an interesting role for evaluation of the profile and potency of drugs with similar chemical structures. The utility of radioligand binding in the search for totally novel chemical structures is promising but not yet demonstrated.

Current research on the isolation of receptors from cell membranes and their subsequent reconstitution into functional biochemical pathways promise to provide insights into the more complex systems in intact cells.

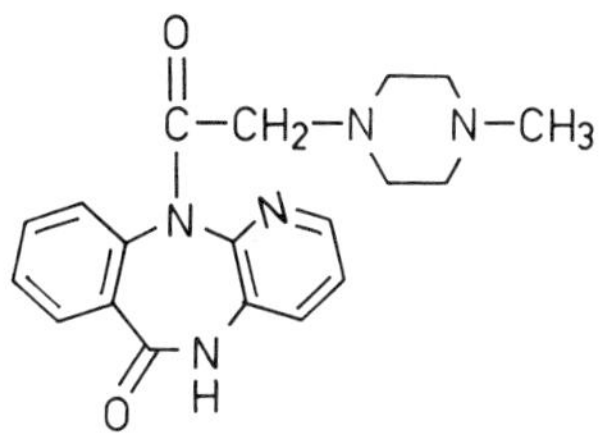

8

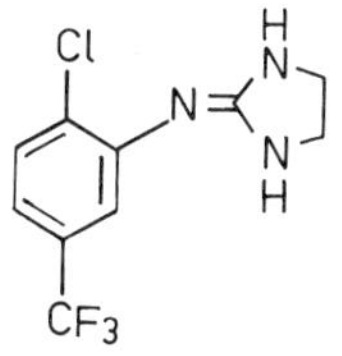

9

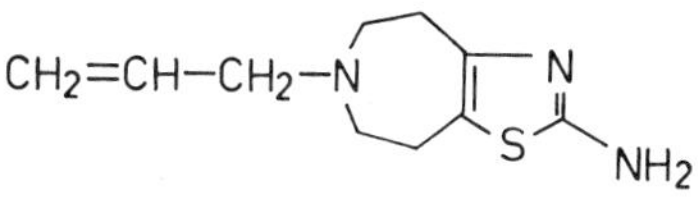

10

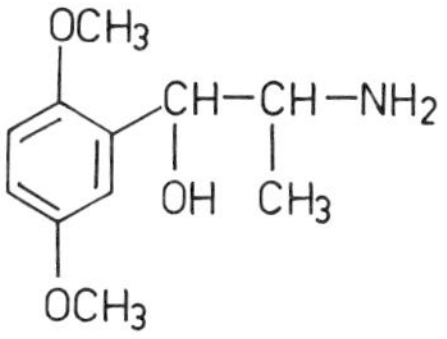

11

C. PREPARATIONS WITH INTACT CELL MEMBRANES

Preparations may be used where receptors are in functional connection with other biochemical systems of the membrane. For example, the binding of an agonist to the D-1 dopamine receptor stimulates the activity of adenyl cylase. Thus one can evaluate the agonist properties of a compound by comparing its effect on adenyl cyclase to that of dopamine. Antagonists would be tested for their ability to antagonize the dopamine-induced changes. In another example erythromycin analogues can be tested for their relative ability to inhibit bacterial protein synthesis.

Such cell-free tests offer the scientist most of the advantages of enzyme and radioligand binding tests but also share many of their disadvantages. The essential differences are the ability to distinguish agonists from antagonists and the ability to establish an effect on an integrated biochemical system.

D. TESTS IN INTACT CELLS

Intact cells represent a higher level of organization of biological test systems as compared to enzymes or cell membranes. This type of experiment may be an alternative to animal experimentation for explorative screening of the biological properties of test compounds for pharmacological properties. Obviously many of the advantages for a rational approach are similar to those cited above for enzyme and radioligand binding assays.

An example of a test in an intact cell is the exchange of Na^+ for K^+ through the membrane of human erythrocytes. This process is inhibited by cardiac glycosides useful in heart disease (Schatzmann, review: Aronson, 1984). The potencies of drugs evaluated with this method parallel their cardiac potencies in man when plasma protein binding is also taken into consideration. Obviously this method is a reliable predictor of potency in humans only for substances that act by the same mechanism as cardiac glycosides.

The test can be performed on cardiac cells from tissue culture. Such cells of primary cultures respond similarly to pharmacological agents as do intact myocardial tissue. Currently such cultures do not maintain the function of contractility and excitability: there is no doubt that establishment of a permanent line of functional cardiac cells would facilitate, improve and rationalize research on cardiac drugs.

Testing in intact cells is often the method of choice for testing antibacterial compounds.

E. EXPERIMENTS WITH ISOLATED ORGANS

The functional responses of isolated organs result from a multiplicity of biochemical events that are highly organized in ways we are only beginning to understand. Thus it is attractive to test compounds in intact organs since they contain intact cells from more than one type of tissue.

Most such preparations result in muscle contraction, formation and conduction of electrical impulses in muscular or neuronal cells, transport or fluxes of ions, or release of hormones. The measured effects may not depend on a single molecular mode of action, but are triggered by a variety of chemical structures. Further advantages of isolated organs for primary screening are the good reproducibility which leads to quantitative comparisons between compounds, the use of only small amounts of drug, and the saving of experimental animals compared to whole-animal experiments by simultaneous use of different organs from one animal or by testing of several doses in one preparation.

Generally there is little variation in the responses of tissues from various species to a particular compound. In some cases specimens of human tissues can be ethically obtained from surgical procedures or rapid post mortem acquisition. This optimizes predictability of effects of test compounds in humans (Kenakin, 1984).

F. EXPERIMENTS IN INTACT ANIMALS

There are a number of therapeutic targets where the primary screening is best achieved by the use of intact animals, especially if small laboratory animals can be used. This approach is necessary if the desired information is complex and not sufficiently determined by single separable mechanisms: this is true for changes in behavior, seizures, blood pressure or diuresis. Another reason for starting evaluation of a test compound with intact animal preparations is the availability of methods that are more economic and save more time or animals than the use of the in vitro methods cited above. For example, the use of intact rats proved more efficient for testing specific bradycardic agents than using isolated spontaneously beating atria. In this context the term specific means that the drug is considerably more potent in lowering heart rate than in exerting other cardiovascular effects. It takes less time to set up the animal experiment than an isolated heart. The whole animal model also has the advantage that in addition to the effects on heart rate it also reveals effects on blood pressure (Kobinger et al., 1979; Stähle et al., 1980).

Sometimes a pharmacokinetic target, such as oral activity or a minimum duration of action, is the primary objective in a project; evaluation of this can only be done in intact animals. As an example, the test introduced by Lawson (1968) for antiarrhythmic drugs can be cited: compounds can be administered by various routes at different times to conscious mice before the test is performed. Compounds that either do not produce an effect orally or that have too short a duration of action would be discarded.

Last, but not least, it should be remembered that testing in intact animals offers the chance of unforeseen

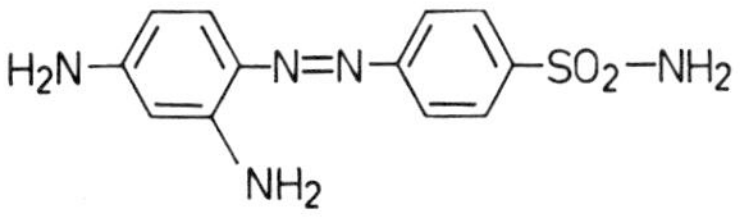

H_2N–$C_6H_3(NH_2)$–N=N–C_6H_4–SO_2–NH_2

12

H_2N–C_6H_4–SO_2–NH_2

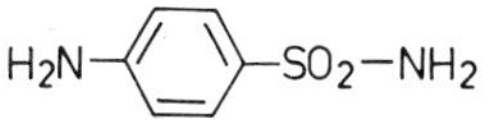

13

discoveries. In 1932 Domagk administered prontosil[R] (12) to intact bacteria-infected mice and thereby discovered the life-saving effect of the first sulfonamide drug. Only later was it discovered that prontosil acted as a prodrug, releasing the active principle sulfanilamide (13) into the organism (review: Weinstein, 1970). A primary test in an isolated bacterial culture at this stage would not have led to this life-saving discovery of Domagk, for which he was awarded the Nobel Prize in Medicine in 1938 and which heralded the era of effective therapy against bacterial infections.

G. SPECIFIC STRATEGIES FOR TESTING POTENTIAL ANTIBIOTICS

The discovery of the sulfonamide antibacterial agents by an in vivo test as described in the last section represents, however, not the usual procedure in the search for new antibiotics. The routine tests for antibiotics are performed in vitro by culturing microorganisms and observing the changes in bacterial growth induced by the test compound. Different concentrations of test substances are added to given amounts of a variety of bacteria that are growing either on a solid medium such as agar plates or in a culture medium. After a given time of incubation the growth of the bacteria is quantified. On the agar plate the effect of the antibiotic appears as a circular translucent area, in solutions the growth is shown as turbidity. Both effects are concentration-dependently inhibited by antibiotics. An important property of an antibiotic is the spectrum of types of bacteria that are inhibited. Usually the optimum therapy is that which eradicates the disease-causing microbe but no others.

Naturally the ultimate aim is effectiveness in vivo with a defined antibacterial spectrum and possibly organ selectivity. These aims can be verified only by studies in intact animals including pharmacokinetics. On the other hand, for many antibiotics molecular modes of action are known and primary screening can be based on very narrow and specific biochemical models. For example, in the search for agents that inhibit cell wall synthesis, a test for inhibition of carboxypeptidase or transpeptidase activity may represent the first step of a screening procedure (Drews, 1983).

VII. PROFILING OF SELECTED COMPOUNDS

During Stage 2 of drug development (Section IV) the goal is to reach a definite conclusion as to the therapeutic usefulness of the substance. As this usually requires sophisticated test models, this advanced testing is started with a model that will reveal compounds that do not fulfill the minimum requirements with respect to activity and selectivity. For example, for cardioprotective ß-adrenoceptor blockers a specified selectivity for $ß_1$-/$ß_2$-receptors may be required; in some antiarrhythmic drugs, it is easy to test for side effects such as sedation or excitation that will eliminate a compound from further development. Usually at this stage experiments in intact animals will predominate.

The following paragraphs give an account of how the pharmacodynamic and pharmacokinetic profile of a compound is determined.

A. PHARMACODYNAMIC CHARACTERIZATION

1. Dose-response curves. The determination of a dose-response curve for the principal effect of the compound is indispenable. It not only establishes the relative potency of a drug, but it may also reveal differences in the mode of action as compared to a reference drug. The parallel dose-response curves of the compounds I, II and III in Figure 2 are compatible with the assumption of an identical mode of action, the different position along the abscissa representing their relative potency: the very left curve (I) is the most potent compound, followed by II and III. It is important to realize that all three compounds, I, II and III, do produce the same maximum effect at an appropriate dose. The relative potencies of I, II and III are approximately 1, 6 and 90.

The experimental results shown in Figure 2 are the urinary excretion of sodium after administration of three thiazide diuretics (bendroflumethiazide I, 14, hydroflumethiazide II, 15, and chlorothiazide III, 16) to rats (Lund and Kobinger, 1960). As was confirmed later in clinical studies, the parallel slopes of the curves not only suggest a similar clinical utility of these drugs, but they also allow one to estimate the relative doses for clinical use. The clinical doses are 2.5 - 5.0 mg for I, 25 - 50 mg for II, and 250 - 500 mg for III (Mudge, 1980).

In contrast to curves I, II and III, the dose-response curve of compound IV has a steeper slope and a higher maximum effect. This different profile not only suggests a different mode of action but also that the compound might be useful in different clinical situations than are I, II and III. Curve IV results from administration of furosemide (17) to rats. In spite of a chemical similarity with saluretic thiazide derivatives, its

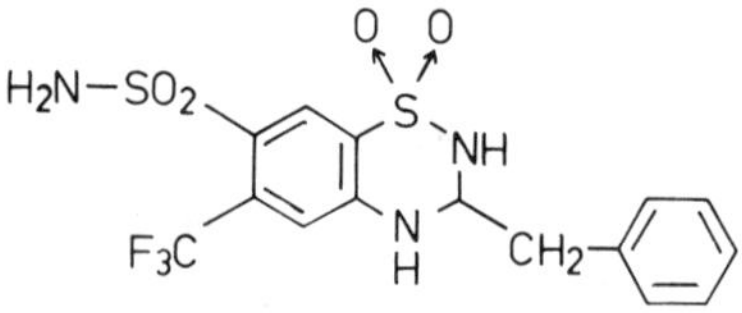

14

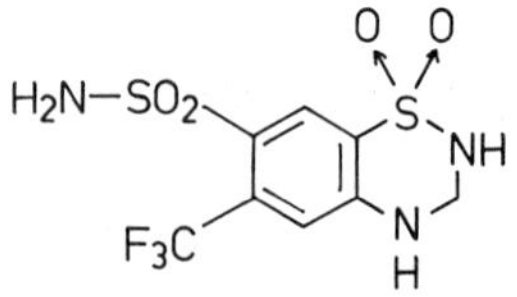

15

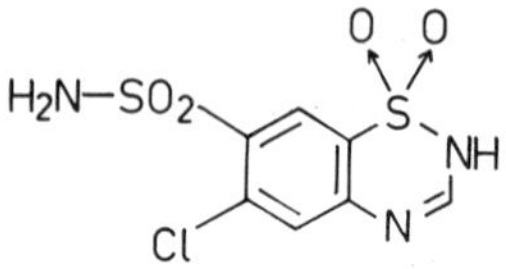

16

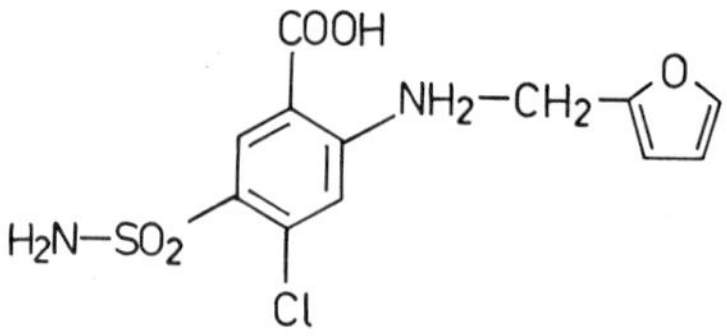

17

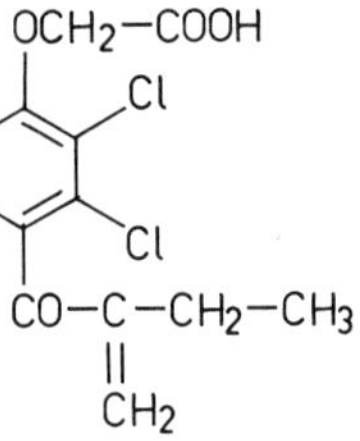

18

higher maximal Na^+ excretion classifies this drug as belonging to the group of high-ceiling diuretics. This group also includes chemically quite different drugs such as ethacrynic acid (18). The high-ceiling activity results in clinical effectiveness in cases of edema that are resistant to thiazides. It may be pointed out that the high ceiling is achieved in spite of a rather low potency of the drug on a dose basis: on the dose-response axis (Figure 2) furosemide is situated rather on the right side of the abscissa; the dose units of 20 or 40 mg tablets for clinical use (Mudge, 1980) are higher than those of the potent thiazide saluretics. Thus the examination of dose-response curves may lead to drugs that are

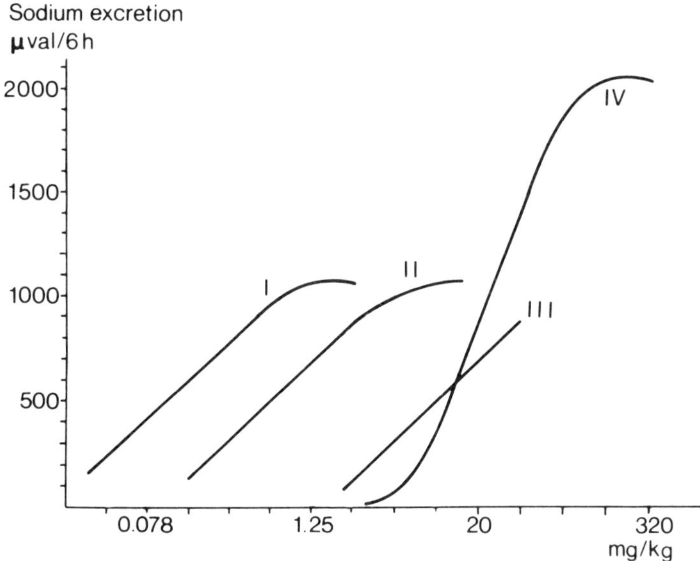

Figure 2 The dose-response curves for several compounds with diuretic action. The straight line portions for bendroflumethiazide (I), hydroflumethiazide (II) and chlorothiazide (III) were calculated from a "6-point assay" (Lund and Kobinger, 1960); the line for furosemide (IV) is based on unpublished results.

more effective in a certain group of patients, but also to drugs that perform a given therapeutic response in a lower dose range and thus allow more economic treatment.
2. Activity profile. Any particular compound may produce many biological effects. To be useful in therapy a compound must possess the appropriate pattern or profile of effects.

Profiling is simplest in the case of potential antibiotics and antiviral compounds. The candidate will be tested in parallel in vitro against a variety of bacteria or viruses and the minimal effective dose vs each organism tabulated. Depending on the particular clinical situation, the physician may choose a very specific or a broad-spectrum compound.

From dose-response curves of various pharmacological activities, doses that produced sub-maximal responses can be determined. These doses are used for further evaluation of the profile of a new compound. An important aspect of drug profiling is to determine the activity pattern of the drug in comparison with drugs that have qualitatively similar effects and whose clinical effect is thought to be understood. The activity profile of a new drug is of central importance for establishing whether a test substance is similar to known compounds or whether it seems to be one of the rare innovations in therapy.

Different biological effects of the compounds are quantified and plotted as in Figure 3. Here two compounds mainly acting upon the heart, mexiletine (19) and sotalol (20), are compared with respect to six cardiac parameters and the arterial blood pressure. As can be seen even on superficial inspection, their profiles are quite different. Both are antiarrhythmics. They differ, however, in electrophysiological properties as measured by intracellular recordings (Vaughan Williams, 1984). In Figure 3 the electrocardiographic (ECG) parameters from spontaneous and electrically induced heart beats were measured in intact anesthetized cats. Obviously the principal effect of mexiletine is the marked increase in diastolic threshold (DT). In contrast DT is not influenced by sotalol; the principal effects are increases in cycle length, CL; effective refractory period, ERP; and the ST-interval in the ECG.

New antiarrhythmic compounds can be precisely characterized by such a profiling procedure (Kobinger and Lillie, 1984 a; Lillie et al., 1985). The profile of each drug is due to a distinct effect on myocardial cell membranes, resulting in a change in transmembrane ion fluxes: mexiletine reduces a fast sodium-inward flux,

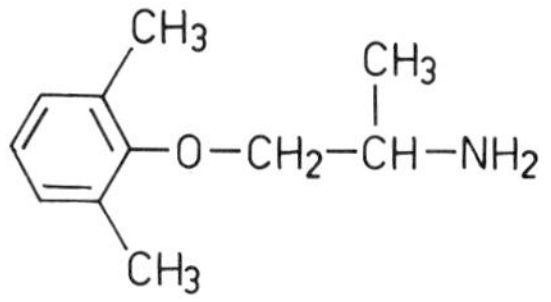

2,6-$(CH_3)_2C_6H_3$–O–CH_2–CH(CH_3)–NH_2

19

H_3C–SO_2–NH–C_6H_4–CH(OH)–CH_2–NH–CH(CH_3)$_2$

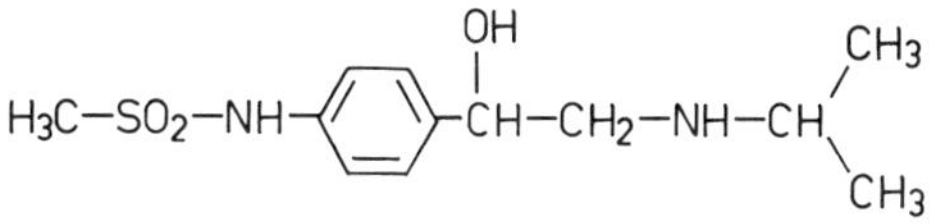

20

whereas sotalol reduces a potassium outward flux. For mexiletine-like substances the therapeutic advantages and limitations are well established. However, for sotalol-like drugs some beneficial clinical effects have been reported (Nademanee et al., 1984), but it is not known whether the predicted advantages of this profile will be verified in long term clinical experience. This example shows that sometimes the pharmacological definition of the desired activity pattern can lead to a new drug that produces a biological effect by a novel mechanism. However, one cannot predict a priori whether the benefit/risk ratio of compounds that act by the novel mechanism will be superior to that of established drugs.

In contrast to the example given in Figure 3, sometimes several drugs may have one important activity in common. In such cases the dose ratio of this activity to other effects is evaluated. Table 2 lists different substances that decrease the rate of spontaneously beating guinea-pig atria. However, quite different concentrations are necessary to achieve the other cardiac effects. From

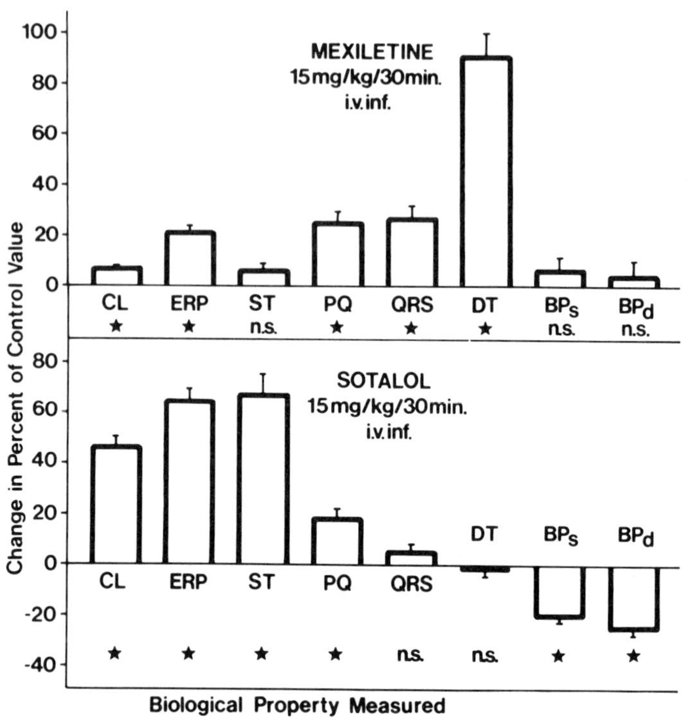

Figure 3 Maximal effects of mexiletine 15 mg/kg and sotalol 15 mg/kg on ECG parameters, ventricular refractory period and blood pressure in anesthetized cats with a stimulating catheter in the right ventricle. BP_s and BP_d = systolic and diastolic blood pressure; ECG parameters: CL = cycle length (RR'-interval); ST = interval from end of S wave to end of T wave; PQ- and QRS intervals; DT = diastolic stimulation threshold; ERP = ventricular effective refractory period. Drug induced changes in percent of control values are shown on the vertical axis (Δ%). Each value represents the mean ± S.E. (n = 6); significance of difference between control (0 min) and drug values (paired t-test):*, $p<0.05$; n.s. = not significant ($p>0.05$) (Lillie et al., 1985).

Table 2 Effects of substances on spontaneous rate, contractility, and maximal driving frequency in isolated guinea-pig atria. From Kobinger et al. (1979).

	Decrease in			Ratio of EC_{30}	
				Contractility	Maximal driving frequency
Substance	Atrial rate (EC_{30} μg/ml)[a]	Contractility (EC_{30} μg/ml)	Maximal driving frequency (EC_{30} μg/ml)	Atrial rate	Atrial rate
Alinidine	2.9 (10)	155 (10)	40 (11)	53	14
Quinidine	7.2 (9)	62 (10)	6.5 (6)	8.6	0.90
Lidocaine	37 (6)	56 (6)	11 (9)	1.5	0.30
Verapamil	0.20 (8)	0.24 (6)	6.5 (7)	1.2	32
Carbachol	0.029 (20)	0.0065 (8)	Increase[b] (6)	0.22	

[a]EC_{30} = Concentration that reduced predrug value by 30%, evaluated from concentration-response curves, No. of experiments in parentheses.
[b]EC_{30} of increase, 0.090 μg/ml.

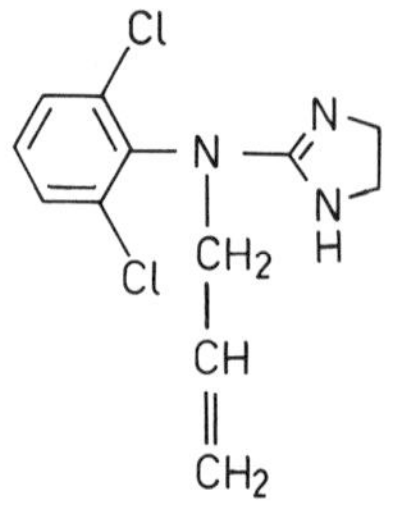

21

the latter concentrations, the ratios given in the last two columns of Table 2 are calculated.

Thus, for alinidine (21) a 53x higher concentration is necessary to decrease atrial contractility than to decrease atrial rate to a comparable extent. Compared with the other substances, this ratio is very high. It constituted one of the essential data to define a novel class of pharmacologically and therapeutically active substances, specific bradycardic agents (Kobinger and Lillie, 1984 b; Kobinger, 1985; Harron and Shanks, 1985). The therapeutic benefit of such substances can be deduced from pathophysiological reasoning: heart rate is one of the major determinants of myocardial oxygen consumption. In pathological conditions such as angina, the oxygen consumption of the heart has to be reduced, for example with ß-adrenoceptor blocking agents or calcium antagonists. Specific bradycardic agents reach this goal by reducing only the heart rate, keeping it within physiological limits. Cardiac contractility and conduction velocity are impaired much less than with ß-blockers and Ca^{++}-antagonists. Because of this improved profile, specific bradycardic agents are expected to be better and safer drugs for the treatment of ischemic heart diseases.

A different pharmacological profile might signal a different molecular mode of action of a compound, but this is not necessarily so: a common mode of action but different tissue distribution may also result in different profiles. An example may be given by the antihypertensive agent clonidine (22), which on a molecular basis stimulates α-adrenoceptors as also does norepinephrine (23). Because of its excellent penetration into the central nervous system, clonidine preferentially stimulates those brain receptors that decrease cardiovascular activity. In contrast, norepinephrine, which is distributed mainly peripherally, increases cardiovascular activity (Kobinger, 1978).

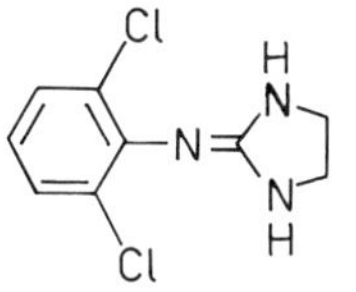

Cl H N N N H Cl

22

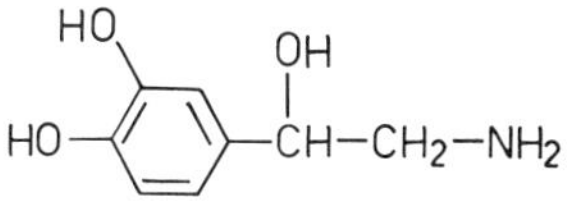

HO OH HO CH—CH_2—NH_2

23

This stage of the investigation of a promising compound might reveal a pharmacodynamic profile that does not fit into the original research concept. It might even happen that the new profile does not permit the prediction of any therapeutic benefit. One of the most vivid examples of a pharmacologic profile of unknown therapeutic utility is the discovery of the usefulness of ß-adrenoceptor blocking drugs for the treatment of angina. In 1959 and 1960 the therapeutic potential of this class of drugs was not widely appreciated, in spite of the availability of potent substances and the knowledge of their specific pharmacological profile (Shanks, 1984). In 1963 J.W. Black and his coworkers suggested that coronary heart disease would be better treated by reducing cardiac oxygen consumption than by trying to increase coronary blood flow, the common therapeutic approach of those days. However, even they did not foresee that ß-adrenoceptor blockers also would be useful for the treatment of hypertension, the most important application of these drugs today.

Therefore, in those cases where testing a compound reveals a novel profile or mode of action, research should be carried beyond that originally planned. The creativity of the whole team concerned with a particular project should be used to propose potential medical uses of such a compound.

3. Side effects, therapeutic range. The pharmacological profile of a compound is not limited to effects that are part of the therapeutic aim, but also covers unwanted actions that might restrict or preclude its clinical use. Again, the ratio between wanted and intolerable effects must be carefully quantified.

Traditionally the therapeutic range of a drug has been defined by the ratio LD_{50}/ED_{50}; the LD_{50} is defined as the dose at which 50 % of animals die and the ED_{50} as the dose which either produces 50 % of the maximum therapeutic effect or where 50 % of animals show the therapeutic effect.

As the aim of drug therapy is to achieve the therapeutic effect in all individuals without any risk of producing toxicity, one would prefer a measure of drug safety that is based on the doses at the lowest toxic and highest therapeutic levels of response. The evaluation of

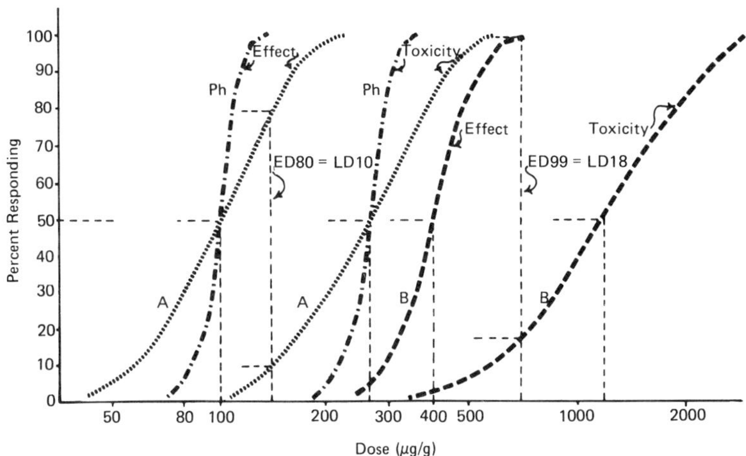

Figure 4 Quantal dose-response curves for effect and toxicity of phenobarbital (Ph), drug A and drug B. Groups of 20 mice were injected with different doses of each of the three drugs. The responses observed were loss of righting reflex (effect) and cessation of respiration (toxicity). The ratio LD_{50}/ED_{50} of phenobarbital and drug A is 2.6; the LD_{50}/ED_{50} of drug B is 3.0 (Levine, 1983).

the "certain safety factor", defined by the ratio LD_1/ED_{99}, expresses the idea of relative safety (Levine, 1983). From this latter textbook Figure 4 is reproduced; it demonstrates the importance of the slopes of dose-response curves. For drug A, efficacy overlaps toxicity because of the low slopes of the two curves. This contrasts with phenobarbital which has steeper slopes. Note that the ratio LD_{50}/ED_{50} is the same for both drugs. For drug B the dose-response curves do not parallel each other, as is usually the case. Again, the ratio LD_{50}/ED_{50} gives false information as to the relative safety of B, and the ratio LD_1/ED_{99} is a more conservative and hence better estimate of its safety. Any estimation of therapeutic ratios in animal experiments cannot be extrapolated directly to humans, but have to be taken as relative measures to be used to compare compounds with each other.

4. Mechanism of action. It is important to investigate the mechanism of action of any compound that shows a new pharmacologic profile. No new drug can be presented to the medical profession without information about its mechanism of action. The description of the biological properties of a drug may be expressed in different terms:

1) for the patient: relief from subjective symptoms and absence of undesirable side effects;
2) for the physician: expected changes in clinical parameters such as blood pressure or clearing of infection as well as no undesirable changes in other parameters;
3) for the physiologist: the dissection of the overall effect into single components, such as interference with endogenous hormones and reflex mechanisms;
4) for the molecular pharmacologist: descriptions of the drug-receptor interactions, enzyme reactions or changes in ionic currents through membranes.

With the development of clinical diagnostic testing the terminology of the physician has approached that of the biochemist. Generally, for the evaluation of a potential new drug, it is of great importance to correlate information from Levels (2) and (3). This usually allows one to distinguish between biologically novel and "me too" compounds. From the viewpoint of the patient or physician information from Level (4) is desirable and of high scientific interest, but it may not always explain the therapeutic advantage of one compound over another.

Aronson has recently reviewed the development of the knowledge on the mode of action of cardiac glycosides (1984). Extracts of foxglove were used in medicine for centuries, until Withering in 1785 described their use for the treatment of dropsy. However, it was not until the beginning of the 20th century that Wenckebach showed that the heart muscle is the primary site of action of digitalis glycosides, the active ingredients of foxglove. It was the discovery by Schatzmann in 1953, that cardiac glycosides inhibit the Na^+/K^+ transport through cell membranes, that opened a wide field of research on the molecular basis of the action of cardiac glycosides. Today, much such detailed information is available; however, we are still unsure about the mode of therapeutic action of cardiac glycosides.

B. PHARMACOKINETICS

As discussed in Chapter 3, pharmacokinetics deals with the factors that determine the magnitude of drug effects or concentrations as a function of time. Three distinct phases of a drug action may be important:

1) the time for onset of action (latency);
2) the time to peak effect;
3) the duration of action.

Figure 5 illustrates these time intervals. The ordinate is the concentration of the drug at the effector organ or tissue; above the minimal effective level one can measure an effect the magnitude of which then usually parallels the tissue concentration.

The pharmacokinetics of a drug can be established by either the analytical determination of drug concentrations at the site of action, or by the measurement of the intensity of the drug action at the target organ. In principle, the time course of the two quantities should be the same, provided that no secondary events occur to change receptor number or sensitivity. For practical

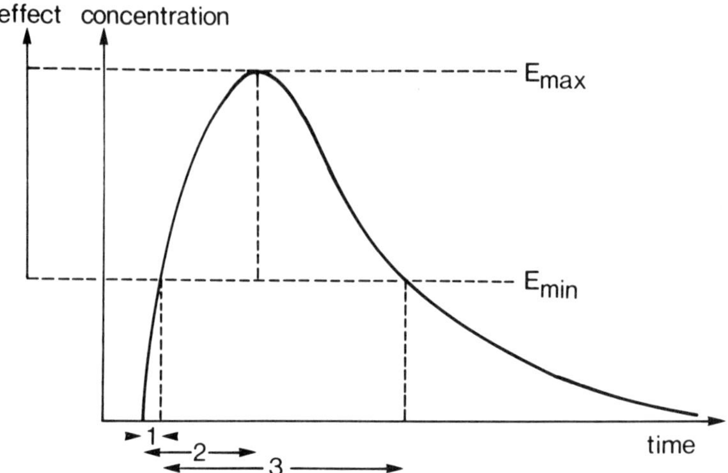

Figure 5 Concentration of a drug at the site of action and intensity of effect as a function of time. Drug is administered at beginning of time period (1). Interval (1) is the time to onset of action. E_{max} is the maximal effect; E_{min} is the minimum level of measurable response; Interval (2) is the time to peak effect; Interval (3) is the duration of measurable effect.

reasons concentration at the site of action usually cannot be measured in humans. Therefore, pharmacokinetics is conveniently determined via the time course of the plasma level of a drug. This procedure leads to relevant conclusions if there is an analytical method that selectively determines concentration of the active drug and if the plasma level parallels the drug concentration at the target site. Misleading results may be obtained if the drug is converted into metabolites that contribute to the biological activity.

Figure 6 is a case for which the plasma level parallels the therapeutic effect and therefore also the concentration at the site of action. Such parallelism does not always occur, as can be seen from Figure 7, which shows the time course of plasma and brain concentrations

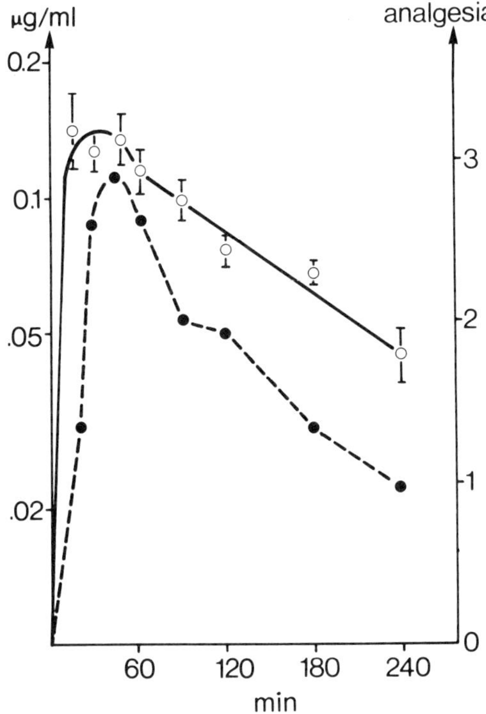

Figure 6 Plasma concentrations (µg/ml, o-o) of pentazocine and its effect on relief of pain (analgesia, ●-●) in humans at various times after intramuscular administration of 45 mg per 70 kg of body weight. Pain evaluation or analgesia: 0 = no pain relief; 1 = some relief; 2 = moderate relief; 3 = high degree of relief; 4 = complete relief, or no pain. The results represent the average (Plasma levels: ± S.E.) of eight patients. Redrawn from Berkowitz et al. (1969).

of phenobarbital in dogs. Due to the slow penetration of the drug into and out of the brain, the concentration in the brain rises even after the plasma concentration starts to fall. As the anesthetic effect of phenobarbital parallels its concentration in the brain, the plasma concentration is not an indicator of the time course of the drug action.

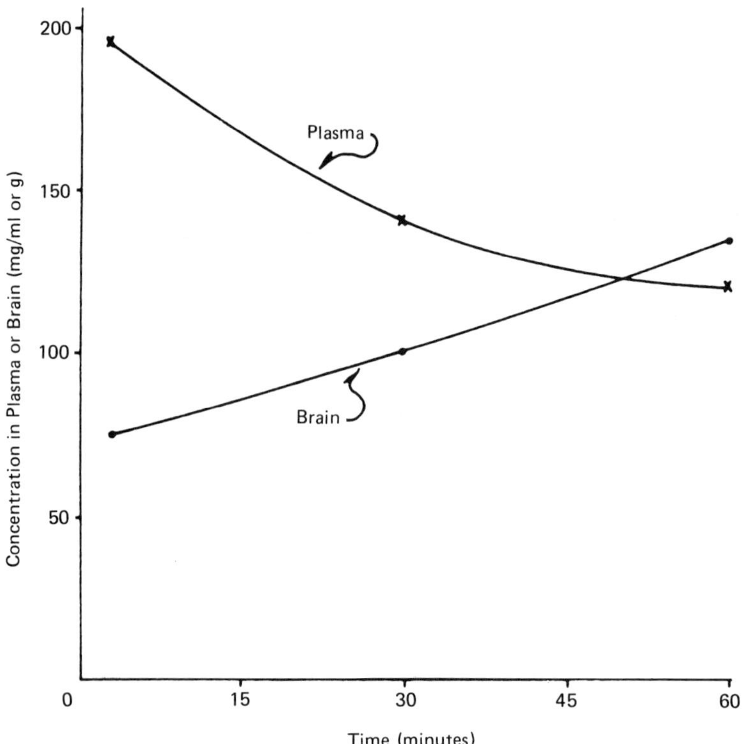

Figure 7 Plasma and brain concentrations of phenobarbital after intravenous administration to dogs. The dose was 100 mg per kilogram. Onset of anesthesia was at 30 minutes after administration. From Levine (1983), based on data from Mark et al. (1958).

Usually the definition of a therapeutic goal includes the idea of a certain duration of action. It is part of the profiling of a new compound to find out whether it fits the desired pharmacokinetic profile. Generally the best way to do this is to measure an effect at the therapeutic target organ. It is advisable to undertake such studies in conscious animals. However, they may not be useful for measuring the desired effect. In such cases the biologist may choose to follow more easily accessible parameters which might even be unwanted side effects. For example, the time course of the action of a ß-adrenoceptor agonist to be developed as an anti-asthma drug might be assessed by following the time course of its effect on heart rate. The latter is easy to monitor continuously. It has to be assumed then that the plasma/heart and plasma/bronchi concentration ratios follow a similar time course. Thus only an approximate conclusion can be drawn from such experimental setups. However they are often of great value during development, if used in comparison with other, well known drugs.

An important criterion for the usefulness of a new drug very often is its biological activity after oral administration. Results gained with oral administration are compared to those with injection. From such experiments the oral bioavailability can be inferred. According to Dost (1968) this can best be done by comparing corresponding areas under the time-concentration or time-effect curves, Figure 5, after oral and intravenous administration respectively. The same methods and criteria for measuring drug levels or drug effects are applied as discussed above. Because of the possible effect of anesthesia on drug absorption, such investigations should be performed in conscious animals or in humans. Furthermore the results will be more reliable if similar maximal effects are achieved after both routes of administration than if the same doses are used, which usually will produce much lower maximal effect after oral administration. In those cases where one cannot continuously monitor the drug effect, an estimate of bioavailability can be made by comparing dose-response relationships after i.v. and after oral administration at selected times, ideally at the time of maximal response. Again, such approximations give valid informations mainly if similar drugs are compared. The pharmacokinetic profile of a drug can be considerably improved by means of controlled drug delivery systems described in detail in Chapter 9.

VIII. EXTRAPOLATION OF RESULTS FROM ANIMAL EXPERIMENTS INTO HUMAN THERAPY

The relevance of a prediction of effects in humans from effects in animal experiments depends on the therapeutic target. Obviously, no animal experiment can provide direct information on the psychological changes produced by a psychoactive drug. In spite of this, today a great number of useful remedies for psychiatric disturbances are available. This is partly due to the serendipidous discovery of unexpected clinical effects such as, for example, the mood-elevating action of the tuberculostatic iproniazide (24). From detailed animal biochemical ex-

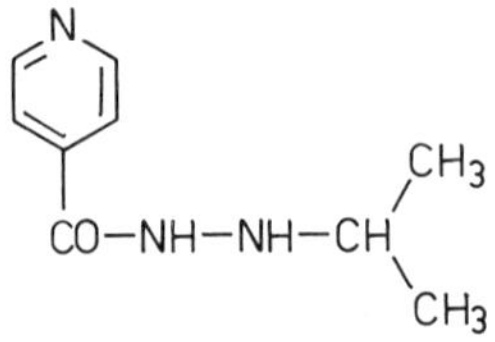

N

CH_3

CO—NH—NH—CH

CH_3

24

periments Zeller et al. then revealed the unique biochemical characteristics of this substance: inhibition of monoamine oxidase (Baldessarini, 1980). Thus a new pharmacologic or biochemical target was now available for the search for more active drugs with similar or analogous modes of action and this search has been successful.

In programs aimed at the discovery of new drugs for action in the central nervous system, drug profiles are constructed from a number of biochemical tests, such as enzyme inhibition or receptor binding, plus observations of effects on animal behavior. Such profiles, in comparison with those of clinically useful drugs, allow a cautious prediction of psychopharmacological actions in humans.

The pharmacologist is in a better situation to make predictions from animal experiments whenever a drug exerts its effect via the autonomic nervous system, some local hormone system such as histamine, neuromuscular functions, or smooth and cardiac muscle. This is because of the similar physiological and biochemical organization of most mammals and because there is usually a good correspondence between the effect measured in the animal and that desired for clinical effect in humans.

However, one must measure the biological effects of potential new drugs in several animal species. For example, cardiac glycosides act in many animal species, but not in rats. The organic mercurial diuretics are barely active in rats, but highly active in dogs; however, activity in rats is highly predictive of human activity for sulfonamide diuretics whether they act by inhibiting carbonic anhydrase or by another mechanism.

As can be appreciated from the discussions of Chapter 3, species differences might be due to different pharmacokinetics or routes of metabolism. This, again points to the importance of testing compounds in several species.

It seems desirable to have models in animals that mimic pathological states in humans. In some indications this is indispensable, as for example, in investigations into anti-infective or anti-inflammatory drugs. For other indications such models might be useful, but not necessary. An example is testing antihypertensive effects in hypertensive animals. Several respective models do exist, but the question remains which, if any, is relevant for the complex of hypertension in man. However, from the viewpoint of consistent test results, an elevated basal blood pressure provides a more sensitive basis for the testing of blood pressure lowering substances than does a normal blood pressure.

As was pointed out in the introduction, for many human diseases we have no adequate animal model. At this stage of our knowledge we should comfort ourselves with the statement of J. Black "that predictions based on the therapeutics of experimental disease are much less reliable than predictions based on pharmacologic actions in normal animals" (Shanks, 1984). However, we must press to understand the biochemical basis of diseases, so that ever more specific drugs can be found with even better predictability.

REFERENCES

Aronson, J.K., Digitalis, in Discoveries in Pharmacology (M.J. Parnham and J. Bruinvels,eds.), Vol. 2, Elsevier, Amsterdam, 1984, pp. 163-184.

Baldessarini, R.J., Drugs and the Treatment of Psychiatric Disorders, in The Pharmacological Basis of Therapeutics (A. Goodman Gilman, L.S. Goodman and A. Gilman, eds.) 6th Ed., Macmillan Publishing Co., Inc., New York, 1980, pp. 391-447.

Bellemann, P., D. Ferry, F. Lübbecke and H. Glossmann, [^{3}H]-Nitrendipine, a Potent Calcium Antagonist, Binds with High Affinity to Cardiac Membranes, Arzneim. Forsch.-Drug Res., 31, 2064-2067 (1981).

Berkowitz, B.A., J.H. Asling, S.M. Shnider and E.L. Way, Relationship of Pentazocine Plasma Levels to Pharmacological Activity in Man, Clin. Pharmacol. Ther., 10, 320-328 (1969).

Bianchine, J.R., Drugs for Parkinson's Disease; Centrally Acting Muscle Relaxants, in The Pharmacological Basis of Therapeutics (A. Goodman Gilman, L.S. Goodman and A. Gilman, eds.) 6th Ed., Macmillan Publishing Co., Inc., New York, 1980, pp. 475-493.

De Jonge, A., P.B.M.W.M. Timmermans and P.A. van Zwieten, Quantitative Aspects of Alpha Adrenergic Effects Induced by Clonidine-Like Imidazolidines. III. Comparison of Central and Peripheral Alpha-1 and Alpha-2 Adrenoceptors, J. Pharmacol. Exp. Ther., 226, 565-571 (1983).

Dost, F.H., Grundlagen der Pharmakokinetik, Georg Thieme Verlag, Stuttgart, 1968.

Drews, J., Experimental Models Relevant for Therapy, in Decision Making in Drug Research (F. Gross, ed.), Raven Press, New York, 1983, pp. 49-55.

Hammer, R., C.P. Berrie, N.J.M. Birdsall, A.S.V. Burgen, and E.C. Hulme, Pirenzepine Distinguishes between Different Subclasses of Muscarinic Receptors, Nature, 283, 90-92 (1980).

Hammer, R. and A. Giachetti, Selective Muscarinic Receptor Antagonists, Trends Pharmacol. Sci., 5, 18-20 (1984).

Harron, D.W.G. and R.G. Shanks, Pharmacology, Clinical Pharmacology and Potential Therapeutic Uses of the Specific Bradycardiac Agent Alinidine, Eur. Heart J., 6, 722-729 (1985).

Hinzen, D., O. Hornykiewicz, W. Kobinger, L. Pichler, C. Pifl and G. Schingnitz, The dopamine autoreceptor agonist B-HT 920 stimulates denervated postsynaptic brain dopamine receptors in rodent and primate models of Parkinsons's disease: a novel approach to treatment, Eur. J. Pharmacol., 131, 75-86 (1986).

Kenakin, T.P., The Classification of Drugs and Drug Receptors in Isolated Tissues, Pharmac. Reviews, 36, 165-222, (1984).

Kobinger, W., Central Alpha Adrenergic Systems as Targets for Hypotensive Drugs, Rev. Physiol. Biochem. Pharmacol., 81, 39-100 (1978).

Kobinger, W., Specific Bradycardic Agents, a New Approach to Therapy in Angina Pectoris?, in Progress in Pharmacology, Vol. 5/4, Gustav Fischer Verlag, Stuttgart, 1985, pp. 89-100.

Kobinger, W. and C. Lillie, Cardiovascular Characterization of UL-FS 49, 1,3,4,5-Tetrahydro- 7,8-dimethoxy-3-[3-[[2-(3,4-dimethoxyphenyl)ethyl]methylimino]propyl]-2H-3-benzazepin-2-on hydrochloride, a New "Specific Bradycardic Agent", Eur. J. Pharmacol., 104, 9-18 (1984 a).

Kobinger, W. and C. Lillie, Alinidine, in New Drugs Annual: Cardiovascular Drugs (A. Scriabine, ed.) Vol. 2, Raven Press, New York, 1984 b, pp. 193-210.

Kobinger, W. and L. Pichler, Alpha-1 Adrenoceptor Blockade by Alpha-2 Adrenoceptor Agonists in the Isolated Perfused Hindquarter of Rats, Eur.J. Pharmacol., 72, 113-115 (1981).

Kobinger, W., C. Lillie and L. Pichler, N-Allyl-Derivative of Clonidine, a Substance with Specific Bradycardic Action at a Cardiac Site, Naunyn-Schmiedeberg's Arch. Pharmacol., 306, 255-262 (1979).

Laduron, P., More Binding, More Fancy, Trends Pharmacol. Sci., 4, 333-335 (1983).

Lawson, J.W., Antiarrhythmic Activity of Some Isoquinoline Derivatives Determined by a Rapid Screening Procedure in the Mouse, J. Pharmacol. Exp. Ther., 160, 22-31 (1968).

Levine, R.R., Pharmacology: Drug Actions and Reactions, 3rd Ed., Little, Brown and Co., Boston, 1983.

Lillie, C., D.W.G. Harron and W. Kobinger, Investigations into the Bradycardic Action of Indoramin, Arzneim. Forsch.-Drug Res., 35, 301-305 (1985).

Lund, F.J. and W. Kobinger, Aromatic Sulphamyl Compounds with Diuretic Actions, Acta Pharmacol. Toxicol., 16, 297-324 (1960).

Mark, L.C., J.J. Burns, L. Brand, C.I. Campomanes, N. Trousof, E.M. Papper and B.B. Brodie, The Passage of Thiobarbiturates and their Oxygen Analogs into Brain, J. Pharmacol. Exp. Ther., 123, 70-73 (1958).

Mudge, G.H., Diuretics and Other Agents Employed in the Mobilization of Edema Fluid, in The Pharmacological Basis of Therapeutics (A. Goodman Gilman, L.S. Goodman and A. Gilman, eds.) 6th Ed., Macmillan Publishing Co., Inc., New York, 1980, pp. 892-915.

Nademanee, K., G. Feld, J. Hendrickson, P.N. Singh and B.N. Singh, Electrophysiologic and Antiarrhythmic Effects of Sotalol in Patients with Life-Threatening Ventricular Tachyarrhythmias, Circulation, 72, 555-564 (1984).

Pichler, L. and W. Kobinger, Alpha-2 Adrenoceptor Blocking Properties of the Alpha-1 Selective Agonist

2-(2-Chloro-5-trifluoromethylphenyl imino)imidazolidine(St 587), Arzneim. Forsch.-Drug Res., 35, 201-205 (1985).

Shanks, R.G., The Discovery of Beta Adrenoceptor Blocking Drugs, in Discoveries in Pharmacology (M.J. Parnham and J. Bruinvels, eds.) Vol. 2, Elsevier, Amsterdam, 1984, pp. 37-72.

Stähle, H., H. Daniel, W. Kobinger, C. Lillie and L. Pichler, Chemistry, Pharmacology and Structure- Activity Relationships with a New Type of Imidazolines Exerting a Specific Bradycardic Action at a Cardiac Site, J. Med. Chem., 23, 1217-1222 (1980).

Timmermans, P.B.M.W.M., F. Karamat Ali, H.Y. Kwa, A.M.C. Schoop, F.P. Slothorst-Grisdijk and P.A. van Zwieten, Identical Antagonist Selectivity of Central and Periheral $Alpha_1$-Adrenoceptors, Mol. Pharmacol., 20, 295-301 (1981).

Van Meel, J.C.A., A. De Jonge, P.B.M.W.M. Timmermans and P.A. van Zwieten, Selectivity of Some Alpha Adrenoceptor Agonists for Peripheral Alpha-1 and Alpha-2 Adrenoceptors in the Normotensive Rat, J. Pharmacol. Exp. Ther., 219, 760-767 (1981).

Vaughan Williams, E.M., A Classification of Antiarrhythmic Actions Reassessed After a Decade of New Drugs, J. Clin. Pharmacol., 24, 129-147 (1984).

Weinstein, L., The Sulfonamides, in The Pharmacological Basis of Therapeutics (L.S.Goodman and A. Gilman, eds.) 4th Ed., Macmillan Co., New York, 1970, pp. 1177-1203.

Williams, M. and D.C. U'Prichard, Drug Discovery at the Molecular Level: a Decade of Radioligand Binding in Retrospect, in Annual Reports in Medicinal Chemistry (D.M. Bailey, ed.) Vol. 19, Academic Press, Inc., Orlando, 1984, pp. 283-292.

8

Safety Evaluation of New Drugs

D. S. Meyer

Department of Experimental Pathology and Toxicology
Boehringer Ingelheim KG
Ingelheim, Federal Republic of Germany

I. INTRODUCTORY REMARKS

The medical use of any drug carries with it certain risks. The aim of toxicological studies in the development a compound to be come a new drug is to provide data for the assessment of its safety for human use. This assessment is the basis for a medical risk-benefit analysis.

A new compound worthy of development and marketing as a therapeutic agent should either be unique or should be safer with respect to adverse side effects than are similar substances already in use. To satisfy these requirements, a number of appropriate toxicological studies has to be carefully performed. Additionally, most governments provide guidelines as to the toxicological tests that must be passed before a compound may be sold in its country. Thus the design of such studies must incorporate these requirements as well as the best science possible.

In general, the use of therapeutic drugs is relatively safe when compared with other risks to life

(Heilmann, 1984). Using a logarithmic scale on populations at risk in the Federal Republic of Germany, the individual risk of death within one year from the use of any drug was calculated. Out of about 250,000 people in total or 5,000 sick persons treated in hospitals, the probability is that one person will die from an effect of the drug. The risks of other activities and personal events are tabulated in Table 1 from data published by Heilmann.

Toxicology integrates three fundamental areas of preclinical drug research: basic science in pharmacology or microbiology, pharmacokinetics and drug metabolism, and morphology. [Klaassen, et al. (1986), Forth et al. (1987), Hayes (1982), Clayson et al. (1985)]

II. ANIMAL WELFARE

In common with the evaluation of efficacy, the evaluation of safety of a compound cannot be gained without animal experiments. The number and the types of these depend upon the current state of the art and science. The Declarations of Helsinki (1964) and Tokyo (1975) state that the use of animals as surrogates for humans is justified morally and ethically to develop safe drugs to cure or to alleviate human or animal diseases.

The particular requirements for the use of animals in scientific experiments are regulated in some countries by national laws on animal welfare. Apart from this, the Swiss Convention of Biomedical Scientists (1983) stated that useless, unnecessary and inhumane experiments on animals should not be tolerated by scientists themselves. Since the legal requirements of the details of the toxicologic studies differ from country to country, more international cooperation in this regard would save animal lives.

Table 1 Individual risk of premature death within one year from certain diseases, use of drugs or voluntary activities (according to Heilmann, 1984)

fatal event	degree of safety (logarithmic scale)	risk 1 : ~
snake bite, poisonous plants	~7	10 000 000
rheumatic fever, tetanus, diphtheria	~7	10 000 000
bee-sting	6.7	5 000 000
antirheumatics, antipyretics, analgesics (FRG 1975)	6.6	4 000 000
lightning-stroke	6.3	2 000 000
malaria, veneral diseases	~6	1 000 000
air travels (USA 1980)	5.9	800 000
tornados (Middle West of USA)	5.7	500 000
butazolidine (single case reports world wide)	5.7	500 000
lanata glycosides (FRG 1982)	5.5	320 000
all drugs (FRG 1975)	5.4	250 000
butazolidine (multiple case reports world wide)	5.25	180 000
the "pill" (non-smokers, 25-34 yrs)	4,4	25 000
arthritis (total population)	~4	10 000
the "pill" (smokers, 25-34 yrs)	3.8	6 300
the "pill" (non-smokers, 35-44 yrs)	3.7	5 000
drug users in hospitals	3.7	5 000
car driver (USA 1980)	3.6	4 000
lung cancer	3.4	2 500
the "pill" (smokers, 35-44 yrs)	3.2	1 600
car driver (FRG)	3.1	1 300
cigarette-smokers (20/day)	2.3	200
arthritis (patients)	~2	100
all cancers (total population)	2	100
myocardial infarction (35 yrs)	1.9	80
all cancers (patients)	1.0	10

A balance must be struck between preventing senseless animal experiments and promoting medical progress through properly designed animal experiments.

III. PREREQUISITES TO TOXICOLOGICAL STUDIES

A. PHARMACOLOGY

As described in Chapter 7, basic information on the effectiveness of a compound is supplied by microbiological or pharmacological testing, including the characterization of side- and adverse effects. These studies identify the pharmacodynamic effects, the effective dose in animals, the slope of the dose-response curve, and the expected human therapeutic dose. This information is essential for establishing the doses to be used in toxicological animal studies.

B. ANALYTICAL CHEMISTRY

To evaluate toxic effects of the test compound one must establish its purity. For acute single-dose toxicity studies it may be acceptable to use a batch with small amounts of possibly more toxic impurities since these impurities will be removed by improving the chemical manufacturing process of the final drug. It would then be less toxic than the batches originally tested. However, long-term toxicity studies must be performed with the compound that is analytically identical to that intended for human use.

In addition, analytical chemists certify the stability and concentration of the active substance in each of its batches and in solutions or in suspensions that will be administered to laboratory animals. Without these analytical controls a continuously equivalent drug supply to the animals over the duration of a multiple-dose toxicological study could not be assured, and one would not know the exact amount of compound administered.

C. DOSAGE FORM DESIGN

Toxicology depends strongly on the pharmaceutical preparation of the drug. A valid comparison of various toxicological studies performed with a particular drug can be made only if the compound administered in all the studies has been formulated identically, since different formulations or pro-drugs may cause different toxic effects due for example to differences in pharmacokinetics or biodistribution. Also, it is of particular importance that preparations for injection, inhalation, or local administration are tested toxicologically in the formulation that contains solvents and/or additives in the same ratio as will be present in the preparations for clinical use.

As will be appreciated from Chapter 3 it is important to control the uniformity of the physical properties of the compound. For compounds tested as solids administered orally, the particle size which may influence the rate of dissolution and absorption, and for compounds administered by inhalation, the droplet size must be controlled. During inhalation only particles with a small diameter reach the bronchi (1 - 5 μm) or alveoli (<1 μm). Larger particles are deposited by impaction within the oropharyngeal area and are then swallowed. In such a case the compound does not exert its intended topical pharmacological effect, but may act systemically following absorption in the gastrointestinal tract.

D. PHARMACOKINETICS AND DRUG METABOLISM

Adequate information on the pharmaco- and toxicokinetic properties and on the metabolism of a compound are also indispensible for the design and the evaluation of chronic toxicity studies. The toxicologist needs to know the rates and route of the absorption, distribution, metabolism and excretion of the compound being tested in species feasible for toxicological studies. This allows the selection of the most suitable species of laboratory animals and the appropriate doses for the chronic toxicity studies in two animal species. Because the appropriate design of chronic animal toxicity studies requires information on the pharmacokinetics and metabolism of a compound in humans, short term animal toxicity studies are performed to demonstrate the safety of a single dose of the compound for man. Then single doses are administered to healthy human volunteers for pharmacokinetic and drug metabolism studies. Finally, the resulting information is used to design long-term animal experiments that in return assess the safety of the compound for repeated administration to humans. Only then is the compound tested in multiple doses in humans.

For practical reasons the rat is usually chosen as the rodent species for chronic toxicity studies. Only rarely are mice or hamsters used. The second species according to regulatory guidelines has to be a non-rodent species. Although theoretically a large number of various non-rodents could be considered, in practice only dogs, mini-pigs and monkeys are suitable for chronic toxicology.

From these non-rodent species, that one should be selected that resembles man most closely in terms of the pharmacokinetics and the metabolism of the drug under investigation. In contrast to the hitherto world-wide existing regulatory demands, this approach should allow to perform chronic toxicity studies only in one, i.e. the pharmacokinetically and metabolically closest species to man.

Pharmacokinetic and drug metabolism studies in animals must be available for other than the oral route of administration if these are intended clinically. Furthermore, pharmacokinetic data are necessary to select the doses to be used in chronic toxicological studies.

Investigations on the plasma-binding capacity of the parent substance and its active metabolite(s) are used in comparisons of species differences of toxic effects. This helps one project toxicity in animals to risk in humans.

Finally, pharmaco- and toxicokinetical studies prevent the toxicologist from applying absurdly high doses in cases where the absorption of a compound is limited. The administration of larger amounts of a compound in such a situation would not produce additional information. Rather it could alter the state of health of the animals, for example, by malnutrition due to disturbances in intestinal function.

IV. VETERINARY CARE OF ANIMALS IN TOXICOLOGY

One of the most crucial aspects of predictive toxicology is to use animals that are in a good state of health and are genetically homogeneous. This allows one to compare results from control and treated animals in the same study as well as results from different studies. This is especially important in studies on non-rodent species, because of the small number of animals (3 - 5 per sex) in each dose group. Strain-specific statistical ranges of various parameters, historical data, can only be achieved by using animals from breeding colonies with a known genetic background.

These conditions are in accord with the legal requirements on animal welfare, for example in the FRG, whereby only animals exclusively bred for that purpose may be used in laboratory experiments.

For long-term toxicity and carcinogenicity studies in rodents, it is advantageous to keep the animals in specified pathogen free conditions. This favors their longevity and reduces the possibilities of infectious diseases producing immunomodulating effects on tumor incidences and growth. A constant diet with continuous analysis of the feed and drinking-water avoids an overload of fats or proteins or a contamination with toxic substances such as chemicals, pesticides, or mycotoxins which may also influence tumorigenesis.

V. TOXICOLOGICAL STUDIES IN LABORATORY ANIMALS

The various countries of the world specify the type and number of toxicity studies to be done for human safety assessment and to be presented to its regulatory authorities. To a large extent these requirements are identical. However, there are differences in some essential details and there are supplementary requirements of some countries that conflict with animal welfare regulations of other countries. Such conflicts and lack of uniformity impedes a standardized drug development world-wide.

A. SINGLE DOSE TOXICITY (ACUTE TOXICITY)

As a prerequisite for chronic multiple dose studies it is necessary to investigate in at least two animal species the effects of single doses of a compound that are high enough to be lethal for some of the animals.

Such tests yield initial information on the degree and possible symptoms of toxicity of that compound in an intact animal. Usually such tests are done in mice and rats by oral and intravenous administration. It is also done by other routes if such routes of administration of the drug are intended for man.

Acute toxicity does not mean the mere generation of a statistically verified number, by counting the dead and surviving animals within a defined period of usually 14 days after administration of the compound. To be statistically valid, such LD_{50} evaluations require a hundred or more animals per experiment. This approach has proved to be obsolete, since the LD_{50} is not constant but rather may depend upon conditions such as the strain, feeding, age, and handling of the animals; the temperature and atmospheric humidity of the laboratory; and the time of day and season of the experiment (Bass et al., 1982).

Today some regulatory agencies (EEC, USA) accept acute toxicity studies on a small number of animals per group with up to 15 - 35 rats or mice per study. If no sex differences are observed with about six additional animals, then the compound may be tested in only one sex of the second rodent species (EEC). This procedure demonstrates the signs and course of toxicity and provides information on the maximum non-lethal dose, the minimum lethal dose and the slope of the dose-response curve. Following necropsy and histopathological examination of macroscopically altered organs and tissues, it indicates possible target organ(s) and allows estimation of the approximate LD_{50}.

If a drug is to be administered by inhalation, it may not be possible for technical reasons to determine even an approximate LC_{50}. This is, for example, the case if the compound is to be tested in its final formulation as an aerosol that contains large amounts of propellant. Under such circumstances the animals die of oxygen deprivation rather than drug toxicity.

Today, there is no consensus amongst regulatory agencies of Japan, the EEC, and the USA as to whether acute toxicity studies in non-rodents are needed, not needed, or are expected but not officially required. From a toxicological point of view, a range-finding study with one or two male and female animals per dose should be sufficient to demonstrate the toxic profile of a compound in non-rodents and to suggest doses for the more important repeated dose toxicity study.

The required acute toxicity tests of industrial chemicals, on the other hand, use the oral LD_{50} values for the classification of chemicals as highly toxic, toxic harmful or non-toxic (Table 2). For this reason the determination of an oral approximate LD_{50} using only a small number of animals is not acceptable. Thus it may happen that if it is transported from one

Table 2 EEC classification for oral toxicity of chemicals

classification	oral LD_{50}, rat
very toxic	25 mg/kg
toxic	25 - 200 mg/kg
harmful	200 - 2 000 mg/kg
practically non-toxic	2 000 mg/kg

site to another, the LD_{50} of an intermediate in the manufacturing of a drug may have to be determined exactly while, for the approval of the final product as a drug, an approximate LD_{50} will do. Of course more total toxicologic information will be provided on the drug.

B. REPEATED DOSE TOXICITY

Each potential new drug for human or animal use is investigated by repeated administration to two or more species of laboratory animals by the intended human route of administration. At least 10 rodents of each sex are used per dose group for studies that last up to three months, and 20 animals per sex and dose are used for longer-term toxicity studies. For non-rodents three to five animals per sex and dose are used. The drug is administered seven days per week.

When designing these repeated-dose studies two decisions must be made. First, the duration of the studies and, second, the doses to be administered.

As shown in Table 3, there are clearly defined standards for the duration of chronic animal toxicity studies depending on the expected length of human therapy with the compound. These regulations differ between countries; this difference may lead to duplication of studies at a cost of human and animal resources. In the EEC, chronic toxicity studies to support an unlimited therapeutic use must be performed for a period of six months; the FDA and the Japanese authorities expect chronic studies to be run for one year in both the rodent and non-rodent species; whereas the Canadian HPB stipulates a period of 1 1/2 year in rodents (Frederick, 1986). Experts have concluded that from a scientific point of view there is no reasonable

Table 3 Requirements for the length of repeated dose toxicity studies in each one rodent and one non-rodent species in relation to the intended duration of human therapeutic drug administration

intended duration of human administration	duration of repeated dose animal toxicity
US FDA:	
several days	2 weeks
up to 2 weeks	3 months
up to 3 months	6 months
unlimited	12 months
EEC:	
once or several times for one day	2 weeks
up to 7 days	4 weeks
up to 30 days	3 months
unlimited	6 months
Japan:	
once	28 days
up to 7 days	90 days
up to 28 days	28 days + 6 months
29 days and more	90 days + 12 months

argument to extend chronic toxicity studies over more than six months (Walker et al. 1984, Lumley and Walker 1986). When carcinogenicity studies of at least 18 months duration in mice and 24 months in rats are also carried out, the latter studies will disclose any possible toxic effect induced by the drug that may not have been seen before in shorter toxicity tests.

Other experts suggest that a single chronic toxicity study in the appropriate animal species, rodent or non-rodent, would be sufficient on the condition that it can be clearly demonstrated that the pharmacokinetics and drug metabolism in that species are similar to those of man. This is especially suggested if all other species of laboratory animals are much less or even not at all comparable to man.

What doses are chosen? The low dose should be a small multiple of the recommended maximum daily therapeutic dose. In addition a plasma concentration of the parent substance and its active metabolite(s) should be reached that is not lower than the therapeutic level in humans. Such data on the human bioavailability and metabolism are lacking at the start of the short-term repeated-dose toxicity tests performed to demonstrate the safety of the first administration of the compound to humans. Thus, the low dose for such toxicity studies has to be set by experience. Usually the total daily low dose relative to body weight should exceed the intended maximum daily relative human dose by about ten times in rats, by about six times in dogs, and should be about double in minipigs and non-human primates. This rough approximation following the appraisals (FDA 1965) should be used only in early toxicologic dose-finding.

The high dose in chronic toxicity but not carcinogenicity studies is expected to reveal any overt sign of general or target-organ toxicity and will best be found by range-finding studies in a few animals of the species to be used. If multi-dose toxicokinetic experiments have shown a limited absorption of a compound, or a saturation of the plasma concentration beyond a certain dose, then the high dose in repeated dose toxicity studies has to be adjusted to these facts. Although it is required by some regulatory agencies (Australia and Japan), we do not see an advantage to the selection of such a high dose that animals die prematurely in toxicity studies of up to 3 month duration. Such premature deaths of animals do not provide better information on the drug effects than does evaluation of animals that reach the termination of a study.

No generally accepted rules exist for the choice of the intermediate dose(s). Some toxicologists propose the intermediate dose(s) to be calculated as the arithmetic mean(s) of, or to be equidistant in terms of a simple multiple to the next lower and higher dose, respectively. The EEC and Japanese guidelines recommend that the mid dose(s) is (are) the geometrical mean(s) of the lower and the higher dose.

For clinical risk assessment, in these toxicity studies it is necessary to assess the reversibility of possibly drug induced toxic effects in the animals after cessation of drug administration. For this reason a high-dose recovery group is also included in the studies. Usually the recovery period is eight weeks. Because of their more rapid growth and body weight gain, it is advantageous with rodents to also include a group of untreated animals that will be sacrificed at the end of the recovery period.

C. TOXICITY STUDIES USING SPECIAL PHARMACEUTICAL PREPARATIONS

If it is intended that a compound will be administered only by a route other than orally, then the toxicity studies must be performed by this route in both a rodent and a non-rodent species. Again the exposure time depends upon the intended duration of the therapeutic application.

If the compound to be used by a non-oral route has already been tested extensively in toxicological studies following oral administration, some supplementary studies should be sufficient: A comparative acute parenteral toxicity study in at least one species and a 4-week to 3-month repeated dose tolerance study using the particular route of administration should be carried out in the species closest to the human with respect to pharmacokinetics and to drug metabolism. It is important to use in these repeated dose tolerance studies the final formulation of the compound, including all additives such as solvents, stabilizers, bactericides, adhesives, propellants etc. A procedure like this should be accepted for risk assessment if there are no crucial differences in toxicokinetics and drug metabolism between the oral and the additional specific route of administration. Otherwise the duration of the special toxicity studies has to be adapted to the intended duration of administration in the therapeutic use of the drug.

D. LOCAL TOLERANCE STUDIES

To prevent unwanted local irritation or tissue damage in humans following parenteral drug administration, the local tolerance of the compound in its final formulation must be investigated.

Drug solutions intended for injection have to be tested locally in animals at least up to the concentration proposed for therapy. Usually one uses rabbits or dogs. A negative control solution is included for comparison. If the compound is to be given intravenously, the local reactions of the blood vessel wall and of the surrounding soft tissue are examined for at least eight to ten days following a single standardized intravenous injection. Because of the possibility of missing the vein when injecting the solution into humans, it is necessary to determine the outcome of unintentional intraarterial and paravascular administration.

In order to preclude hemolytic events following therapeutic administration a possible hemolytic property of a drug is investigated by in vitro-tests on citrated human blood.

The irritation potential of compounds that may come into contact, even unintentionally, with the skin or mucous membranes must be tested. The primary skin irritation test is usually performed on albino rabbits by scoring the erythema, oedema, inflammation, or necrosis produced by the compound. The rabbit skin is more susceptible to irritants than is the skin of humans. This is considered "ideal for the pharmaceutical manufacturer who needed a margin of safety for the protection of the more sensitive members (i.e. high risk groups) of the human population" (Mershon and Callahan, 1975).

The Draize test is the classic tests for the potential of a compound to irritate mucous membranes. (Draize 1944). A solution of the compound is put into the rabbit eye, and the effects on the cornea, iris and conjunctiva are graded. Despite the facts that there are differences between rabbits and humans in ophthalmic anatomy and physiology, and that the response of the rhesus monkey is closer to that of the human (Beckley et al., 1969), until now the rabbit eye is the model of choice in regulatory toxicology for practical reasons. Agents that are erosive or are severe irritants to the skin should not be examined on the rabbit eye. It is expected that validations now in progress will support the use of the chorio-allantoic membrane of hen-eggs after 10 days of hatching to replace the rabbit eye test (Luepke, 1985).

E. SKIN SENSITIZATION AND PHOTOTOXICITY

In addition to systemic and local toxicological testing compounds to be administered by the dermal route must be tested for their ability to produce allergic sensitization or light sensitivity. The potency of these compounds is assessed either in their final formulation or using each of the constituents separately.

Even though the guinea pig is less capable of being sensitized by weak or moderate sensitizers than is the human, of all small laboratory animals the guinea pig has proved to provide the most reliable prediction of human sensitization. Currently the mostly recommended assay is the maximization-test according to Magnusson and Kligman (1969).

If exposure to light is suspected to induce dermal photoallergy or dermal phototoxicity, then the possible response to photoactivation of a compound by wavelengths above 320 nm is determind in guinea pigs, rabbits or hairless mice.

F. TOXICITY STUDIES WITH FIXED DRUG COMBINATIONS

For the regulatory agencies of most countries, pharmaceutical preparations of two or more previously approved compounds in a new quantitatively fixed combination must be tested in the final formulation in acute single-dose toxicity tests. This will reveal possible additive or potentiating toxic effects. In addition, toxicity studies of 1 to 6 months duration have to be carried out. Preferably this will be done in the species that shows a pharmacokinetic and metabolic pattern of the single compounds closest to that of the human. The clinically intended route(s) of administration are used.

If the drug combination contains one unapproved new compound, all toxicity studies required for the approval of new drugs have to be performed.

However, the regulatory agencies of Canada and Australia consider any new drug combination, whether of unapproved substances or of compounds previously approved, to be a new drug. They therefore expect the complete package of toxicity studies to be submitted for any new fixed combination.

The composition of fixed drug combinations should be such that the active components are similar in pharmacokinetic properties. Otherwise an efficient bioavailibility of the combined drugs would not be attained, thus questioning the significance of the fixed drug combination.

G. REPRODUCTIVE TOXICOLOGY

The first guidelines for the assessment of the possible effect of a potential new drug were published by the FDA in 1966. They were formulated in response to the thalidomide tragedy in which birth defects were attributed to the ingestion of the drug by pregnant women. This event taught the public that chemicals are not always safe and it ushered in an era of increasing government regulation of drugs and other substances.

However, not every prospective parent is disease free, so new drugs must be evaluated for their possible effects on fertility and on the developing fetus.

Fertility, Segment I, studies are usually performed in one rodent species, most often the rat. The compound is given in at least three doses to the males for at least 60 days (Canada: 80 days) and to the females for at least 14 days before mating and then continued in the females during gestation. This regimen allows the evaluation of a possible influence of the compound on the germ cells and on the processes of spermatogenesis, ovulation, fecundation or conception. In addition, one can detect embryotoxic or teratogenetic lesions caused by damage to the germ cells prior to or following conception. The Japanese Guidelines (1984) request that treatment of the females is continued until day 6 or day 7 of gestation, in mice or rats respectively, in order to also cover the early phase of embryogenesis.

The main concern of reproductive toxicology subsequent to the thalidomide disaster has been to find experimental models that would allow prediction of a possible teratogenic potential of chemical agents. There are, however, fundamental structural and functional differences in placentation in rodents and humans (Beck, 1976). In addition, scientists understand the "incapability of any one model to accurately predict the response of humans in a consistent manner" (Harbison, 1980). Therefore the regulatory agencies decided that testing for teratogenicity should be done in at least two mammalian species. They recommend rodents (rats, mice or hamsters) and rabbits as suitable species.

In the teratogenicity, Segment II, studies the compound usually is administered orally in at least three dosages. If the main route of therapeutic administration is expected to be inhalation, intravenous, or subcutaneous, or if there are differences in drug metabolism or pharmacokinetics in comparison with the oral route of administration, then the experimental design has to address these particularities. The low dose should not exceed a small multiple of the projected maximum daily human dose if the absorption rate and the blood concentrations of the drug are comparable in the human and animal species. The high dose is recommended to be subtoxic without inducing evident general or target organ toxicity in the females, but high enough, for example, to reduce somewhat the body weight gain. This minimal toxic dose has to be determined in pilot studies, since doses that are frankly maternotoxic are apt to generally affect embryo- or organogenesis. The result would be false positive findings of embryotoxicity, embryolethality, or anomalies in the pups (Khera, 1985). The Australian Guidelines (1983) are the only ones that expect the high dose to cause embryolethality.

The drug has to be administered during the most sensitive phase of embryo- and organogenesis: in mice or rats this is from day 7 through day 15 or 17 of gestation; in rabbits, from day 6 or 7 through day 18; and in hamsters, from day 4 through day 14 of gestation. Some pups are delivered by caesarian section one day before expected parturition and others by spontaneous birth. They are examined immediately after birth for external malformations, and later on for visceral and skeletal anomalies.

Fully aware that "no experimentally available placental model is identical to man" (Beck, 1967), Calabrese (1983) stated "that the rodent and rabbit models, which are so commonly used in teratological evaluations, may be highly sensitive models for a variety of teratogenic agents". However, he also suggested that "the principal findings of the present assessment suggest that nonhuman primates would probably make the best model for predicting human susceptibility to potential teratogenic agents."

It is, however, ethically questionable to use large numbers of monkeys routinely in teratology. Thus the development and validation of new biologic tests for teratogenicity would be highly appreciated. Unfortunately, no biologic test system using cell cultures, or invertebrate, amphibian or bird embryos, or embryonic limb buds has convincingly shown predictive properties (Neubert, 1980).

The perinatal and postnatal, Segment III, reproductive toxicology studies involve the exposure of pregnant rodents to drugs during the last third of gestation and the lactation period up to weaning. This reveals possible fetotoxic effects of the compound on the late development of the fetus; on the labor, delivery, and lactation of the dams; on the neonatal viability, growth, and bodily and behavioral development of the pups; and on the reproductive capability of subsequent generations. For example, reduced fetal body weight gain, retarded opening of the eye lids, lack of auditory reflexes or reduced capability of learning and memory may reflect adverse drug effects.

H. GENETIC TOXICITY

The evidence that there is a strong relationship between mutagenic and carcinogenic effects on cell systems due to chemical or physical actions on the DNA and the knowledge that chromosomal anomalies are responsible for some human diseases, clearly requires a strategy to test any potential drug for a possible chemical or physical effect on cellular DNA.

Toxically induced genetic lesions may manifest themselves on different end-points: gene mutations, structural or numerical chromosomal aberrations, or toxic effects on the cellular apparatus for cell division. Each of these is elucidated by different experimental protocols; thus a battery of short-term genotoxic assays has been recommended (EEC 1983, JAPAN 1984, EPA 1985). There are a total of probably about 100 in vitro and in vivo tests; only some of them that are commonly used and well validated shall be mentioned here.

Point-mutations within a single gene, due to substitution or addition of base-pairs or to deletion of small sequences of base pairs in the DNA-strand, are best investigated in the bacterial/mammalian-microsome mutagenicity test (Ames et al., 1975). Various strains of Salmonella typhimurium and Escherichia coli are used. Depending upon the chemical class of the agent, the predictive association between mutagenicity and carcinogenicity in the AMES-test is reported to be 63 - 92 % (Calabrese, 1983). In these bacteria, however, the DNA is not arranged in a nucleus, so the barrier of the nuclear membrane is absent. Because of this, the gene mutation test should also be carried out in at least one eukaryotic cell system, such as fungi, yeasts or cultured mammalian cells. For the latter cell lines from the Chinese hamster, or mouse lymphoma cells with a defined genetic deficiency on the HGPRT (hypoxanthine-guanine phosphoribosyl transferase)- or TK (thymidine kinase activity)-locus, respectively are frequently used.

The clastogenic potential of a compound is its

capability to induce aberrations in chromosomal structure such as achromatic lesions or gaps, breaks, deletions and displacements of pieces of chromosomes and their exchange or re-arrangement within or between chromosomes. Common tests to evaluate such effects are in vitro tests with cultured mammalian cells, the in vivo cytogenetic metaphase assay on rodent bone marrow cells, or the micronucleus test in polychromatic erythrocytes from mouse bone marrow. These tests are also suitable for the detection of the gain or loss of one or more complete chromosomes, aneuploidization.

The genetic toxicity of a compound may be a property of the parent substance itself or it may require metabolic activation. Alternatively, metabolism may inactivate the intrinsic genetic toxicity of a compound. For these reasons, in vitro assay systems that lack the mammalian metabolic capacity are performed both with and without the addition of metabolically activated cell fractions. Usually the metabolically induced liver microsome fraction of rodents (S9-mix) is used.

There are a number of additional genotoxicity assays that may help to answer particular questions arising from perhaps contradictory results obtained when testing a compound. Unscheduled DNA synthesis, the sister chromatid exchange assay, the dominant lethal test, the mouse spot test, and the cell transformation test are some of these.

I. CARCINOGENICITY STUDIES

Since the first clinical observation by the London surgeon Percival Pott (1713-1788) that chimney sweeps suffered from cancer of the scrotum more often than do other men and the suggestion that this might be due to their professional contact with soot and coal tars, many other chemical carcinogens (Table 4) have been detected (Williams and Weisburger, 1986).

In modern drug development all compounds intended to be given to humans for longer than a short therapeutic or diagnostic administration have to undergo long-term studies in laboratory animals to evaluate their carcinogenic potential. As recommended by several guidelines (HPB 1981, EEC 1983, Japan 1984, EPA 1985), at least two carcinogenicity studies in rodents are required. The possible carcinogenic potential of synthetic hormones is determined in dogs or monkeys with studies that extend over several years.

It is a scientific issue whether it is relevant to perform carcinogenicity studies in more than one rodent species if only one of them is comparable to the human in pharmacokinetics and drug metabolism.

The compound is administered daily by the oral route, usually in the food or drinking water, at 3 dose levels for a minimum of 24 months to rats, and 18 months to mice or hamsters (EEC 1983, FDA 1984). The Japanese guidelines (1984) require a premature discontinuance of the study if there is a mortality rate in the control or low dose group of 75 %. The EEC-guidelines propose two parallel control groups to assist in the interpretation of random variations of tumor incidences. Each group, including the controls, consists of 50 male und 50 female animals.

Pharmacodynamically, pharmacokinetically and toxicologically effective doses, as well as the intended therapeutical doses must be considered in choosing the doses for carcinogenicity studies. However, there is no international agreement on dose selection. In particular, the selection of the highest dose is a sophisticated decision. Some investigators establish the

Table 4 Classification of carcinogenic chemicals (from: WILLIAMS and WEISBURGER, 1986)

Category and Class	Example
A. DNA-Reactive (Genotoxic) Carcinogens	
1. Activation-independent	alkylating agent
2. Activation dependent	polycyclic aromatic hydrocarbon, nitrosamine
3. Inorganic*	metal
B. Epigenetic Carcinogens	
1. Promoter	organochlorine pesticide saccharin
2. Cytotoxic	nitrilotriacetic acid
3. Hormone-modifying	estrogen, amitrole
4. Immunosuppressor	purine analog
5. Solid state	plastics
C. Unclassified	
1. Peroxisome proliferators	clofibrate, phthalate esters
2. Miscellaneous	dioxane

*Some are tentatively categorized as genotoxic because of evidence for damage of DNA; others may operate through epigenetic mechanisms such as alterations in fidelity of DNA polymerases.

highest dose as that dose that is projected to produce a 10 % reduction of body weight in the animals by the end of the experiment. A more reasonable approach is to choose as the highest dose the minimal toxic dose (MTD) where, in preceding studies, toxic effects first appeared, regardless of the kind of effect, but only if such effects were definitely induced by the compound.

For non-toxic compounds another standard is needed to avoid false positive neoplastic incidences caused, for example, by chronic enzyme induction and/or glutathion deficiency due to chemical overload of target cells. According to the EEC-guidelines (1983) it is acceptable that the high dose should not exceed 100-fold the intended maximum daily human dose, if the toxic dose is a still larger multiple of the therapeutic dose.

The official guidelines for the selection of the lowest dose also vary by country. The FDA recommends that the low dose be one-fourth the high dose; whereas the EEC-guidelines propose that the low dose should be two to three times the maximal daily human dose.

The middle dose is expected to be one half of the high dose or the arithmetic mean of the low and high doses (HPB 1981, FDA 1984), or to correspond to the geometrical mean of the low and high doses (EEC 1983, Japan 1984).

An increased incidence of neoplastic lesions may also be attributed to substances other than potential new drugs. There is good evidence of altered rates of tumorigenesis from dietary influences (protein, fat, carbohydrates, fiber); disturbances of hormonal balance (prolactin, thyroxine, estrogen or other steroids); immunological effects following viral, chronic bacterial or protozoal infections; and repeated or prolonged stress that may affect hormonal and/or immunological regulation (Newberne, 1985).

Early in the investigation of a new compound there may be data that suggest that additional information on its possible carcinogenicity should be obtained. For example, one or two mutagenicity tests of different types may have given equivocal results. It may then be advisable to test the compound in an in vitro cell transformation test, using Syrian hamster embryonic cells, that shows atypical colony formation in the presence of a mutagenic agent. Other established short-term carcinogenicity bioassays are the mouse skin or lung tumor induction assay, the altered foci induction assay in rodent liver, or the breast cancer induction assay in female rats (Williams and Weisburger, 1986). If one of these tests and a mutagenicity test is positive, the compound is highly suspect as a carcinogen. Further development should be stopped unless the compound is life-saving with overwhelming clinical benefit. A prompt decision not only spares the lives of hundreds of animals, but also time and money. On the other hand, one has to keep in mind that none of the short-term carcinogenicity assays definitly proves the potency of a substance to initiate or promote cancer (Kayser, 1986).

VI. TEST PARAMETERS IN TOXICOLOGICAL STUDIES

What are the main issues in assessing the risk of using a new drug in humans?

First, a detailed protocol of each intended study has to be worked out by the study director. It must follow written standard operational procedures. Deviations from, and supplements to, this protocol have to be detailed in addenda. These documents form the basis of the work of the quality assurance unit. This is an independent internal control group whose function is to observe and to testify that, in the course of any toxicity study up to the final report, the rules of Good Laboratory Practice have been observed. Good Laboratory Practice, an internal laboratory standard, was enacted in 1979 by the FDA. It has been adopted voluntarily by most European pharmaceutical companies and has been officially established in Japan and in the FRG.

During toxicity studies, the clinical state of the animals is monitored by cage-side observation under surveillance of veterinarians. Observations are made of feed and water intake, excretion, body weight development, behavior and microbiological status of the animals and palpation of the body. Premature deaths, whether animals die or are killed for humane reasons, establish the mortality rate. These animals, as well as those sacrificed at the end of the study, are subjected to complete necropsy in order to observe all gross pathology.

Before starting treatment, at intervals during the test-period and at the end of the study, a series of biochemical parameters is monitored. These are indicators of possible organ toxicity; for example, serum enzyme activities monitor heart, liver, or muscle function; endogenous metabolites or substrates monitor liver, pancreas, or kidney function; serum electrolytes monitor renal or metabolic function; and total plasma protein and protein fractions monitor liver, kidney, and immune system function. In addition, one may investigate possible drug-induced changes in hormonal or immunological balance in the animals by using electrophoresis, radioimmuno- or enzyme-linked assays of serum proteins. To detect glomerular or tubular renal lesions or dysfunction or metabolic disturbances, one performs repeated urinalysis of protein(s), electrolytes, substrates and, on occasion, enzymes.

Hematological data obtained at various times are used to reveal possible drug-induced disorders of the peripheral blood, of the blood clotting system, or of bone marrow function. Such disorders are often sensitive indicators of immunologic reactions or the onset of neoplastic proliferation of blood-forming tissue.

Ophthalmologic surveillance is very important (Weisse and Hockwin, 1985). This is done not only where the topical administration of a ophthalmic preparation is being investigated, but also in systemic toxicity studies when the compound is administered orally or parenterally. The eyes are monitored by direct or indirect ophthalmoscopy, the slit-lamp, pupillometry, tonometry, the Schirmer-test and, at the end of the study, serial sections of the eyes for histopathological investigation. These procedures detect possible damage to specific parts of the eye, for example induced by cholinergic or adrenergic compounds or by cataractogenic substances.

When studying compounds for cardiovascular use, it is of great advantage to use electrocardiographic measurements to monitor the possible effects of the compound on the systemic blood pressure or on cardiac activity in toxicologic long-term studies.

It is crucial to investigate the possible effects of repeated drug administration on the morphology of various organs and tissues, at cellular and subcellular levels, by light- or electronmicroscopy. Samples of all organs and tissues from all treated and control animals are processed for histopathologic examination.

Considering that about 40 tissue samples per animal are analyzed and that, for example, a carcinogenicity study includes about 500 animals, the pathologist has to scrutinize some 20,000 histological specimens per study. Usually more than one diagnosis or diagnostic remark per specimen are necessary, and this data must be tabulated and statistically analyzed. Thus it is easy to understand that the histopathology data cannot be managed correctly without a computerised data processing system.

The data obtained in toxicological and carcinogenicity studies are evaluated statistically to verify the possible significance of inter-group and inter-sex differences. However, statistical significance does not necessarily mean biological relevance. The decision how to interpret the findings is up to the scientist, who, for example, considers the influence of individual values on the statistical results or compares actual experimental data with historical data obtained under identical experimental conditions.

VII. EVALUATION OF TOXICITY STUDIES

The data from animal toxicity studies have to be evaluated carefully to establish their relevance with respect to the intended human use.

In single dose acute toxicity studies the clinical signs of intoxication and the degree of toxicity are the main points of interest. Since Paracelsus (1493 - 1541), we have been aware that the dose is decisive in the classification of a substance as more or less toxic. Knowledge of both the pharmacodynamically effective and the non-lethal and lethal toxic doses is important in estimating the width of the therapeutic window of a compound. The careful monitoring of clinical symptoms following a single administration of the compound may provide information on the organ(s) or organ-system(s) most sensitive to the compound. Special attention should be paid to cases of delayed animal mortality. This suggests metabolic transformation to a toxic substance or slowly acting neurotoxicity. Such acute toxicity studies may also be helpful for screening a variety of related compounds or in assessing a possible change of toxicity of different formulations or of different batches.

The repeated administration of a compound reveals, on the one hand, the possible toxic effects on target organs, and characterizes, on the other hand, the "no observed effect level" (NOEL). Considering a necessary safety margin and species differences in drug kinetics and metabolism, one can approximate the maximum allowed therapeutic dose for humans. Moreover, these repeated dose toxicity studies demonstrate predictable adverse reactions at higher doses. This information enables the clinician to prevent organ damage by careful monitoring during drug administration. Thus the risk to humans will be smaller the more comprehensive and unequivocal the data are that have been obtained from experimental toxicology. Additionally, repeated dose toxicity studies may show whether a toxic effect of a compound depends upon exceeding a threshold dose or is a function of the total cumulative dose.

The question is often raised whether the findings in toxicological animal experiments can be extrapolated to humans and, if so, to what extent. The British Royal Society of Medicine concluded that there was a rather good concordance of adverse or side effects that were initially observed in laboratory animals and later endorsed by clinical drug surveillance (Fletcher, 1978).

However, adverse drug reactions that concern sensory phenomena or general feelings of well-being (e.g. headache, dizziness, fatigue, sleeplessness, nausea etc.) cannot be predicted from animal experiments. Lumley and Walker (1986) showed that, in repeated dose toxicity studies in rats and dogs with 29 compounds, there was a tendency to underpredict effects on the cardiovascular, central nervous and gastrointestinal systems, but a tendency to overpredict effects on the liver and urinary systems. Griffin (1985a) compared adverse effects of antiarrhythmic drugs in animal toxicity studies and in clinical use: In more recently developed substances there was a good concordance.

It is difficult to interpret experiments in rodents and rabbits when teratogenic effects were observed in only one species and only in higher dose groups, and where the dams showed signs of maternal toxicity. Are those effects induced by direct or primary teratogenicity, or by secondary mechanisms due to maternotoxicity? In contrast to mutagens, true teratogens are characterized by a threshold dose and a clear dose-effect relationship (Harbison, 1980).

According to Brent (1986), we have to differentiate between: (1) teratogens, which produce permanent pathological alterations in the offspring at exposures or in circumstances that commonly occur; (2) potential teratogens, which can alter fetal or embryonic development if exposures is increased a 100- or 1,000-fold; and (3) nonteratogens, which have no teratogenetic potential at any exposure, or have equal or greater maternal toxicity and, therefore, affect the mother before the embryo or fetus.

From 391 reproductive toxicity studies on rodents

with various chemical agents and drugs, Khera (1985) suggested that 56 % of the malformations observed in the pups were due to maternotoxicity. Secondary embryolethality or teratogenicity may result from bursts of high blood levels of a compound that severely affects cardiovascular regulation, as is true of caffeine and antihypertensive agents. The amount of caffeine that induced skeletal lesions in rabbits when given once a day did not produce the same effects when administered in split doses three times a day (Sullivan, 1982). Similarly the oral embryolethal dose of clonidine hydrochloride in rats is elevated dramatically when the drug is supplied continuously with a subcutaneous mini-osmotic pump to avoid high peaks in plasma concentration (Niggeschulze, personal communication).

There is no general agreement on the predictive value of in vitro and in vivo mutagenicity studies for the carcinogenic potential of a compound in humans (Clive, 1985). The EPA's (1980) position is that "positive findings in as few as two mutagenicity tests would provide a reasonable evidence to conclude that exposure to the substance presents a hazard" (Calabrese, 1983). OSHA does not accept that positive results in short-term tests identify potential carcinogens, but considers them to be confirmatory of a positive result in a long-term bioassay (Brusick, 1980). This is in accordance with the statement of Rinkus and Legator (1979) that "the fact that the host's immunological competency, about which mutagenicity (in vitro-) testing makes no evaluation, plays a role in the course of carcinogenesis". This argues against equating mutagenic potency with carcinogenic potency.

In a battery of genotoxic assays, a positive result in an in vivo test deserves more weight than a positive bacterial test. The overall risk-benefit assessment has to consider the whole pharmacokinetic, metabolic, toxicity and clinical profile of the compound (Griffin, 1985b).

Moreover, none of the short-term carcinogenicity tests has been validated as a test of high predictability of carcinogenicity (Kayser, 1986). Therefore, each compound intended to be administered to humans for a long period has to be investigated in two rodent species for its possible carcinogenic potential. A large number of data must be assessed: tumor incidences with respect to their site, type, benign or malignant growth, degree of proliferation (grading), capacity to metastasise, latency of onset; and the numbers of tumor bearing animals and animals with more than one tumor. In addition all non-neoplastic lesions must be evaluated. The tumor data must be compared statistically for heterogeneity and possible trends (PETO et al. 1980) to the untreated controls in the same study and to historical data obtained in identically conducted studies carried out during the previous 4 - 5 years (Haseman et al. 1984).

To evaluate animal carcinogenicity studies in terms of the neoplastic hazard of a compound to humans is a highly sophisticated procedure, as has been shown recently for formaldehyde (Albert et al. 1982) and methylendichloride (ECETOC 1987). The problem is that no experimental study in rodents can unequivocally prove a chemical agent to be definitely non-carcinogenic and only very extensive epidemiologic studies can establish a substance as a human carcinogen. Thus the objective of experimental mutagenicity and carcinogenicity studies is to position the compound somewhere between these extremes. Therefore, a medical benefit risk assessment includes a number of issues such as the chemical class, the dose-effect relationship, the pathways of metabolism, the pharmacokinetics, the clinical indication, the duration of administration, and the maximal therapeutic dose of the compound as well as the age, sex, and general vitality of the patient to be treated and the characteristics of alternative therapies.

VIII. TOXICOLOGY OF BIOTECHNOLOGICAL PRODUCTS

It is a challenge to the toxicologist to design tests to assess the safety of the peptide products of biotechnology. According to Bass et al. (1986) these are (1) natural human proteins, (2) analogues of human proteins, (3) novel oligopeptides, (4) isosteric peptides, and (5) monoclonal antibodies.

Irrespective of the method of production of these compounds, their common denominator is that they are peptides of various molecular weight. This means that, a priori, they are capable of generating antibodies following parenteral administration.

In contrast to the toxicology of small molecules, for such products we have very little historical experience. While conventional toxicological tests are often inadequate (Teelmann et al. 1986), especially for long-term studies, we need to establish preclinical test procedures for the estimation of the safety of the use of such products in human therapy.

The proposals of the French (1984) and Japanese (1984) regulatory agencies are considered to be too rigid and only a first step in defining the ultimate safety regulations. For the time being, the conflict between our incomplete knowledge and experience and the necessity for a preclinical proof of quality may be bridged best by choosing toxicological tests on a case-by-case decision. This conforms to the FDA proposal that studies should be carried out in order to determine the biologic and presumably toxic properties of these agents "with the intent of learning about the compounds under study, not just to demonstrate the lack of harmful effects" (Galbraith 1986).

As stated by Zbinden (1986), there are three main areas of toxicological concern for the design of protocols for biotechnological products: (1) the intrinsic toxicity of the product, (2) the toxicity related to pharmacodynamic effects of the product and (3) the biologic toxicity of the product.

In addition to problems that may arise from lack of homogeneity and the presence of small amounts of impurities or of contaminants due to fermentation and isolation, products of biotechnology can cause intrinsic toxic reactions in man. This has been shown with interferons (Scott 1983) and interleukin-2 (Rosenberg et al. 1985). These intrinsic reactions have only partly been reproduced in animal studies (Teelmann et al. 1986). The development of more predictive in vitro and in vivo tests, therefore, is necessary to improve the preclinical safety assessment of this class of agent.

Due to their high biologic specificity and potency, endogenous human proteins such as insulin, growth hormone, or human placental lactogen may exert excessive pharmacodynamic effects when administered in doses that produce higher than physiological concentrations or that affect other than their primary target receptors. This may lead to unexpected effects in various organ systems (Galbraith 1986). Making such currently unpredictable reactions predictable by test models is an important step in increasing the safety for patients.

It is difficult to assess the influences of biotechnology products on the immune system, including the induction of autoimmune diseases. This is another serious problem in evaluating biotechnological products experimentally as well as clinically. It is known that proteins can alter the hosts' susceptibility to and the course of infectious disease by immunomodulation. Proteins may also influence the manifestation or growth of tumors by hormonal effects (Anonymous 1986): Another strong argument for the need to develop predictive preclinical tests.

Finally, the biologic toxicity of the these products results from their ability to induce antibody formation. This may neutralize the therapeutic effect or again have effects on the immune system. For example, one might observe allergic hypersensitivity responses, microthrombotic events by platelet aggregation, or the blockade of the cell system that is responsible for the prevention of bacterial, viral or endotoxin induced diseases.

As suggested by Galbraith (1986), preclinical animal studies on monoclonal antibodies should be carried out with single and multiple dose treatment in rodents and in at least one non-rodent species, such as the dog or sub-human primate. The lowest dose should be the highest expected human dose. Additionally at least two doses which are multiples of the high human dose should be used. This range of doses allows the estimation of the highest non-toxic dose and of the dose capable of inducing pathological alterations in clinical animal observation, hematology, urinalysis, biochemistry and histopathology. To evaluate the preclinical toxicity of new endogenous proteins, Galbraith proposes to "select an animal model in which the pharmacologic activity of the protein can be demonstrated to be the same as for man" (1986).

The immunogenicity of human proteins to test animals jeopardizes interpretation of chronic toxicity studies. Repeated dose studies of 2 - 4 weeks duration are sufficient if the formation of neutralizing antibodies in the test animals can be assessed. Clinical observation during animal studies must include the surveillance of blood pressure, body temperature and cardio-pulmonary function.

The genotoxic potential of biotechnical products should be evaluated in the Ames-test for point-mutations, and in a cytogenetic test in a mammalian cell line. If the drug is to be used in women of childbearing potential for the therapy of a non-life threatening disease, reproductive toxicity studies are necessary. If there is any reason to suggest that the compound may cause abortion, as could be shown with interferons that interact with prostaglandins, reproductive toxicology studies are indispensible. Sub-human primates would be used because of the immunological and physiological similarity of this species to man. The amino acid sequence of proteins in rhesus and cynomolgus monkeys, both members of the family of macaques, is about 90 % compared to man. This is exceeded only by chimpanzees with a homology of 99 % (Hobson, 1986). From an ethical point of view, however, only the macaques are acceptable for reproductive toxicological studies of biotechnological products in sub-human primates. Animal welfare considerations make it urgent to search for alternative laboratory methods for such investigations.

IX. EXPENDITURE IN TOXICOLOGY

Toxicological studies necessary to elucidate safety of

a new drug are one of the major expenses of drug research and development. The costs for one compound are calculated today to reach about sixty million dollars before marketing.

In addition to the expenses of laboratory work and animal husbandry, the costs, for example, of animals necessary for one long-term toxicity study are approximately $1,000 for rodents and up to $100,000 for monkeys. The expenses for one complete carcinogenicity study in a rodent species amount to approximately one half million dollars.

Considerations such as these clearly show that, in addition to ethical factors, there are important economic arguments against unnecessary animal studies in industry or academic drug research and development. The sole justification for performing animal studies is the need to minimize any predictable drug-induced hazard to the patient.

Thus, it is important to have very strict planning and control of the toxicological studies. Step by step decisions on the progression from short-term studies to long-term toxicity and carcinogenicity tests ought to be practised in order to avoid wastage of animals and money. Teratology and genotoxic tests should be finished before starting long-term animal and clinical phase-II studies. If one of the teratogenicity studies or in vivo genotoxicity assays yield unequivocally positive teratogenic or mutagenic results, then there will probably be insurmountable difficulties in testing the compound in humans unless it is an anti-cancer or absolutely life-saving drug. Thus long-term studies would only waste resources and should not be initiated.

Consequently, the toxicologist in a pharmaceutical company is responsible not only for scientific planning, precise realization and careful evaluation of toxicological studies, but also for the best possible financial planning in drug development. This will lower the producer's expenditure and so reduce the cost of medical care.

X. SUMMARY

Animal toxicology studies are decisive in the development of better and safer drugs by allowing the evaluation in humans only of those compounds that are expected to be safe. These studies also provide information that allows the physician to monitor the patient for the first possible signs of drug-related toxicity.

The toxicologist uses a battery of tests, including acute single dose, short-term und long-term repeated dose toxicity, reproductive toxicology, genotoxicity and carcinogenicity studies. These are done to meet the ethical responsibilities of the drug producing companies and the requests for the projected human safety of new drugs by the regulatory agencies of various countries. Every effort has to be made to save animal lives by reduction in the number of animals used, by refinement of techniques and, if possible, by replacement of in vivo tests. However, there must be no decrease in the projections of safety of the compounds. International agreements on the test strategies and details would save animal lives and patient costs.

New challenges in experimental toxicology have been brought about by biotechnologically manufactured drugs. These need to be solved in the near future in order to maintain or even improve the standard of safety of drugs developed by conventional means.

REFERENCES

a. Official publications:

Australian Department of Health, Guidelines for Preparing Applications for the General Market or Clinical Investigational Use of a Therapeutic Substance - NDF 4. Amended May 1980 and Corrigendum September 1983

Bundesgesetzblatt, Neufassung des Tierschutzgesetzes vom 18.08.1986, Teil I, 22. Aug. 1986, Federal Republic of Germany

EEC-Guidelines - Official Journal of the European Communities, L 332, 28 November 1983; L 73, 16 March 1987

Environmental Protection Agency: Toxic Substances Control Act Test Guidelines, Final Rules, Part. II. Fed. Reg. 50, 27 Sept., 1985, USA

Ethische Grundsätze und Richtlinien für wissenschaftliche Tierversuche, Bern, 6. u. 7. Mai 1983, Schweiz. Ärztezeitung 64 (Heft 24) vom 15.06.1983, Switzerland

FDA, Dept. of Health, Education and Welfare: Appraisal of the Safety of Chemicals in Foods, Drugs and Cosmetics, 2nd Printing 1965, USA

FDA Guidelines Fed. Reg. 22.05.1984 Part II Chem. Carcinogens, USA

Health Protection Branch - Health and Welfare Canada, Bureau of Human Prescription Drugs, Preclinical Toxicologic Guidelines 1981, Canada

Ministry of Health and Welfare, Japan, Notification 118, 15.02.1984

Ministry of Health und Welfare, Japan, Preparation of data required for approval applications for drugs manufactured by the application of recombinant DNA technology. Notification No. 243, Pharmaceutical Affairs Bureau, March 30, 1984

Nordic Council on Medicine, Nordiska Läkemedelsnämnden NLN Publication No. 12, Uppsala 1983, Sweden

Syndicat National de l'Industrie Pharmaceutique, Rue de la Faisandrie 88, F-75782 Paris 16, Recommendation concernant le protocole toxicologique des interferons pour l'obtention d'une autorisation de mise sur le marche. Mars 16, 1984, France.

b. Publications by specified authors:

Albert, R.E., A.R. Sellakumar, M. Kuschner, N. Nelson, and C.A. Snyder, Gaseous formaldehyde and hydrogen chloride induction of nasal cancer in the rat, J. Nat. Cancer Inst. 68, 597 - 603 (1982)

Ames, B.N., J. McCann, and E. Yamasaki, Methods for detecting carcinogens and mutagenes with the Salmonella/ mammalian-microsome mutagenicity test, Mutat. Res. 31, 347 - 364 (1975)

Anonymous, Growth factors and malignancy, The Lancet, 1986, 317 - 319

Bass, R., P. Günzel, D. Henschler, J. König, D. Lorke, D. Neubert, E. Schütz, E. Schuppan, and G. Zbinden, LD_{50} versus acute toxicity: critical assessment of the methodology currently in use, Arch. Toxicol. 51, 183 - 186 (1982)

Bass, R., E. Scheibner and B. Schnieders, Preclinical safety of biotechnology products intended for human use: Government Regulatory Position in the Federal Republic of Germany. Satellite Symposion to IV International Congress of Toxicology, Tokyo 1986 (Abstracts)

Beck, F., Comparative placental morphology and function, Environ. Health Perspect. 18, 5 - 12 (1976)

Beckley, J.H., T.J. Russell, and L.F. Rubin, Use of rhesus monkey for predicting human response to eye irritants, Toxicol. Appl. Pharmacol. 15, 1 - 9 (1969)

Brent, R.L., Definition of a teratogen and the relationship of teratogenicity to carcinogenicity, Teratology 34, 359 - 360 (1986)

Brusick, D.J., Testimony at the OSHA carcinogen hearings, Fed. Reg. 45(15), 5112 (1980)

Calabrese, E.J., "Principles of Animal Extrapolation", John Wiley & Sons, Inc., New York 1983

Clayson, D.B., D. Krewski, I. Munro, "Toxicological Risk Assessment", CRC Press, Inc., Boca Raton, Florida, 1985

Clive, D., Mutagenicity in drug development: Interpretation and significance of test results, Regul. Toxicol. Pharmacol. 5, 79 - 100 (1985)

Draize, J.H., G. Woodard, and H.O. Calvary, Methods for the study of irritation and toxicity of substances applied topically to the skin and mucous membranes, J. Pharmacol. Exp. Ther. 82, 377 - 390 (1944)

ECETOC. "Joint Assessment of Commodity Chemicals" No. 4, Methylene chloride, CAS: 75-09-2, 17. January 1984

Fletcher, A.P., Drug safety tests and subsequent clinical experience, J. Royal Soc. Med. 71, 693 - 696 (1978)

Forth, W., D. Henschler, W. Rummel, "Allgemeine und Spezielle Pharmakologie und Toxikologie", 5. Auflage. Wissenschaftsverlag Mannheim, Wien, Zürich 1987

Frederick, G.L., The necessary minimal duration of final long-term toxicologic tests of drugs, Fund. Appl. Toxicol. 6, 385 - 394 (1986)

Galbraith, W.M., New challenges in the safety evaluation of drugs and biologics, Satellite Symposium to the IV International Congress of Toxicology, Tokyo 1986 (Abstract)

Griffin, J.P., Predictive value of animal toxicity studies, ATLA 12, 163 - 170 (1985a)

Griffin, J.P., The Committee on Proprietary Medicinal Products Safety Working Party's recommendations to minimise the unnecessary use of animals in the evaluation of new medicines, ATLA 12, 251 - 257 (1985b)

Harbison, R.D., Teratogens, in "Casarett and Doull's Toxicology", (J. Doull, C.D. Klaassen, M.O.Amdur, eds.) 2nd edition, Macmillan Publishing Co., Inc., New York pp. 158 - 175, (1980)

Haseman, J.K., J. Huff, and G.A. Boorman, Use of historical control data in carcinogenicity studies in rodents, Toxicol. Pathol. 12, 126 - 135 (1984)

Hayes, A.W., Principles and Methods of Toxicology, Raven Press, New York, 1982

Heilmann, K., Das Arzneimittel als Risiko, Pharma dialog 82, Hsg. Bundesverband der Pharmazeutischen Industrie, Frankfurt/Main, Mai 1984

Hobson, W.C., Species selection. Satellite Symposium to the IV. International Congress of Toxicology, Tokyo 1986 (Abstract)

Kayser, D., Kurzzeittests zum Nachweis von Kanzerogenen, BGA-Schriften 1/86. MMV Medizin Verlag München, 1986

Khera, K.S., Maternal toxicity: A possible etiological factor in embryo-fetal deaths and fetal malformations of rodent-rabbit species, Teratology 31, 129 - 153 (1985)

Klaassen, C.D., M.O. Amdur, J. Doull, Casarett and Doull's Toxicology, 3rd edition, Macmillan Publishing Co., Inc., New York 1986

Luepke, N.P., Hen's egg chorioallantoic membrane test for irritation potential, Fd. Chem. Tox. 23, 287 - 291 (1985)

Lumley, C.E. and S.R. Walker, Repeat-dose animal toxicity tests - are studies longer than 6 months necessary? Poster: Satellite Symposium to the IV International Congress of Toxicology, Tokyo 1986.

Magnusson, B., and A.M. Kligman, The identification of contact allergens by animal assay. The guinea pig maximization test, J. Invest. Dermatol. 52, 268 - 276 (1969)

Mershon, M.M. and J.F. Callahan, Exploiting the differences. in "Animal Models in Dermatology", (H.I. Maibach ed.). Churchill Livingstone, New York, 1975, pp. 36 - 52

Neubert, D., Übertragbarkeit von Tierversuchen. Prospektive in-vitro-Modelle als Ersatz für Langzeituntersuchungen. AMI-Berichte 1/80, 277 - 287 (1980)

Newberne, P.M., The influence of nutrition, immunologic status, and other factors on the development of cancer, in "Toxicological Risk Assessment",(D.B. Clayson, D. Krewski, I. Munro, eds), Vol. II. CRC Press, Inc., Boca Raton, Florida, 1985, pp. 43 - 87

Peto, R., M.C. Pike, N.E. Day, R.G. Gray, P.N. Lee, S. Parish, J. Peto, S. Richards, and J. Wahrendorf, Guidelines for simple, sensitive significance tests for carcinogencity effects in long-term animal experiments, IARC Monographs on the Evaluation of the Carcinogenic Risk of Chemicals to Humans, Suppl. 2. Lyon, International Agency for Research on Cancer, 1980

Rinkus, S.J. and M.S. Legator, Chemical characterization of 465 known or suspected carcinogens and their correlation with mutagenic activity in the Salmonella typhimurium system, Cancer Res. 39, 3289 - 3318 (1979)

Rosenberg, S.A., M.T. Lotze, L.M. Muul, S. Leitman, A.E. Chang, S.E. Ettinghausen, Y.L. Matory, J.M. Skibber, E. Shiloni, J.T. Vetto, C.A. Seipp, C. Simpson, C.M. Reichert, Observations on the systemic administration of autologous lymphokine-activated cells and recombinant interleukin-2 to patients with metastatic cancer, N. Engl. J. Med. 313, 1485 - 1492 (1985)

Scott, G.M., The toxic effect of interferons in man, Interferon 5, 85 - 114 (1983)

Sullivan, F., Effects of various dosing regimens of caffeine, IV. International Caffeine Workshop, Athens, Oct 17 - 21, 1982

Teelmann, K., Chr. Hohbach, H. Lehmann, and the International Working Group, Preclinical safety testing of species-specific proteins produced with recombinant DNA-techniques, Arch. Toxicol. 59, 195 - 200 (1986)

Walker, S.R., E. Schuetz, D. Schuppan, J. Gelzer, A comparative retrospective analysis of data from short- and long-term animal toxicity studies on 40 pharmaceutical compounds, Arch. Toxicol. Suppl. 7, 485 - 487 (1984)

Weisse, I., and O. Hockwin, Auswirkungen von Spezies-Unterschieden für die Okulotoxizitäts-Prüfung von Arzneimitteln, in "Biochemie des Auges" (O. Hockwin ed.), Ferdinand Enke Verlag, Stuttgart 1985

Williams, G.M., and J.H. Weisburger, Chemical carcinogens, in "Casarett and Doull's Toxicology", (C.D. Klaassen, M.O. Amdur and J. Doull, eds.) 3rd edition, Macmillan Publishing Company, N.Y. (1986)

Zbinden, G., Biotechnology products intended for human use, toxicological targets and research strategies. Satellite Symposium to IV International Congress of Toxicology, Tokyo 1986 (Abstract).

9

The Contribution of Pharmaceutics to Improvements in Drug Therapy

B. Zierenberg

Department of Pharmaceutics
Boehringer Ingelheim KG
Ingelheim, Federal Republic of Germany

I. GENERAL ASPECTS OF PHARMACEUTICS IN DRUG DEVELOPMENT

A. HISTORICAL DEVELOPMENT OF PHARMACEUTICS

Until the beginning of this century the pharmaceutical contribution to medical care came more or less from pharmacies. The pharmacist's task here was to isolate the active ingredient, render it into a form suitable for use by the patient and establish its concentration in the dosage form.

Since then this situation has changed. Today research, development and production of dosage forms are done almost exclusively by the pharmaceutical industry. The principal benefit of these changes for the patient is a steady improvement of dosage forms resulting from the better organization and flow of information in industry compared to pharmacies.

The old art of manufacturing dosage forms has been transformed into a science. In the 1950's and 1960's pharmaceutics described the dosage forms scientifically. Such techniques as tabletting and coating were investigated intensively and ultimately successfully described. The fate of the drug in the body became a research emphasis after that time. Current research involves a closer integration of these two areas into the design of dosage forms that provide an even more optimal delivery of the active ingredient to its site of action. These subjects will form the subject of this chapter.

B. REQUIRED PROPERTIES OF A DOSAGE FORM

Contributions of pharmaceutics to the development of a medication are needed even in phases where the drug is tested in animals: pharmaceutics are always involved when human efficacy trials are performed. What tasks should pharmaceutics fulfill in converting an active ingredient into a dosage form?

The patient must receive the drug in a form appropriate for the administration route. For oral administration a suitably designed tablet, capsule, solution or suspension has to be prepared and manufactured; for parenteral, intravenous injection a stable tolerable solution is required. The dosage form has to fulfill the following three technical requirements: (1) The dosage form must contain the specified quantity of the active ingredient. The active ingredient must not degrade during the shelf life, nor must it react with the additional ingredients used in the dosage form. (2) The dosage form must be producible on an industrial scale. This means one must devise and acquire the production machinery as well as create the necessary production process. (3) The dosage form must be of reproducible quality. This entails adherence to a quality standard governing the manufacturing process. The elements of this standard are laid down in the Good Manufacturing Practice (GMP) requirements of different government agencies.

Additionally, the dosage form must be such that the active ingredient reaches the target site at the necessary rate and in sufficient concentration. Depending on the therapeutic aim this may mean either that the active ingredient is released as rapidly as possible after intake into the body or that the rate of release of the active ingredient is controlled to a slower rate.

Finally, the dosage form is required to be as convenient and palatable as possible for the patient. Thus for example bitter substances must not reach the taste buds, the dosage form should not cause gastric or intestinal upsets, and the dosage frequency should not interfere with normal sleep nor be complicated to remember. From the patient's point of view a dosage form that is superior to others in these respects is a better drug.

C. CONTROLLED DRUG DELIVERY

In designing dosage forms that deliver the drug to the right place at the right time one needs to understand the behavior of the drug with respect to the concepts of drug absorption, distribution, metabolism and elimination discussed in Chapter 3. It took several decades to introduce these pharmacokinetic concepts into development and evaluation of dosage forms. This time span was needed to create an appropriate experimental basis, that is, to develop methods for the measurements of levels of drugs in human blood and also for the analysis of such data in a convenient manner. The availability of chromatography, radioimmunoassays as well as of computers paved the way to these ends.

The concepts of pharmacokinetics led to the consideration of the effect of the dosage form on the biological distribution of the drug. This can be done by treating the administered dosage form as a further compartment in pharmacokinetic models. This man made compartment is crucially important for the systemic action of the drug. Besides the intrinsic pharmacological action of the chemical entity, the dynamics of drug release from the dosage form and subsequent availability are important in determining the safety and efficacy of therapy.

Although for centuries pharmacists had known that the dose of a drug is correlated with the pharmacodynamic effect, it has only recently become possible to correlate blood levels of a drug with its effects. An example is given in Figure 1 for the drug theophylline. We can distinguish between no effect levels, therapeutic levels and toxic levels.

New materials had to be developed for the implementation of controlled drug delivery. The excipients for traditional dosage forms stem more or less from natural sources. Such materials possess serious shortcomings: first the quality may change from batch to batch; and second, the physico-chemical properties of such materials impose limits on the type of use. Today one can design

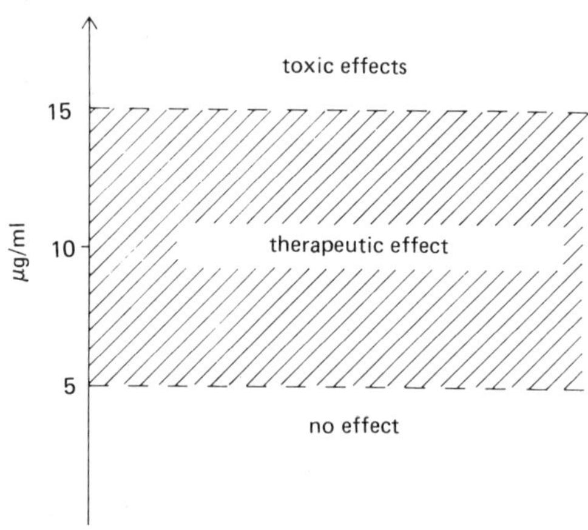

Figure 1 Plasma level and pharmacodynamic response for theophylline.

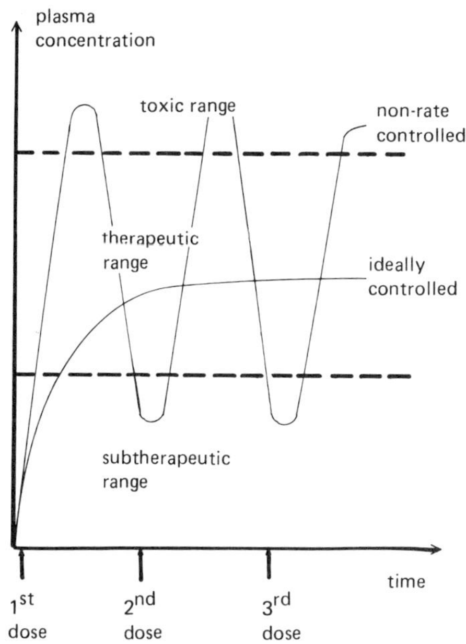

Figure 2 Comparison of plasma levels of repeated doses of a non rate controlled and a single dose of an ideally controlled release form.

polymers that enable one to overcome these deficiencies. The availability of such polymers opened the way to producing new dosage forms that release drugs in a completely controlled manner.

Figure 2 compares a conventional dosage form with an ideally controlled release form in terms of plasma levels of the active ingredient. The aim of medical treatment in many cases is to maintain the drug level at steady-state. Kinetically, this means that the input rate from the drug delivery system, V_{input}, is the same as the rate of elimination from the body, $V_{elim.}$.

$$V_{input} = V_{elim.} \text{ (in amount/time)} \tag{1}$$

What are the benefits for the patient of a controlled delivery system that fulfills Equation 1? For acute illnesses such as infectious diseases the total time of medical treatment will be reduced, because the periods in the treatment of both acute and chronic illnesses of subtherapeutic blood levels will be eliminated. Furthermore, the toxic range of blood levels will not be reached; this will diminish the possibility of undesirable side effects.

So far we have discussed the blood as the site of measuring the concentration of the drug. In many cases the drug's site of action is not the blood but specific organs or cells. A sufficiently high concentration of the drug at its target must be built up. Controlled drug delivery can accomplish this also. Such target-controlled delivery is a main objective of the design of dosage forms for cancer therapy because they would help reduce the severe side effects of current cytostatics.

II. PRINCIPAL WORKING AREAS IN PHARMACEUTICAL RESEARCH

A. GENERAL ASPECTS

1. Dosage forms for drugs. Before a drug is tested in living organisms we must decide which route of administration should be selected. The most common and hence most widely used sites of administration are the gastrointestinal tract, the lungs, the rectum or direct injection into the bloodstream. The absorption route should be selected after consultation with pharmacologists and physicians. Which route is to be preferred depends on the pharmacologic, chemical and metabolic properties of the substance; the medical indication for its use; and the patient's habits.

The dosage form must be so designed that the active substance reaches the target site of absorption and is released in the desired manner. An example is the administration of antiasthmatic drugs to the lung as an aerosol. If a substance is designed for inhalation, the particle size distribution must lie between 1 and 10 μm. Particles with a diameter of less than 1 μm reach the bronchial region, but cannot be not deposited there since they are immediately breathed out again. Particles greater than 10 μm in diameter are separated off in the nose/throat area and thus never reach the lung at all.

The dosage form often protects the premature degradation of the active compound. Some substances administered orally are hydrolyzed in the acid pH of the gastric juice and therefore have to be protected during passage through this region. Tablets or capsules may be coated with enteric polymers that prevent any flux of material at the acid pH of the stomach. Once the tablet or capsule enters the intestine, a pH range of approximately 7.0, the coating dissolves to release the active substance. However in the case of the peptide insulin, oral administration has so far been unsuccessful because the enzymatic degradation of the insulin in the gastrointestinal tract is faster than its absorption through the gastrointestinal wall. An oral form of insulin would have to be designed so as to increase the absorption rate to such an extent that the rate of metabolism is negligible in comparison. Alternatively, the metabolism of insulin during absorption would have to be prevented by suitable chemical or physical means. All attempts to do this have so far failed.

Another function of the dosage form is to ensure that a sufficient quantity of the substance is dissolved at the site of administration. The rate of passive diffusion of active substance passes through biological membranes into the bloodstream is directly proportional to the concentration of dissolved substance. Thus insufficient solubility of drugs can lead to poor absorption. Such difficulties can be overcome by:

(1) changing the salt and/or crystal form of the molecule,
(2) adjusting the pH by adding organic acids or bases to the dosage form,
(3) adding cosolvents, or
(4) increasing the surface area of the substance in order to promote formation of supersaturated solutions.

The effects and success of such measures must be demonstrated with in vitro studies before being tested in vivo. Examples supported by the relevant laws of physical chemistry can be found in books on physical pharmacy (Martin et al., 1970).

The relative amount of drug available for the therapeutic effect in the organism is described quantitatively with the bioavailability parameter described in Chapter 3. In addition to the bioavailability, the time course of the blood level is an important criterion for the absolute and comparative assessment of dosage forms.

In order to turn an active compound into a dosage form that is simple for both physician and patient to handle, excipients must be used. A few commonly used excipients are sugars, such as lactose; polymers, such as ethylcellulose; and solvents, such as ethanol. Excipients may also maintain the chemical, physical, and microbiological stability of the active substance. The excipients must be non-toxic and should not cause intolerance at the site of administration.

To sum up, we can say that the pharmacist must incorporate the active molecule into a dosage form that is so designed that it releases the active substance in therapeutic doses at the site of absorption, that is adapted to conditions at the site of absorption, that uses only safe excipients, and that is acceptable to the patient.

2. Stability of drugs and dosage forms. Therapeutically used substances usually exist in a crystalline form. After combination with excipients, the drug may be crystalline, amorphous, or in solution. The dosage form must be designed so as to ensure that both the substance and the form remain stable. This means that formulation produces no change in the chemical structure of the drug or its physical properties, such as crystal structure and particle size distribution. In the case of peptides the three-dimensional structure must be retained. The shelf-life, which is the time from manufacture to use by the patient, should range from two to five years. As drugs are often marketed worldwide and hence exposed to different ranges of temperature and humidity, the shelf-life must be sufficiently independent of climatic conditions.

Short-term stress tests and long-term stability tests are performed to obtain information about the stability of a substance in a particular dosage form. The stress test is a type of screening. It is used to select those excipients or pH conditions that stabilize the substance or at least have no detrimental effect. For the stress test, the substance is combined with potential excipients in a tablet or in solution and is stored at high temperatures (from 50 - 80 °C). Samples are taken at appropriate intervals and the concentration determined by suitable methods. Thus substance content c is measured as a function of time t, c(t). The two simplest equations for interpreting the results are zero (Equation 2) and first order decreases (Equation 3) respectively

$$c(t) = c(0) - k_o \cdot t \tag{2}$$

$$c(t) = c(0) \cdot e^{-k_1 \cdot t} \tag{3}$$

c(0) is the initial concentration, and k_o and k_1 are the zero and first order decomposition rate constants.

The selection of excipients that gives sufficiently low k_o or k_1 values is expected to lead to stable dosage forms. Long-term stability can be predicted by extrapolating stability data obtained at several higher temperatures to the temperatures of storage.

Stress tests enable one to recognize immediately any factors that might affect stability, and thus to confine long-term stability tests to a few near optimal dosage forms. Only if a dosage form passes this second test successfully can it be released for use in patients.

One contribution of pharmaceutics to drug efficiency and safety is therefore the incorporation of the drug into dosage forms in which it remains stable until use (Martin et al., 1970).

3. The manufacturing of dosage forms of drugs. Nowadays drugs are manufactured almost exclusively at industrial plants. Production processes must ensure that the dosage forms are of constant quality. This is the only way to guarantee consistent drug therapy for patients.

For consistent therapy the physician and the patient must be assured that, in addition to the stability discussed above, both the quantity and the timing of release of the drug from the dosage form also remain constant. In order to ascertain this in quantitative terms, several variables are used to describe the properties of the dosage form. Examples of such variables include: content uniformity, compressibility, and thickness of film-coating.

The observed values of the variables depend on certain manufacturing factors. For example, content uniformity may be determined by the mixing time; compressibility by the amount of excipients, and the thickness of film-coating by the viscosity of the coating substance. The extent to which the variables depend on these potential manufacturing factors is thoroughly and systematically examined via experiments designed with the aid of factorial schemes. Such schemes are described in Chapter 6. Numerical evaluation of the design reveals which of the manufacturing factors studied correlate to a statistically significant degree with the dosage form variables.

The next stage in the experimentation is to optimize the values of the factors with the aid of established optimization methods, such as the Nelder-Mead method (Nelder, 1965). This method uses an iterative approach in which the factors are varied systematically and the variables of interest measured. For a selected new combination of factors the values of the variables are calculated. The values thus obtained are used as a basis for selecting a new combination of factors to be tested experimentally. This search for the optimum ends when the values of the variables cannot be further improved. The factors obtained are then used for manufacturing the dosage form. Often the values for several variables point in opposite directions. For example, the stability of a parenteral dosage form of a drug might increase with pH whereas the physiological compatibility decreases with pH. In such cases one establishes those factor values at which each of the variables still meets the minimum requirements. Computer-assisted methods help to keep the cost and duration of this search for the optimum within reasonable limits.

Once a drug has been incorporated into a dosage form that meets the medical and pharmaceutical requirements, it is produced in quantity. A prerequisite is of course the appropriate production equipment and scaling-up of the dosage form; this is to ensure that the results obtained at laboratory level on small machines can be transferred to production level.

Once the technical requirements of the dosage form have been met, then appropriate managerial methods must be taken to guarantee the end product is of perfect quality. The basic conditions and regulations are laid down in the Good Manufacturing Practice specifications. They help to ensure that the dosage form produced has no contamination by other drugs, will not be confused with other dosage forms, and is microbiologically safe. These regulations also require that there is a documentation available that reflects every step in production. This catalogue of measures is the final activity in the long development process of a drug before it is brought onto the market and is used by patients.

B. SPECIAL ASPECTS OF CONTROLLED DRUG DELIVERY

1. Some general remarks on polymers. Polymer science and technology have been among the most important areas of discovery during the twentieth century. Polymers have a very broad range of physical, biological and chemical properties. The main factors relevant for these properties are the constituent monomers, the molecular weight distribution and the method of polymerization. In many cases the polymer scientist can tailor polymers to the desired properties. It is, therefore, not surprising that polymers are increasingly being used in pharmaceu-

tics, especially in controlled drug release, polymer implants, prosthetic materials and polymeric drugs. Biocompatibility and non-toxicity of the polymer is a fundamental prerequisite. This property is easier to establish when the polymer is used in an external or oral application. Expenditure of cost and time is far greater for implanted polymer systems loaded with drugs. The ratio of benefit to risk has to be considered in detail before starting with an implanted system.

The principal polymers used in controlled drug delivery are silicones, acrylates, cellulose derivates, pyrrolidones and lactides. One common feature of pharmaceutical formulations is the judicious selection of a polymer material to act as a rate-controlling device, container or carrier for the agent to be released. In some cases the choice of polymer is the result of sophisticated considerations while in others evolution from previous successes dictates the selection. One of the central problems in designing a controlled drug release dosage form is to find a method for achieving a release profile that best fits the targeted in vivo levels. One can dissolve or disperse the drug in the polymer or bind it chemically to the polymer.

2. Chemically controlled drug release. In chemically controlled drug delivery systems, the release of the pharmacologically active agent takes place in the biological environment either by gradual degradation of the polymer with dispersed drug or by breaking the bond that links the drug to the polymer chain. The degradation process can be induced either chemically or by enzymes that are naturally present in the biological fluids. In practice, chemical approaches are preferred because degradation based on enzymatic processes often leads to greater interindividual differences in the release profile.

There are a large number of potentially biodegradable polymer systems that can be used as carriers for dispersed or dissolved drugs: activated carbon-carbon polymers, polyurethanes, polyamides, polyesters and polyacetals, etc. Since the monomers resulting from the in vivo cleavage process must be biocompatible, poly(lactides) and the copolymers with glycolic acid are especially suitable. The poly(lactides) are prepared by ring opening of the dimer of the lactic acid, according to the following formula (1).

$$2n \times \text{(lactide: } H_3C{-}HC{<}\text{ ring with two } C{=}O \text{ and two O, } CH{-}CH_3) \xrightarrow[\text{e.g. Sn-octanoate}]{\text{catalyst}} HO{-}\overset{H}{\underset{H_3C}{C}}{-}\left[\underset{\|}{\overset{}{C}}{-}O{-}\overset{H}{\underset{H_3C}{C}}\right]_{2n-1}{-}C(=O)OH$$

The physical parameters that influence the in vivo degradation of poly(lactides) are the molecular weight average and the distribution as well as the type and configuration of the α-hydroxy acid used. In the pharmaceutical step the polymer is loaded with the drug by using either a solution or a melting process. The resulting material is then transformed to a defined geometric shape. This shape influences the release pattern.

The rate of release is, of course, also dependent on the initial concentration of the drug in the dosage form and on the rate of degradation of that form. The latter property can be expressed in terms of an erosion constant K_0 that is characteristic for the particular polymer. An indirect proof of the dependency of K_0 on the molecular weight of the polymer is demonstrated in Figure 3. It shows the release of clonidine from two poly(L-lactides) of different molecular weight. The higher the number of bonds that need to be hydrolyzed, the slower

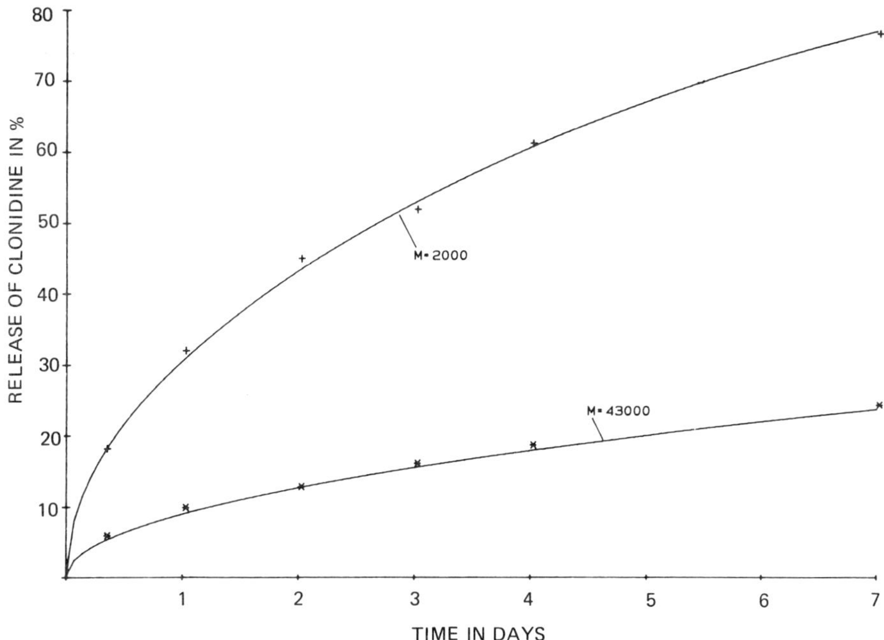

Figure 3 Release profiles of clonidine from two poly(L-lactides) carriers with different molecular weights.

the erosion of the host polymer and therefore the slower the release of drug.

Long-acting contraceptives would be an important use of chemically controlled drug release using poly(lactides) (Zatuchni, 1983). The aim is to develop an intramuscularly injected dosage form that maintains an effective serum level of a contraceptive such as norethisterone for several months. If this goal can be reached it will be an important contribution to safe contraceptive treatment and, ultimately, to better regulation of worldwide birth rates. The success in development of such a dosage form so far achieved is shown in Figure 4 in which the mean serum level of norethisterone is plotted as a function of time. The minimal effective blood level of norethisterone is about 1 ng/ml; a single dose of 100 mg has a duration of contraceptive action of approximately 5 months.

A second use of poly(lactides) as a carrier will probably be for the administration of peptides.

3. Diffusional systems for controlled drug release. In diffusional systems the embedded drug diffuses out through the polymer leaving the latter intact. Diffusional systems are of interest for oral or external forms of controlled release because for these the carrier is not absorbed by the body. The incorporation of the drug into a polymer carrier can result in either a two phase or a single phase system. In a two phase system the drug is dispersed as discrete particles in the polymer carrier. Two phase systems belong to the oldest release systems described in the pharmaceutical literature. Diffusion of the drug from the carrier is then governed mainly by three factors: the size distribution of the drug particles, the loading dose of the drug and the diffusion coefficient of the drug in the polymer. The amount of drug released from a two phase system was shown to be proportional to the square root of time (Higuchi, 1961). Knowing the equation and constant for the release rate of the drug from the polymer carrier it is possible to develop a controlled release two phase system in a straightforward manner.

In single phase systems the drug is completely dissolved in the polymer, that is the drug is physically not discernable from the carrier. Such a system is therefore called a solid solution. The amount of drug released from a solid solution in a unit of time depends on three principal parameters: the loading dose, the diffusion coefficient of the drug in the polymer and the length (or thickness) of the system (Crank, 1975).

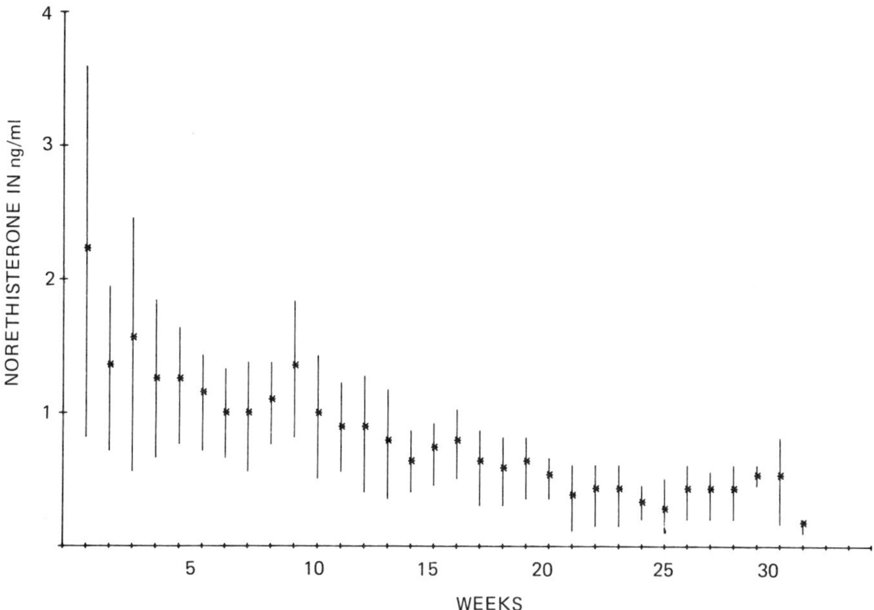

Figure 4 Mean serum levels of immune reactive norethisterone (NET) in 13 women treated by intramuscular injection of sterilized poly(lactic acid) microcapsules (mean dose = 2.45 + 0.44 mg NET/kg). (Zatuchni et al., 1983)

A two phase system results if the concentration of the incorporated drug exceeds its solubility in the polymer. At drug concentrations below the solubility a single phase system prevails. This situation applies especially to highly potent drugs.

For a solid solution we shall now demonstrate in two examples how one can vary the properties of the polymer so as to specifically change the diffusion coefficient and thereby the release profile of a drug. One possibility consists in altering the chemical properties of the polymer carrier. An example of a chemically controlled mechanism is shown in Figure 5 for the drug clonidine embedded in emulsion polymerized poly(acrylate) containing varying percentages of additional carboxyl-groups in the polymer chain. The drug, being partially present as a protonated base, forms an ionic interaction with the carboxyl-group. This decreases the diffusion rate.

A second way to modify a release profile is to vary the physical properties of the polymer as illustrated in Figure 6. In this case two emulsion polymerized polymer carriers that differ in glass temperature were used. At temperatures below the glass temperature the polymer chains are immobilized and rigid. Therefore high glass temperatures correspond to a low diffusion coefficient of the drug.

4. Membrane systems for controlled drug release. For membrane controlled drug delivery systems, polymers are

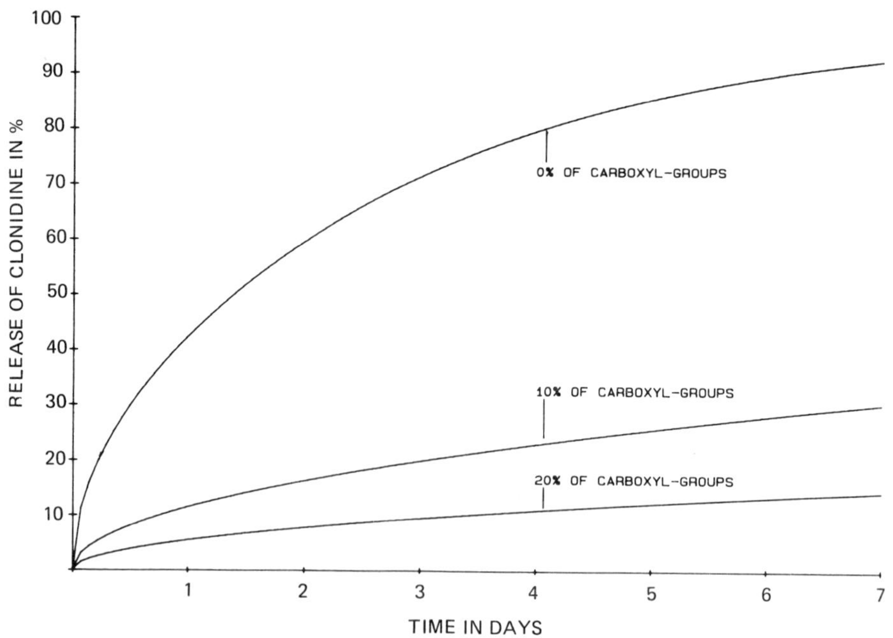

Figure 5 Example of how the diffusion coefficient D of the drug changes in response to modifying the percentage of carboxyl groups on the polymer carrier.

again the fundamental construction material. But in this case the polymer forms a membrane barrier between the compartment that contains the drug and that into which it is to be delivered.

Membrane systems with passive diffusion are the oldest systems for producing controlled release forms of drugs. The rate of drug release depends on the diffusion coefficient of the drug, the difference in drug concentration across the membrane and the thickness of the membrane. The diffusion coefficient can be varied by the choice of the polymer. The concentration difference depends on the solubility of the drug in the phase enclosed by the membrane and the transport rate of the drug from

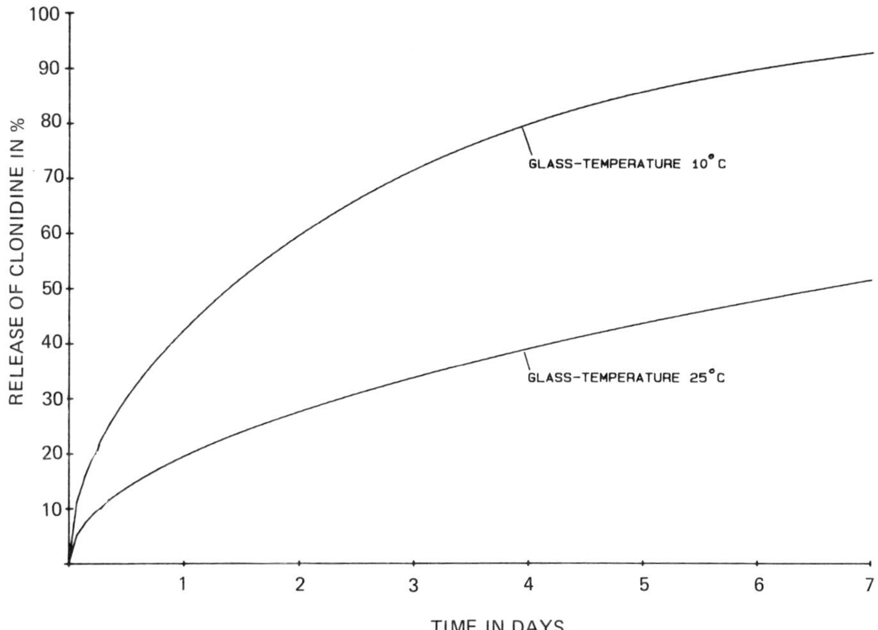

Figure 6 Example of how the diffusion coefficient D of the drug changes in response to modifying the glass temperature of the polymer carrier.

the outer surface into the surrounding phase. The membranes are manufactured mainly by coating processes. The drug can either be coated directly or after being transformed to a solid dosage form e.g. a tablet.

Besides the diffusional properties other properties of the membrane can become important for drug delivery. Especially significant are partial pH-dependent permeability and the ability to adhere to gastric or intestinal walls. Both factors can be exploited for the local targeting of the drug.

The coating polymers used cover a broad spectrum of natural and man-made polymers (Park et al., 1984). The membrane properties and the coating processes have to be adapted to the specific requirements of the drug investigated; some typical cases will be discussed later.

An internal driving force such as osmosis can also be used to control the release of a drug from a polymer carrier. The key features of an osmotically driven drug delivery system is a mechanically rigid semipermeable membrane, see Figure 7. Within the drug delivery device is an osmotic agent that, upon contact with water exerts an osmotic force on the flexible reservoir that contains

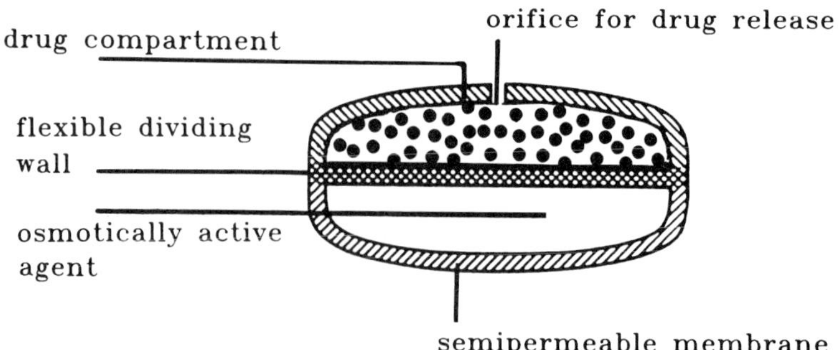

Figure 7 Schematic drawing of an osmotically driven drug delivery system.

a solution of the drug. This osmotic force then pushes the dissolved drug molecules into the external environment. In such devices the flux of drug is constant and depends only on the osmotic pressure, the thickness of the membrane and the drug concentration within the system.

Delivery systems that use the osmotic principle are oral, rectal and subcutaneous minipumps (e.g. Alzet®). Of these, the subcutaneous administration of drugs with osmotically driven minipumps is an important tool in pharmacology for measuring the dose dependency of drug effects, and, in pharmacokinetics for determining the input rate that maintains a required constant serum concentration (see Equation 1). These kinds of measurements are prerequisite to the successful development of controlled release forms and for substantiating an in vitro/ in vivo correlation of the dosage form.

In modern therapeutics the demand for dosage forms with variable delivery rates will increase. Such dosage forms would be useful for the treatment of diseases that are caused by deficiencies in the production of natural hormones and transmitters. The first approaches to systems with controllable rates of drug delivery are implantable mechanical pumps that are triggered manually by the patient or by an appropriate sensor. This approach has been applied to the administration of insulin discussed later in this chapter (Sefton, 1984).

III. SPECIFIC EXAMPLES OF NEW DRUG DELIVERY SYSTEMS

A. TRANSDERMAL SYSTEMS

1. General remarks about the use of the transdermal route. The traditional way of administering drugs is the

oral route because this is the natural way for the uptake of substances into the human body. However, the topical application has also been used throughout the centuries although usually it has been reserved for treating localized skin diseases. In such local treatment the drug should penetrate only into the outer layer of the skin. The concept of using the skin as a point of entry into the systemic circulation is relatively new. The first transdermal systems, those involving scopolamine and nitroglycerin have been on the market only since 1980. Their acceptance and potential for standardizing the time course of blood levels have resulted in the pharmaceutical industry and academia displaying broad interest in this dosage form. The principal advantages of administering drugs via the transdermal route can be summarized as follows: (1) The passage of the drug molecule through the skin is equivalent to an i.v.-injection; a first-pass effect is therefore avoided. (2) In contrast to classical dosage forms the delivery of a drug from a transdermal system will not be in terms of hours but rather days. This makes it easier to control blood levels and to keep them constant which in turn lowers the incidence of toxic side effects. (3) Patient compliance will be improved as the drug needs to be taken only once a day or even once a week instead of several times daily. (4) In principle, the amount of the drug remaining in the transdermal system worn by the patient enables us to determine the amount of drug that has entered the body. From this information the physician can accurately titrate the patient's blood level. This requires simple analytical methods compared to those for blood level measurements.

Enumeration of all these advantages raises the question: why don't we administer all drugs transdermally? If we are to do so the drug has to fulfill certain requirements: (1) The drug must be used for an indication where prolonged treatment is necessary, for example hypertension, asthma, coronary heart disease. (2) The drug must be highly potent, less than 5 mg per day. Otherwise the transdermal system becomes too large. (3) The drug must penetrate the skin at such a rate that the necessary therapeutic dose, released from the transdermal system, reaches the systemic circulation within the appropriate time.

Normally, the rate of penetration can either be roughly estimated from physico-chemical data or experimentally determined by diffusion measurements. In general drug molecules that are reasonably water-soluble and at the same time sufficiently lipophilic should penetrate the skin best.

To measure the flux, one uses excised human skin in an appropriate cell. For example, the amount of clonidine that appears in the receptor phase as a function of time is shown in Figure 8. From the slope of the line, the penetration rate is calculated to be 9.3 $\mu g/cm^2$ per h. To complete this example, the daily therapeutic dose for clonidine is 150 μg per 24 hours. In order to deliver this dose a system that covers a skin area of 10 cm^2 requires a minimum penetration rate of 0.6 $\mu g/cm^2$ per h. Since this is well below that observed, clonidine is therefore a promising candidate for transdermal administration.

2. Practical examples of transdermal systems. Whenever a drug appears to be suitable for transdermal administration, the pharmacist is faced with the task of developing the actual delivery system. From the discussion above, one would consider diffusional or membrane systems. Figure 9 is a diagram that illustrates the two release

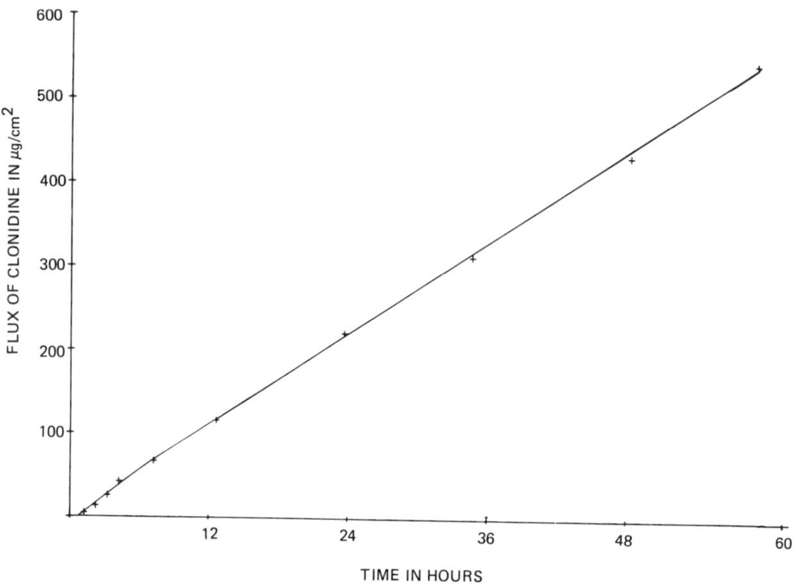

Figure 8 Plot of the flux I of the drug clonidine through excised human skin from a saturated aqueous solution at T = 32 °C.

principles. These have since been put into practice for the drugs clonidine and nitroglycerin respectively. The transdermal system is affixed to the patient's skin, usually the upper arm with a skin adhesive. Due to the concentration gradient the drug diffuses out of the system through the skin into the blood circulation. The amount of drug to be released from the system is adjusted in such a way that the appropriate concentration/time course of the blood level results. The parameters that control the rate of drug release from diffusional and membrane systems have already been discussed above. In the case of transdermal systems a further parameter is the area of contact between the system and the skin. In the systems already on the market this lies in the range from 2 to 20 cm^2. Covering larger areas of skin leads

diffusional system

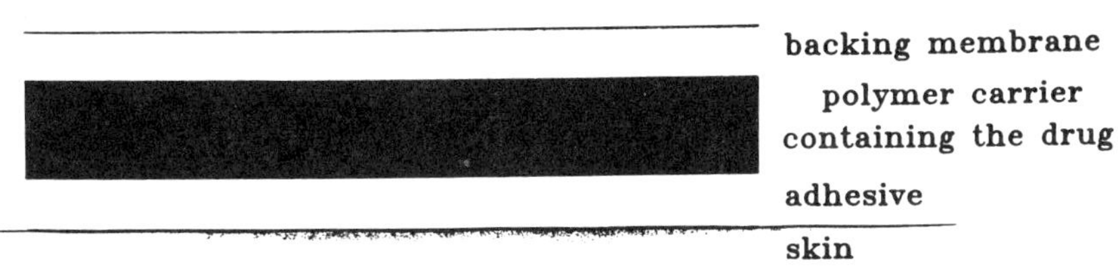

membrane system

backing membrane
suspension/
solution of the drug
rate controlling
membrane
adhesive
skin

Figure 9 Principal setups of transdermal systems.

to skin reactions. Aesthetic considerations also speak against applying excessively large patches.

How have transdermal release systems so far contributed to improvements in drug therapy? As an illustration the two examples will be discussed in more detail.

Nitroglycerin, which serves as an antianginal drug, has a high first-pass effect and a biological half-life of only a few minutes. Oral forms therefore demand very

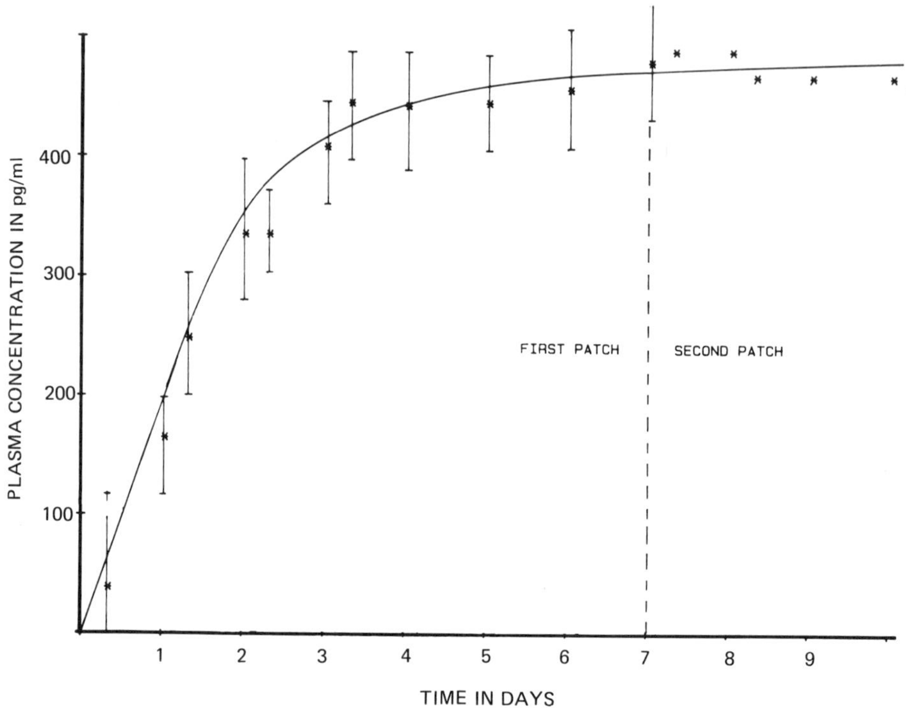

Figure 10 Mean plasma clonidine levels in 17 subjects wearing a transdermal system being replaced at 7 days.

high doses at frequent intervals. By using sublingual or buccal forms (i.e. dosage forms from which drug is absorbed from the mouth) one can circumvent the first-pass effect and lower the dose somewhat; the problem of the short half-life, however, remains. Prolonged protection against aginal attacks is not possible. Both problems were solved by developing a transdermal system for this drug. Via controlled absorption of nitroglycerin from such a delivery form, the protective antianginal effect of nitroglycerin persists for over 24 hours or longer, if the patch were to be replaced every day (Shaw, 1984).

In the case of clonidine an antihypertensive drug, the goal is to keep blood levels as constant as possible as this minimizes of side effects such as sedation and dryness of the mouth. Figure 10 shows the extent to which constant blood levels can be maintained with a transdermal system. The figure shows that plasma levels reach a steady state three days after the first patch has been affixed. This plasma level is maintained also after the first patch has been replaced by a new one on Day 7. Further changes of the patches at one week intervals maintain this level for as long as treatment is required. The time courses of the effects in 32 hypertensive patients are shown in Figure 11. These patients received a fresh patch of a transdermal form of clonidine every week. Their blood pressure was monitored for over a year. Obviously blood pressure returned to normal values and remained there in all the patients. These results are also indicative of a high patient compliance with the transdermal therapy, that is patients are willing and able to follow the dosage regimen.

What other drugs may be candidates for transdermal administration? Generally speaking, all potent drugs that can permeate the skin at sufficiently high rates and that do not exert sensitizing effects could be considered. The number of such pharmaceutical preparations will certainly increase in future.

B. ORAL SYSTEMS

1. General remarks on the use of the oral route. The oral administration of drugs is the most commonly used route. As daily food is taken in by this way it is understandable that patients prefer this route for drug administration too. On using this route for introducing a chemical entity into the body, one has to consider three factors that have a decisive influence on the potential success of this route of administration. A key factor is

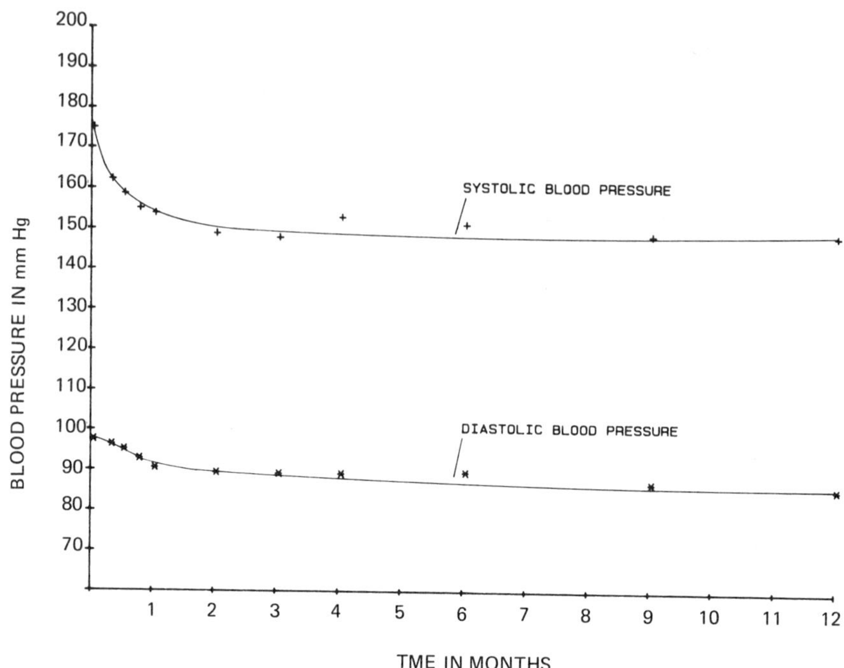

Figure 11 Representation of the course of the blood pressure in hypertensive patients using clonidine in a transdermal system over 12 months (average of 32 patients).

the transit-time of a drug, which is the time interval between the dosage form intake and its appearance in the colon. Transit-time can vary considerably depending on the amount and kind of food taken in by the patient. A second factor is that during the passage through the gastrointestinal tract the drug molecules pass through a pronounced pH-gradient, ranging from pH 1 - 2 in the stomach to pH 7.5 in the colon. In many cases only the non-ionized form of a drug can diffuse through the intestinal wall. Therefore the absorption rate of ionizable drugs depends on their pKa-values and on the pH-value of the environment. The final factor is that absorption for each drug might be different in different sections of the GI.

In order to be able to design proper oral dosage forms one therefore first must study gastrointestinal transit. An established method for obtaining information about the transit of a dosage form through the GI is the technique of gamma scintigraphy (Davis, 1985). For this purpose the formulation is labelled with an appropriate radionuclide such as technetium 99 m. After the labelled dosage form has been administered to the patient, its position can be monitored with a gamma camera that detects the radiation coming from the radio nucleotide label. From the data obtained by a large number of such studies, the following general conclusions have been drawn (Davis, 1985): (1) Solutions and pellet systems (size < 1.2 mm) empty quite rapidly (< 1 h) from the stomach. The gastric emptying of pellet systems is delayed by the presence of food up to 6 h. (2) For single unit systems (size $\gg 1.2$ mm) the nutritional state has a pronounced effect on retention time in the stomach. The system may remain there for up to 10 h after a heavy meal. (3) Transit in the small intestine is remarkably constant (in the order of 3 hours); it is independent of the size and kind of dosage form and the nutritional state of the subject. (4) Single units can be held for periods of 4 to 12 h at the ileocaecal valve before being moved into the colon. (5) The transit pattern for old and young patients does not differ.

In order to minimize the intra- and interindividual variability of blood levels that result from different transition times of oral drug delivery systems, the patient has to coordinate times of food and drug intake.

The total residence time of oral systems in the gastrointestinal tract is in the order 8 to 12 h and this cannot presently be extended routinely. However, given the general interest in oral controlled release systems and the associated basic research in GI physiology these limitations will be overcome in the near future. Interesting results are emerging from floating (Muller-Lissner, 1981) and mucoadhesive (Park, 1984) devices that remain in the stomach and deliver the drug for a fixed time. So far, however the clinical data in support of such approaches are still limited.

In contrast to the other routes, oral administration exposes the drug to quite acidic environments. Drugs that are unstable in acid have to be administered in a dosage form that protects the drug from the acidic environment but releases it at the proper site. For stable drugs with a pH-dependent solubility a shift in the blood level will result. This shift can be predicted by using model calculations (Zierenberg, 1980). The acid sensitivity can also be partially corrected in release systems by adding a buffer substance that creates a microenvironment with a favorable pH-value (Brickl, 1984).

A further factor, unknown to the pharmacists at the time of the original studies, is the absorption rate constant k_i of the drug and its possible dependence on the absorption site. The latter information is important mainly for the development of sustained release formulations. With this knowledge the release profile of a drug from the formulation can be optimized to improve the bioavailability and blood level profile. The influence of different k_i-profiles over the GI tract can be demonstrated numerically with pharmacokinetic models (Stricker, 1980).

In one experimental setup, to obtain the k_i-profile for a drug, one uses a release system such as an osmotic pump whose in vivo release profile is known.

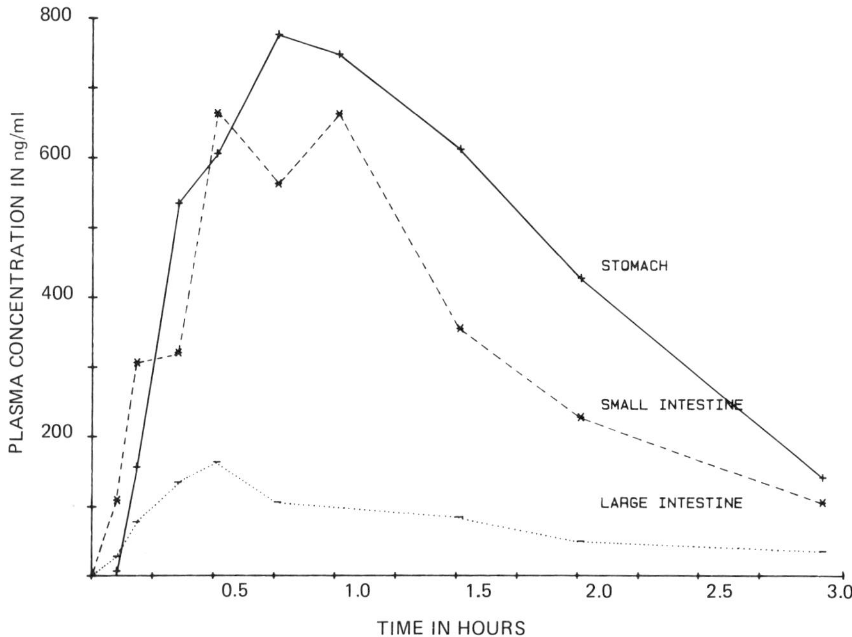

Figure 12 Plasma level concentration of the drug AR-L 115 BS following the opening of the capsule at 3 various sites in the gut. (Zimmer et al., 1981)

From the measured blood level concentration and the known pharmacokinetic properties of the drug under investigation the k_i-profile can be deduced mathematically. Alternatively the data can be examined with the convolution technique (Stepanek, 1976). The latter procedure has the advantage that it does not depend on a previous detailed pharmacokinetic analysis of the drug.

Another experimental procedure evaluating the k_i-value in defined sections of the GI, uses a capsule that releases its drug content in response to an external signal (Zimmer et al., 1981). An X-ray camera is used to follow the movement of the capsule; when it reaches the desired site the signal is activated. An example of the plasma levels obtained on release at different sites with this kind of technical setup is shown in Figure 12. Again there are mathematical procedures available that allow k_i-values to be derived from such data.

Both methods are important tools to facilitate the development of oral dosage forms and shorten development time because they decrease the amount of experimental work compared to a pure trial and error approach of dosage form design and testing.

2. Sustained release forms of oral systems. The achievable transit time for a drug may not be fully realized because it may be released from the dosage form in a much shorter time period. It is this discrepancy between release time and transit time that provides the general impetus for developing sustained release forms. Such forms aim at using the transit time completely and to improving the drug concentration/time profile. Many such systems are described in the literature; some are commercially available.

The many sustained release forms known exploit one or another of all the physical principles discussed in section II B for controlling the release rate. For example, the Perlongets® oral sustained release form consists of differently coated tablets in a hard gelatin capsule. Since a capsule can be filled with differently coated tablets, and since one can use various combinations of such tablets, a large number of release profiles can be generated. For example, one can generate 126 different release profiles if one uses five different tablets in one capsule. Again, the fundamental question is which release profile is needed to generate a desired blood level profile. Currently, this question is answered by a trial and error procedure guided by theoretical considerations, i.e. the compartment models and/or convolution techniques mentioned previously.

Figures 13 and 14 show the result of such an investigation for three different release formulations (Perlongets®) with the drug mexiletine. Figure 13 shows the in vitro release profiles. Figure 14 shows a comparison of the observed human blood levels and the levels predicted from the release profiles by the convolution technique. From the good agreement we are confident that there is a correspondence between in vitro and in vivo release profiles. With this knowledge at hand one can now use in vitro studies to select that combination of coated tablets that will give the optimal blood level course obtainable from an oral form.

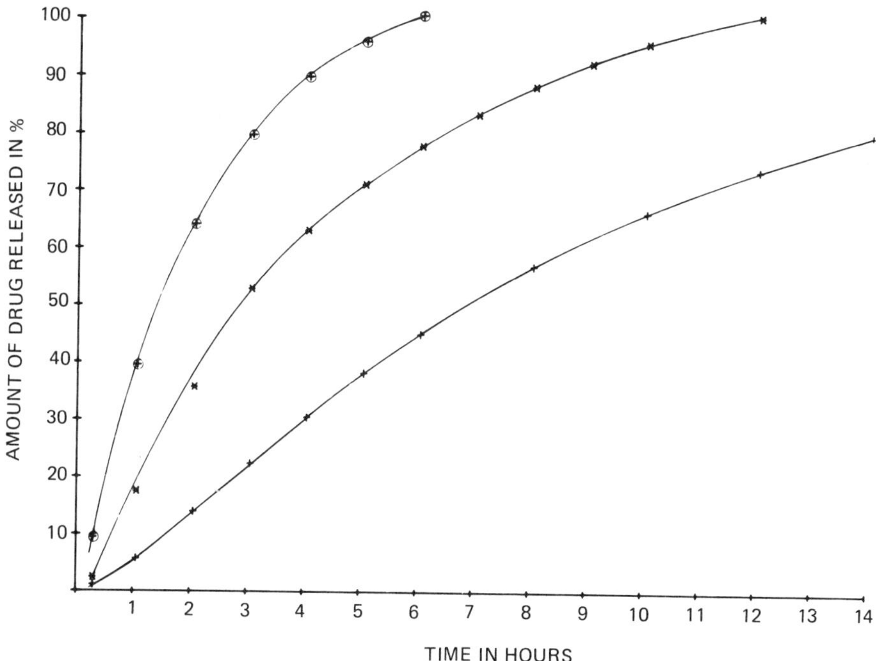

Figure 13 In vitro release profiles for three different release formulations (Perlongets®) with the drug mexiletine.
⊕ type 1, * type 2, + type 3

The improvement in drug delivery that is to be expected from sustained release forms can be summarized as follows: (1) The whole resorption surface available in the GI can be used for drug absorption. (2) Local drug concentrations are generally lower than with nonsustained forms. The potential for local irritation is thus reduced. (3) By smoothing and prolonging the blood level course, side effects can again be reduced. In addition the intervals between drug administration can be extended.

IV. OUTLOOK FOR THE CONTRIBUTION OF PHARMACEUTICS TO DRUG DEVELOPMENT

A. THE DISCOVERY OF NEW AND THE BETTER USE OF CURRENTLY USED ROUTES OF DRUG ADMINISTRATION

Progress in drug delivery will come from a better understanding of the barrier functions of membranes and from technical improvement in delivery systems themselves. For instance, the initial success of transdermal systems has awakened fundamental interest in studying the skin as a barrier and a metabolizing organ. The growth of knowledge based on the application of sophisticated analysis techniques in studying the skin will help us to understand the transport mechanisms at the molecular level. The final goal will be to lower the barrier function of natural membranes such as the skin in order to regulate the input rate of drugs from dosage forms more precisely. The lowering of the barrier function must be reversible, locally limited and non-irritating. The last requirement must be fulfilled for long-term treatment. First approaches for lowering the diffusional barrier are under investigation using penetration enhancers (Woodford, 1986). Others have suggested using ion pair

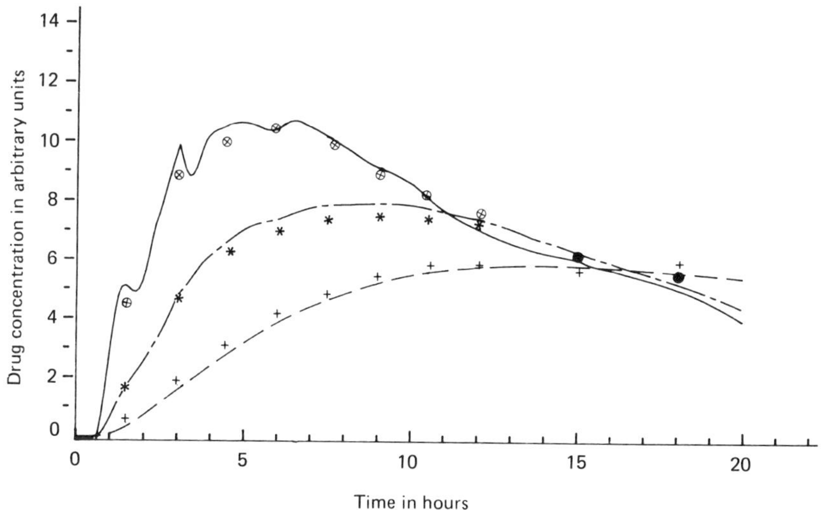

Figure 14 Comparison of measured (⊕ type 1, * type 2, + type 3) and calculated (by the convolution method) blood level curves for mexiletine with different in vitro profiles.
——————— type 1, —— - —— type 2, --------- type 3

mechanisms in order to facilitate transport. Thus suitable amines may function as carriers for acidic drugs (Barker, 1981).

An alternative to passive penetration of a drug through a membrane is diffusion driven by external power sources. For example, by applying additionally an electrical field to passive diffusion it is possible to regulate the flux of a drug more precisely and independently of the constitution of the individual membranes. Specific transport could also be based on using molecules that provide energy for the transport of drugs through membranes. This is a working principle frequently used in nature, e.g. for sodium transport across cell membranes. The problem in application to dosage forms is to find appropriate carrier systems.

In addition to such basic studies, efforts are also being made to use other routes of systemic drug administration such as intranasal administration (Chien, 1985). For this route the advantages over oral administration are similar to those of transdermal administration. However, compared with the skin, the diffusion resistance of the nasal mucosa is much lower. Therefore, this route appears promising in particular for protein-based drugs such as products of biotechnology.

At the same time others are designing dispensing systems adapted to specific biochemical or physiological situations within the body, for example bioadhesive materials that adhere to a predetermined site of administration. Based on our knowledge of membrane structure, one perhaps could design corresponding polymer systems which interact with the surface structure. Such carrier systems might be targeted for buccal, nasal or oral administration. The buccal absorption route is being investigated in great depth since it also resembles the transdermal route and appears suitable for peptide-type substances with a major first-pass effect. Taken together, we have acquired a broad knowledge concerning the fundamentals of the physiological conditions and the appropriate interpretation and adaptation of drug delivery systems. This should lead us to an increasingly specific drug therapy which is less prone to interindividual variation and which is more agreeable to the patient.

B. THE AIMS OF INNOVATION IN DRUG DELIVERY

The new developments in the field of drug delivery systems mentioned in the previous paragraph do not solve all the remaining problems of drug administration. In order to solve these problems a close cooperation between various scientific disciplines is required.

For certain substances the therapeutic goal is to use pulsed drug administration driven by physiological demand. An example is insulin. The problem of designing a feedback release system that determines the required insulin doses from the actual blood sugar levels can be solved only with an interdisciplinary approach: we have to develop a sensor to constantly monitoring blood sugar. The signal from this sensor would serve to drive a pump that administers the insulin (Sefton, 1984).

We can assume that, in the future, feedback or demand-oriented drug delivery systems, will become more widely used in human therapy. This is because with such delivery systems we would be able to specifically influence the interplay between the course of a disease and drug requirement. Such an advance requires the development of biocompatible sensors. For delivery of the drug in addition to the mechanically driven micropumps used today, we expect to see novel materials based on fundamentally different principles. For instance polypyrrole films may be loaded with a drug which, in response to an electric current, is released (Zinger, 1984). Since the rate of administration is adjusted by varying the current, these films offer a fresh perspective for the pulsed administration of active ingredients.

These examples show that we need to combine our knowledge from medicine, biochemistry, electrochemistry and the engineering sciences to solve the problem of feedback drug administration using the appropriate administration system. The complexity of such projects requires not only committed researchers but, increasingly, efficient management methods as well. One approach here could be the multidisciplinary project management method.

Organ or cell-specific drug administration is another area of active research. The aim is to deliver the active ingredient directly to the site of action in the diseased organ and thus minimize side effects. This approach is particularly significant for the treatment of malignant tumours with cytostatics. The basic design of organ or cell specific delivery systems involves coating the active ingredient with a particular material or linking it to a high-molecular carrier. For example, one may coat with phospholipids that are converted into liposomes that contain the drug (Schneider, 1985). These liposomes are made organ or cell-specific, by adjusting there surface properties, such as charge. Alternatively monoclonal antibodies are incorporated into the phospholipid liposomal membrane and serve to direct the liposome to the appropriate cell. The active ingredient can also be linked directly to the antibody and as such be directed to the organ or diseased cell.

All of these new developments open up new perspectives for novel methods for drug administration. Expectations are high. However, translating results into industrial production will involve a major deployment of technical and financial resources. Results will only come from pooling the individual contributions made by committed scientists in academia and industry. A commitment is required to exploring every avenue that may render therapy more efficient. After all, we are patients ourselves sometimes. We then expect the most efficient help from physicians, the pharmaceutical industry and its products.

Acknowledgement

I would like to thank Mr. Taylor for his help with the English version and Ms. Lieth for typing the manuscript.

REFERENCES

Barker, N. and J. Hadgraft, Facilitated Percutaneous Absorption: a Model System, Int. J. Pharm. 8, 193 - 202 (1981)

Brickl, R. (Dr. Karl Thomae GmbH), US-Patent 4427648 (1984)

Chien, Y. W., Transnasal Systemic Medication, Elsevier, Amsterdam - Oxford - New York - Tokyo (1985)

Crank, J., The Mathematics of Diffusion, Clarendon Press, Oxford (1975)

Davis, S. S., The Design and Evaluation of Controlled Release Systems for the Gastrointestinal Tract, Journal of Controlled Release 2, 27 - 38 (1985)

Higuchi, T., Rate of Release of Medicaments from Ointment Bases Containing Drugs in Suspension, J. Pharm. Sci. 50, 10, 874 - 876 (1961)

Martin, N., J. Swarbrick and A. Commarata, Physical Pharmacy, Lea & Fabiger, Philadelphia, Pennsylvania (1970)

Muller-Lissner, S. R. and A. L. Blum, The Effect of Specific Gravity and Eating on the Gastric Emptying of Slow-Release Capsules, N. Engl. J. Med. 304, 1365 - 1366 (1981)

Nelder, J. A. and R. Mead, A Simplex Method for Function Minimization, Computer J. 7, 308 - 313 (1965)

Park, K. and J. R. Robinson, Bioadhesive Polymers as Platforms for Oral-Controlled Drug Delivery: Method to Study Bioadhesion, Int. J. Pharm. 19, 107 - 127 (1984)

Park, K., R. W. Wood and J. R. Robinson, Medical Applications of Controlled Release, Volume I, (R. S. Langer and D. L. Wise, ed.), CRC Press, Inc. Boca Raton, Florida, pp. 159 - 201 (1984)

Schneider, M., Liposomes as Drug Carriers: 10 Years of Research, Drug Targeting, Proc. Symp. 1984 (Burri, P. and A. Gumma, ed.), Elsevier, Amsterdam, Neth., pp. 119 - 134 (1985)

Sefton, M. V., Medical Applications of Controlled Release, Volume I, (R. S. Langer and D. L. Wise, ed.), CRC Press, Inc. Boca Raton, Florida, pp. 129 - 158 (1984)

Shaw, J. E., Pharmacokinetics of Nitroglycerin and Clonidine Delivered by the Transdermal Route, Amer. Heart J., 108, 217 - 223 (1984)

Stepanek, E., Praktische Analyse Linearer Systeme durch Faltungsoperationen, Geest & Portig K.-G., Leipzig (1976)

Stricker, H. and B. Zierenberg, Modellbetrachtungen zur Optimierung der relativen Bioverfügbarkeit von Arzneistoffen in peroralen Retardformen, Pharm. Ind. 42, 637 - 642 (1980)

Woodford, R. and B. W. Barry, Penetration Enhancers and the Percutaneous Absorption of Drugs: an Update, J. Toxicol. Cutaneous Ocul. Toxicol., 5, 165 - 175 (1986)

Zatuchni G. I., A. Goldsmith, J. D. Shelton and J. J. Sciarra, Long-Acting Contraceptive Delivery Systems, Harper & Row, Publishers, Philadelphia (1983)

Zierenberg, B., Mathematical Studies on the Relationship of the Size Distribution and Bioavailability of Solid Drugs, 2nd International Conference on Pharmaceutical Technology, Paris, pp. 105 - 112 (1980)

Zimmer, A., W. Roth, B. Hugemann, W. Spieth, and F. W. Koss, A Novel Method to Study Drug Absorption, C.-R.-Congr. Europ. Biopharm. Pharmacocinet., 1st, 2, 211 - 214 (1981)

Zinger, B. and L. L. Miller, Timed Release of Chemicals from Polypyrrole Films, J. Am. Chem. Soc., 106, 6861 - 6863 (1984)

10

Search for Better and Safer Drugs: Impact of External Factors—The Social, Financial, and Working Environment

E. Kutter

Department of Research and Development
Boehringer Ingelheim KG
Ingelheim, Federal Republic of Germany

A. Kieser

Institute of Organizational Behavior
University of Mannheim
Mannheim, Federal Republic of Germany

There is no question that there are diseases to be cured and research still necessary to find new drugs. However, there are other factors to be considered. Times have changed since the golden age of drug discovery. In the 1950's the scientist's creativity was the major prerequisite for successful drug research and development. Raising funds and getting public support was not a major problem since there was a general atmosphere of trust in drug therapy and research. Since then this situation has changed profoundly. Public opinion is often critical of the costs of drug therapy and skeptical of its benefits. In addition, drug research has reached a high level of scientific sophistication and drug development is heavily regulated. This increases the costs to such an extent that drug research is now a high risk investment. Today we understand that the search for better and safer drugs can only be performed successfully in an innovative climate. Nowadays the climate of drug research has three components:

1) public opinion of its value
2) the availability of money to do the necessary research and development
3) the organizational framework and working climate (the "corporate culture") in which the work is done.

The following sections in this chapter will discuss how each of these components may be used to optimize the climate for innovative discovery of better and safer drugs. Only when these components function properly will scientists be able to discover better and safer drugs using the type of information described in the earlier chapters.

I. DRUG THERAPY AND PUBLIC OPINION

The pharmaceutical industry operates within the context of society. Because drugs may literally control the life or death of an individual, this industry is heavily regulated in most countries. In addition, since health care is considered to be a basic human right, society bears much of the financial costs of health care, whether it be by insurance or government. Thus every citizen has an ethical and financial interest in the pharmaceutical industry: effective drugs prevent or alleviate human disability and enhance the productive lives of the members of society, whereas bad ones do the opposite. Public opinion directly affects the search for better and safer drugs because it influences the availability and morale of productive employees; private and government funds to support basic research; investment capital to support industrial research and development; production and marketing of better and safer agents; and reasonable government regulation and patent protection of its products. If a society does not want or cannot afford drug research, its only source of better and safer drugs will be those developed elsewhere and without its influence.

There are those who believe that synthetic drugs require countless, unnecessary and dangerous animal and human experiments; that manufacturing them pollutes the environment; that they increase costs of health care, harm patients and, at best, merely serve the corporate profit motive. These or similar rejectionist formulae sum up the way fanatical supporters of the natural lifestyle in some countries of Europe criticize institutions involved in drug research, manufacture and marketing. Although this attitude is rarer in the U.S., there, too, the number of skeptics is increasing.

Another aspect of public disillusionment with drug research is seen in the animal rights movement. A society that sees no value to humans from its drug industry is less tolerant of any use of animals for research purposes. This attitude directly affects individual innovators because they have to use energy and creativity to convince family and friends that such use is ethical.

But what are the causes of this growing negative, emotional attitude towards modern drug therapy and the pharmaceutical industry? Is twentieth century drug research unable to demonstrate any convincing success? Has it been a complete failure? By no means!

Only in this century have we had drugs that are sufficiently effective to prevent or to alleviate a great deal of human suffering. Drugs and vaccines have significantly reduced the frequency of infectious diseases such as pneumonia, tuberculosis, typhoid, whooping cough, polio, measles, diphtheria and tetanus. In the 1930's and 1940's, the development of sulphonamides and antibiotics was the beginning of a dramatic battle in the fight against infectious bacterial diseases. Today no longer do children die of ear infections!

Recent decades have been marked by the development of drugs for the therapy of chronic diseases such as hypertension, coronary insufficiency, ulcer, cardiac rhythm disturbances, arthritis, diabetes, asthma, chronic bronchitis and numerous diseases of the central nervous system such as epilepsy, Parkinson's disease and endogenous depression. A further branch of medical therapy, namely, surgery, would not be possible in today's perfected form without modern narcotics, anaesthetics and analgesics. The carefree use of individual modes of transportation and the pursuit of popular sports with their sometimes extremely high accident rates can only be rationalized against a background of unconscious trust in the achievements of modern surgery. Thanks to suitable drugs, surgery can restore health without causing excessive pain.

It can be argued that drugs are one of the most cost effective aspects of medical care. Consider the highly effective anti-ulcer agents: by taking one of these a person does not have to be hospitalized. Drugs have made a vital contribution to the virtual doubling of life expectancy during this century.

In spite of the proud record in some quarters public opinion is highly critical of drugs. Why? We can understand this paradox if, instead of looking at drugs in isolation, we consider the double-edged sword of technical progress in general. We live in a time of disillusionment about what can be expected from science and technology. This disillusionment has gained ground more rapidly in Europe compared with the strong belief in progress that still prevails in the U.S., but it exists everywhere.

The extent of this change in attitude with respect to technology becomes particularly evident if one considers the situation in the 1950's and 1960's. During that time people had blind faith in the value of technical and scientific progress. The idea that "if it's technically possible, do it because it's progress" became a kind of substitute religion. Nuclear research opened up the microcosm, space travel set out to conquer the macrocosm, and it appeared that defects in the biological machinery of the human organism could be corrected or at least controlled with drugs. The opportunities thrown open by modern technology to secure our energy supply, to expand our living space, and to improve the quality of life appeared to presage the dawn of a bright new scientific and technical era. This optimism thrust into the background any anxieties and misgivings about the risks of modern technology.

Opinion changed forever with dramatic evidence of the hazards of technology: the thalidomide tragedy (1961); the Seveso (1976), Bophal (1984) and Challenger accidents (1986); the Tschernobyl reactor (1986) catastrophe; and the Rhine pollution of Basel (1986) were serious warnings of the risks inherent in modern technological achievements. In some parts of society, these disasters converted the original extreme enthusiasm into an equally extreme distrust in the merits of technical innovation: possible risks induced a fear that obscured any realistic assessment of the overall benefits of technological progress for mankind. However, all elements of society share the concern to some degree.

The dream is over that man may be totally liberated by modern technology from the inadequacies of human existence. It is now clear to all that technological progress has its price in the form of certain risks. This is uncomfortable and painful. Therefore, in the future, the significance of scientific findings and of technical progress for human society must be carefully considered in terms of the expected benefit-to-risk ratio. Blind faith in technology and uncritical acceptance of anything new must be replaced by sober and careful analysis of the consequences of innovation. No matter how admirable an innovation may be scientifically, it must be evaluated in terms of all of the risks associated with its industrial use.

The risks we face today should only be discussed against a background of the risks of yesterday. We can only hope that this will also become recognized by the general public in due course and that responsible risk/benefit analysis within the historical dimension will be accepted as the only appropriate instrument for evaluating technical and, of course, also pharmaceutical innovations.

From this point of view there appears to be an excellent chance that public acceptance of drug research will improve. The need for better and safer drugs cannot be denied. The importance of an effective drug therapy and thus drug research as well is at present very clearly exemplified by the problems connected with AIDS. Following the triumph of drugs over infectious bacterial diseases and the mainly symptomatic alleviation of complaints caused by numerous cardiovascular, metabolic, respiratory and psychiatric diseases, the scientific groundwork has been laid for highly promising research into yet unsolved problems of therapy. Many of these diseases became increasingly significant due to the increase in human lifespan. Pertinent examples are chronic diseases and diseases of the immune system, various forms of viral diseases and cancer as well as diseases of the central nervous system ranging across to senile dementia of the Alzheimer type. For the patient, development of new drugs has obvious benefits. It is to be expected that average life expectancy will continue to rise as soon as a specific and more causal therapy of cancer and viral diseases becomes possible. A further increase in life expectancy is, however, desirable only if it is linked to an acceptable level of quality of life, both physical and mental. Together with a health-conscious, sensible lifestyle, drugs can make an essential contribution to further improving the quality of life. In particular, along with other measures to preserve health, drugs should help people to adopt a positive attitude to life even at an advanced age: "Die young, but as late as possible" (Doll, 1983).

The previous chapters in this book document the fact that within biomedical and pharmaceutical science there is currently an explosion of knowledge. This constitutes the basis for the development of better and safer drugs. The coming decades will see a synergism of molecular pharmacology, structural biology, gene technology, polymer chemistry, stereoselective synthesis, theoretical chemistry and computer graphics. As a result we may usher in a second pharmacological revolution and see new therapies for currently devastating diseases. Industry now must turn these opportunities to good account by offering patients considerably better and safer drugs in future than we can at present.

When such drugs are available we can expect again a gradual change in public opinion in favor of the products yielded by research.

II. DRUG RESEARCH AND FINANCIAL CONSTRAINTS

All of us, and thus also the public, are at least potential consumers of drugs. We purchase them either directly, by contributions to health insurance systems or as taxpayers that support government assisted medical care. Therefore, only a belief in the benefit of drug therapy among the public can guarantee the finances for pharmaceutical research in the long term. This research is expensive, so much so that doubts are raised in many quarters as to whether research and development in this sector will make commercial sense in future.

We all desire greater drug safety and efficacy. It is obvious to us all that this has its price, because drug research has become considerably more expensive. In addition, because of greater safety testing, the usable remaining patent life of a newly marketed drug has declined dramatically. This severely detracts from market exclusivity as a commercial incentive to investment in research. On the basis of scientific and technical progress already achieved, industry could produce considerably improved drugs in the years to come, including drugs for diseases currently untreatable with drug therapy. However, realizing this potential will require a powerful international pharmaceutical industry with sufficient financial resources. This in turn requires the appropriate incentives for investment in innovative research which will depend on supranational agreements to improve the general economic conditions for such research. For the smaller European countries in particular such agreements are of vital importance. They would also be of advantage for such major world markets as the USA and Japan.

Developing a new active ingredient into a drug and registering it in the most important European countries, the USA, Japan and Australia costs on average US$ 100 million. This figure, however, includes only the direct expenses for successful products. If we also account for failed projects, we arrive at an overall figure of approximately US$ 130 - 180 million for each new drug. According to a calculation model of J. Drews (1985), a new drug which has cost US$ 160 million for development must show gross integrated sales worth around US$ 800 million within ten years of introduction in order to achieve a 10% repayment of interest on the capital used. Beyond that time the competition of generic or better agents decreases sales significantly. However, the reality is that only one in twenty-five new products hits this target. A bare 4% of new drugs have achieved annual sales of greater than US$ 120 million. Investment in pharmaceutical research is now a high-risk venture. The chances of achieving an acceptable level of returns on research and development in this sector have fallen. For products introduced in the US market in the period 1955 - 1958, the return on R & D capital invested was 11 - 18%: for 1966 - 1972, the figures were down to 3.3 - 7.5%. Thus pharmaceutical research in the 1950's and 1960's was highly profitable but since then its profitability has been declining steadily. For many companies drug research is now only moderately profitable. In fact, the interest borne by capital invested in pharmaceutical research has dropped far below the yields of other investments.

A major reason why drug research has become so much more costly is undoubtedly the demands for more drug safety voiced in the wake of the thalidomide tragedy in the early 1960's. These demands led to rapid development in the methods and models used to examine drug safety and thus to an enormous rise in the level of scientific knowledge. This in turn was reflected in the more stringent legal standards adopted. However, more drug safety means more animal experiments, more clinical trials, more bureaucracy, more official influence and so, of course, substantially higher costs. Figures 1-3 convincingly show the resulting longer development times, higher research expenditures and shorter patent life times for new drugs that have to be borne by research-based companies.

The increased amount of time required for developing new drugs is as much of a burden on financing research as is the cost pressure exerted by additional development studies. This is because each additional year of development time is one year less that the company has exclusive patent protection. With novel and original substances the development time can be even longer than the period of validity of the patent! This means that the more time investigations into drug efficacy and safety require, the earlier in the marketing cycle pharmaceutical companies have to fear generics. The innovative companies bear all the risks since, in spite of all the development work, only a certain percentage of the new drugs are a market

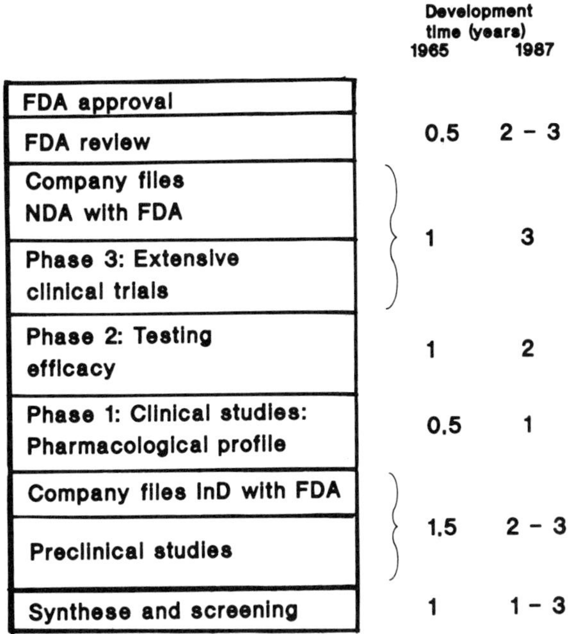

Figure 1 Increase in development time (Comparison 1965/ 1987 estimated values)

success. However, the generic companies of course are interested only in products with high sales and large markets.

The current conflict between the governmental requirements for lengthly testing of new drugs without the traditional period of market exclusivity is hostile to innovation and as such serves neither therapeutic progress nor ultimately the desire for greater drug safety. This is clearly shown by the experience of Canada in which, in an attempt to control prices, it was required that any manufacturer license rights to sell a drug to others. Prices indeed fell, but drug research stopped also. The incentive to put up capital for pharmaceutical research and development will vanish if appropriate measures are not taken to adapt the

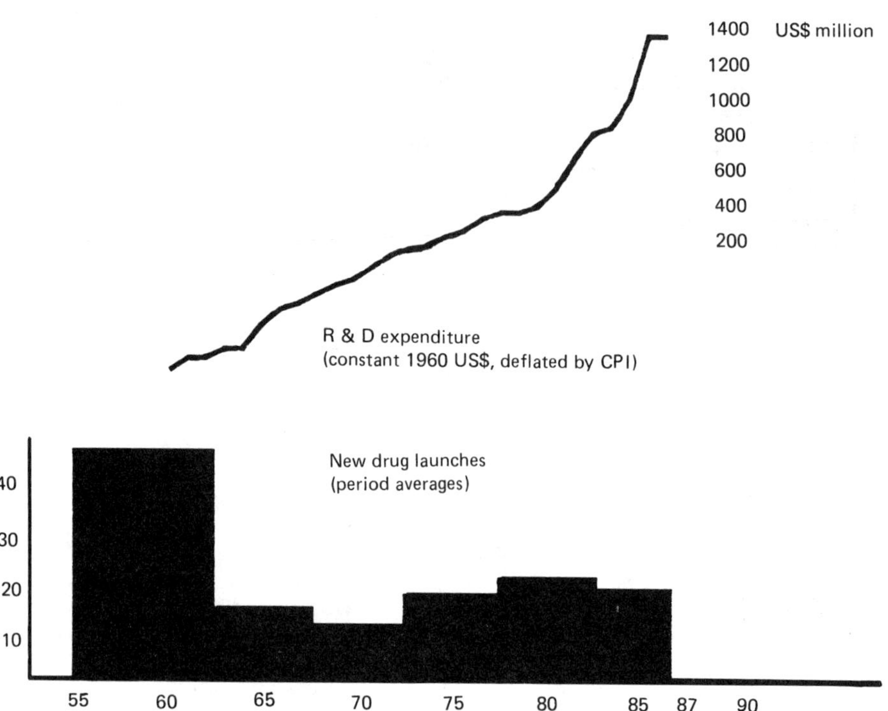

Figure 2 New drug launches and R+D expenditure (CPI = consumer price index)
Source: Pharmaceutical Manufacturers Association

length of patents for pharmaceutical products to development times. In some countries with a strong pharmaceutical research base, attempts have been made to alleviate this problem by extending patents by the length of governmental reviews (USA), barring registration for generics (Japan) or the introduction of a user bar for particular periods (EEC, including the Federal Republic of Germany). In the long term, however, broader international solutions to this problem are indispensable to guarantee survival of research-based pharmaceutical companies dependent on international marketing.

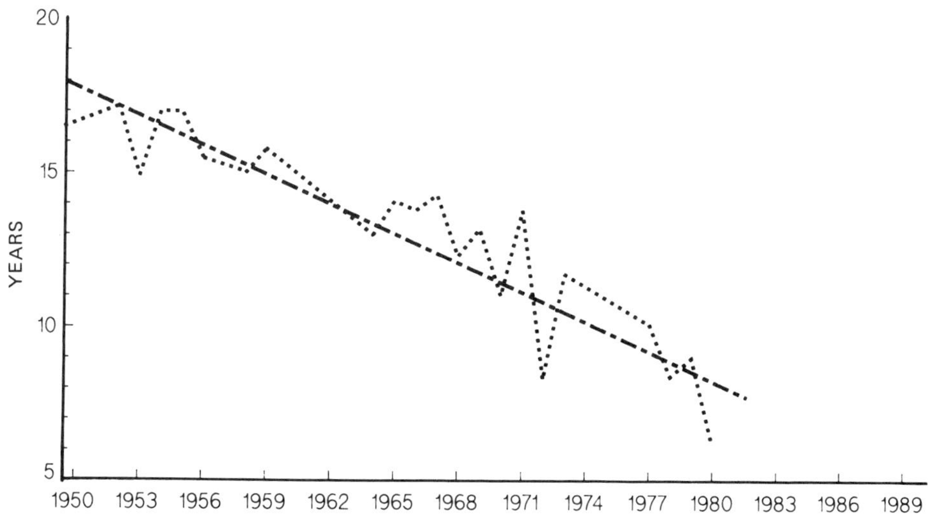

Figure 3 Decrease in effective patent life time for newly introduced drugs in Germany
.-.-. Regression line
Source: Medizinisch Pharmazeutische Studiengesellschaft., Germany

The registration authorities are often severely criticized for impeding pharmaceutical innovation and increasing development costs by being unduly bureaucratic. Critics have claimed that patients have been deprived of vital therapeutic help because of registration delays. The United States' FDA, for instance, was blamed for the premature death of many American coronary patients because the use of newer beta-blockers successfully used in Europe had been delayed. Therefore some countries have changed their practice in that they now use standard registration criteria. The latest level of scientific knowledge has become largely internationalized, yet many governmental regulations and guidelines are based on outdated scientific information or the scientifically untenable view that more preclinical tests inevitably result in more drug safety. As discussed in detail in Chapter 8, a carefully planned test, one that takes into account the latest scientific findings and is appropriate to the problem in hand, can be more valuable for effectively calculating the risk-to-benefit ratio than a whole series of routine investigations in all the animal species imaginable. On the other hand, the authorities also come under fire because of their desire and the public's expectation to achieve unrealistic, absolute safety standards. These goals are unrealistic because they are impossible to attain. In 1980 Sir Derrik Dunlop described the situation trenchantly: "Officials charged with approving a new drug can make two kinds of mistakes; they can approve a new drug which turns out to be unexpectedly toxic or refuse one that could have been life-saving with few adverse effects. If they make the first mistake their folly will be emblazoned in the public media and disgrace will follow; if the second, few will know of it, and those whose lives might have been saved, will not be there to protest."

Certainly, bureaucracy, excessive rules and regulations, and officiousness have by no means accelerated the development of drugs and the financial return to the investors. In individual cases the patience of research-based companies has been stretched almost to breaking point. Even so, criticism can be leveled only with extreme caution. It is indeed true that official bureaucracy has led to many drugs appearing on the market considerably later than would have otherwise been the case, thus preventing patients from receiving the most advanced drug therapy. However, it is also the case that many substances have not come onto the market, or have done so only with special precautions, because of critical official analysis of the safety data. It must also be said that we have so far been spared a second pharmaceutical disaster of the magnitude of the thalidomide tragedy. A disaster of such dimensions would have far more negative repercussions on the rate of future innovation and money invested in the drug sector than all the bureaucratic delays put together. Therefore, pharmaceutical companies are forced to consider the quite legitimate question of whether they should not perhaps be grateful to these state institutions as protection for the consumer. If they function properly, the authorities can assist in gaining and increasing public trust in drugs and drug therapy.

There have also been criticisms of the inconsistency of policy-makers that are responsible for ensuring the profitability and hence the continuation of future research efforts. Regulation of development times and inventor protection are necessary. However, development times that are already necessarily from 8 to 15 years long should not be extended even more because of registration requirements that differ from country to country. In Europe we urgently need to harmonize the national legal and administrative provisions on drug specialties; in the USA obligatory deadlines for examining registration documentation are necessary; in Japan the international standard should be accepted and special, national regulations reduced to the essential minimum. Reciprocal recognition of registration assessments and registration data is also urgently needed. These measures would reduce costs and so attract investment and also protect health by making new therapies available to all.

Whereas the development time for a new drug has greatly increased in recent years, length of patent protection has remained more or less the same. This has shortened the period of market exclusivity to the innovative company. Inventor protection should give innovators a period of protection from generics to provide research-based pharmaceutical companies with the incentive to invest revenues from drug sales in research and development. At present, worldwide patents afford protection of from 17 to 20 years, depending on the country, starting from the time of application or issue respectively. However, one must apply for patent protection at a time when the product's properties are still far from known. Particularly deficient at the time of patent application are data on efficacy in man and the results of in depth safety investigations in animals. To get this information can take 8 to 15 years. This period, though, is completely at the expense of the period of validity of the patent. Therefore, a drug for which efficacy and safety investigations require especially great effort, and which is perhaps more valuable to the patient for precisely that reason, may not enjoy any patent protection whatever following market introduction! On the other hand, a product with a marginal medical advantage may enjoy a longer patent protection following introduction. This paradoxical situation is in glaring contradiction to the spirit of invention and inhibits medical progress. Hopefully, policy-makers will find a sensible way out of this dilemma. A basis for this could be the worldwide guaranteeing of a protection period of at least ten years for the registration documents of a new product following introduction, as has been proposed by the EEC Commission for Europe.

Money for research must be made available to those most able to achieve results. These are not only the countries with the largest markets: in the past, European countries have produced most of the innovative new drugs. To ensure continued vitality of research in relatively small countries and thus the availability of innovative therapy for all, the free movement of pharmaceuticals between countries will be vital. Only a large domestic European market can counterbalance the large US and Japanese home markets. We need to create the general conditions enabling companies to pursue their own development in a common market. Governmental regulation of drug prices and encouragement by price of imitators stifles investment in research and development. This type of policy is short-sighted, because it jeopardizes domestic drug research and spin-offs for other sectors of the economy such as agriculture. The drug industry in France, Italy and Canada has already been hard hit, and, without prompt action at international level, it will also be at risk in the United Kingdom, the Federal Republic of Germany and Switzerland.

The pharmaceutical industry is one of the most research-intensive of all industries. This is chiefly because, when it comes to drugs, enormous improvements are still conceivable. Those interested in therapeutic progress at the fastest possible rate must cooperate in bringing about overall conditions in which drug companies will be prepared to risk financial resources for R & D. It is especially in Europe where the appropriate measures have to be taken in order to retain this traditionally successful research capacity.

III. WORKING CLIMATE FOR INNOVATIVE DRUG RESEARCH

Public opinion that is favorable to research, novel scientific and technical findings, and sufficient capital to finance research and development efforts - all these factors are vital to successful research. Yet they are by no means enough. A further, essential pillar of a pharmaceutical company's innovative capacity is a working atmosphere conducive to the creation of new ideas and the translation of these ideas into usable products. A climate that favors risk has to be created within a company. It is not the function of research management to plan and implement innovations rationally from above, but to bring about the conditions in which ideas can be evaluated. The driving force behind these creative processes is the researchers and their staff. They are motivated to experiment when their company is guided by the principles of innovation and creativity, if it tolerates failures, if staff policy lessens the personal risks involved in organizational changes, and if efforts to innovate are rewarded.

Popular studies such as "In Search of Excellence" (Peters and Waterman, 1982) and the follow-up study "A Passion for Excellence" (Peters and Austin, 1985) emphasize the attitudes of sucessful research-based organizations. Central to successful research is the champion who, with just the right amount of fanaticism, pushes forward with his innovation as research management clears the necessary path. The significance of people-orientation is also expressed in the emphasis on the principle of management by wandering around (MBWA).

Innovative projects always involve risk. Failures are unavoidable. If you don't try a shot you won't score a goal. Of course the more shots you have the higher your chances of scoring, but then the more that will also go wide. Successful innovative companies see and accept this connection. In sucessful companies management recognizes successful and unsuccessful innovations because of the initiative shown and the exemplary efforts made.

Kieser predicts that in the future, pharmaceutical companies with the appropriate corporate culture will attain a strong competitive edge over rigidly hierarchical, uniformly structured rivals (1985). Most major research-based pharmaceutical companies still have enough capital to create a competitive technical infrastructure; they all have access to the technologies that accelerate innovation discussed in the other chapters of this book; they all have connections to institutes of higher education and hospitals with the appropriate facilities; qualified scientists and technicians are available. Research-based pharmaceutical companies vary in one essential point, however. Not all have a corporate culture that favors innovation. The culture of a company is usually rooted in its tradition and stems from the founders themselves although later leaders change emphasis. The culture fosters general and stabilizing principles according to which the company operates. This corporate culture embodies a company's vitality and its ability to maintain its position in a competitive environment. Both the blossoming and decay of research-based companies have their roots in company culture. And it is precisely in company culture that companies differ.

Successful drug research, development, production and marketing require particularly close cooperation between the most varied disciplines. This cooperation has to take place between organizational units that have diverse roles to play in the process. Because their roles differ, the organization and management of those organizational units are usually structured differently. Thus research and development, for instance, require an organizational structure that encourages independent thought and entrepreneural activity. However, production requires more regulation in the shape of officially authorized manufacturing specifications and guidelines, and for planning targets. A general call to abolish bureaucracy is therefore not appropriate. Management's problem is to coordinate the various areas. Coordination is hampered because managerial attitudes and behavior are molded by the varying functions and structures. For example, production managers are often cost-oriented and tend to cultivate an immediately practical and somewhat authoritarian style of leadership. Research and development managers, on the other hand, recognize that their success arises from a certain level of autonomy of their staff members. They usually prefer a more people-oriented leadership style. Failures may be disasters in production or quality control but unavoidable and to a certain extent essential for innovative research. When one wishes to ensure cooperation between representatives of the various organizational units to bring about product innovation and development, the differing values, orientations and behavior of these representatives can cause considerable conflicts. These conflicts arise not only from technical problems but also from different ways of looking at them.

To prevent and resolve such conflicts, organizational measures, such as setting up efficient project management or department-specific organizations, are not enough. It is also crucially important that the managerial system enables all those involved to be geared to the company's objectives and strategies and that it also promotes a method of conflict resolution that encourages rather than obstructs innovation. Therefore vital for success or the lack of it is the company's system of values, the way it is conveyed and the motivation and coordination of those involved in innovation - in short, corporate culture.

But what are the characteristics of corporate cultures that foster innovation? The easiest way to answer this question is to think about the salient characteristics of those cultures that hinder innovation. Companies where innovation is inhibited are characterized by a heavy orientation towards departmental goals. Communication between departments is suppressed. The company becomes a conglomeration of scarcely connected subcultures; there is not an identification with the company as a whole or special corporate goals. People from other departments are seen more as adversaries than cooperation partners. For this reason, teamwork decreed from on high is not usually very fruitful because it is not rewarded in middle management.

A second major feature of cultures opposed to innovation is a predominance of hierarchy. Top management is constantly at pains to monitor closely the progress of innovative projects and reserve the right to make relatively minor decisions. Following the procedures laid down appears to be more important than results. The climate is one of control and tutelage; leadership is not based on trusting individuals or teams seeking to innovate. A parallel to this is that information is generally traded as a scarce commodity. People are not informed about company strategy and current projects. In such companies it is difficult to obtain general support for projects that are out of the ordinary: internal venture capital is in extremely short supply. In these companies with a detrimental climate for innovation, moral support for projects not embedded in the "research plan" can only be drummed up with difficulty. There the hierarchical atmosphere thus discourages individual innovation.

A corporate culture that fosters innovation is quite the opposite. The foundation of such companies is the champion, the highly motivated individual who has recognized a central problem, the solution of which he has elevated to his own pressing concern. Successful, innovative companies back champions in many ways.

Firstly, champions are backed with information since information is not a scarce commodity. Every employee has the responsibility to use whatever means possible, including calling meetings with other parts of the organization, to gather needed information. This kind of information policy is successful only if top management sets the example with open-door methods, comprehensive information to all scientists on strategies and results, frequent coordination meetings and "face-to-face" communication in preference to written communication.

The champion is also encouraged by the availability of resources. Personnel, material, and financial resources for risky projects are easier to obtain in innovative companies. Projects that require the cooperation of several people are relatively easy to get off the ground. Some companies set up innovation banks to release "venture capital" for internal projects. Since research budgets are not unlimited, project activities must be integrated in an internal, results-oriented control system and exposed to continuous peer review and internal competition. This competition limits the extent and duration of projects based on performance. Peer-review committees in which experienced innovation champions have a central role are used to evaluate a project's potential for success in the light of the company's long-term objectives.

It is easier to ensure information transfer and easy shifting of resources in research units that include just a few hundred members than in the mammoth organizations that are still usual in many places. Smaller units are more efficient than larger ones. Quality takes precedence over quantity. The astonishing success of small but competent and highly motivated gene technology companies underlines the correctness of this assumption. Innovative companies need a highly decentralized organizational structure with simple, transparent managerial systems. A corporate culture that favors innovation by ensuring communication and cooperation toward common goals allows middle management to feel comfortable with decentralization.

It is for these reasons that repeated calls for research cooperation or research in one field to be conducted at one center instead of at many research based companies should be rejected. Competition and motivation in small, easily graspable units are the best guarantee of success. This makes occasional duplication of research activities inevitable. But the final results reached are rarely the same. Moreover, the pharmaceutical research world is so small and patents have to be applied for so early that virtually everyone knows who is researching what in which direction. In research, state influence on communication is superfluous.

Apart from providing the essential structures, most companies have to take further measures to create a climate conducive to innovation. First, the top management must credibly demonstrate that it attaches pivotal importance to innovation in its corporate system of objectives: official statements such as business reports, talks delivered to the workforce, inhouse publications, the written corporate concept and other official announcements must all give a clear indication that innovation is absolutely vital to the company's future. The role of innovation must be highlighted in interpretations of the company's historical development and its current situation, as is required from day to day in one-to-one talks or group discussions with top management. However, such an aggressive approach to innovation must not be allowed to decline into an exercise in mere rhetoric; it must be reflected in changes in the system of management. The value system, as Peters and Waterman put it, must be "lived".

The system of strategic management offers a starting point for consideration. In the strategic management systems of most companies, more scope is generally allowed for current business rather than innovative activities - not least of all because many different, tried and tested devices are available for strategic analysis and planning of ongoing business: devices for forecasting market potential, analysing the competition, assessing alternative strategies, implementing these strategies, etc.

The strategic plan covering innovative activities which may open up new fields of business is often not specific enough as far as the necessary budget for these activities is concerned. In the pharmaceutical business quite frequently only the overall total budget for research is defined and not the budget for a particular project.

The result of proceeding in this way is that if we face problems in sales in existing fields of business, some of the research resources earmarked for long-term projects may be rechanneled into me-too type improvements. Therefore, without the back-up of the strategic plan, it is difficult to put risky long-term projects through against apparently sure profits from minor improvements, especially if sales are sticky.

Incorporating innovative activities from outside the existing fields of business into the strategic plan first of all means strategically determining the proportion of the research budget a company is willing to spend on research and development activities in these new fields. The appropriate research budget for these activities should be based upon the future opportunities of these new fields and one's own research potential there. Of course, ongoing research projects should be assessed from a strategic standpoint on the basis of the results achieved in comparison to the competitors' position. A number of analytical tools have been developed recently which may prove useful for the strategic management of innovative activities (see e.g. Foster 1986, Servatius 1986).

Attracting first class scientists, identifying and selecting these from among the other candidates and allowing them to develop in accordance with their own development potential is probably the most important task of research and development management. Nothing determines the fate of research and development enterprise more than the quality of its people. Therefore, research and development managers should spend most of their energy on finding good people and keeping them productive once they have been hired (see Wolff, 1987). However, in order to keep scientists productive, research and development management should do everything to keep their scientist motivated. Three rules should be strictly followed in order to achieve this end (Wolff, 1987):

1) The importance and value of the scientists' work should be emphasized again and again.
2) Management should focus on the quality of the scientists' ideas, not on their position in the organization, their personality or anything else.
3) Research and development management should give them a chance to practice their craft as they rise in the organization, promote them to project leaders and allowing even presidents and vice-presidents to be scientists on a regular basis.

In order to motivate scientists properly however, their performance has to be thoroughly evaluated. This can only be done by an expert peer-review committee. But how can the individual creative researchers be motivated without detracting from the general effectiveness of the group work as a whole? Often it is only the "heroes" who are given any recognition and those who have worked away doggedly are completely overlooked. Committees that evaluate innovative work have to identify very carefully the group of researchers involved in that particular innovative exercise. If only "heroes of innovation" are picked out for special attention, other members of the team may well be frustrated. If too large a group is praised, the sense of motivation disappears into thin air. The expert committee therefore has to move very carefully between these two extremes.

An important aspect for the motivation of scientists is also to give them a chance to climb up in the company's hierarchy. This is particularly difficult because a creative researcher need not necessarily be a good research manager too. On the other hand non-researchers are usually not capable of managing a research and development unit properly because they lack the scientific knowledge which is essential for such a position. Management career decisions are therefore particularly difficult to make. However, they are of great importance for the research climate, the quality of the research performed and therefore important for the future fate of a research-based company. In order to fulfill their tasks properly, research managers need to be good leaders. Good leaders need to have a vision of which direction to take. They connect with people - people understand them and they understand people; and - they enjoy helping others grow. By doing this they contribute significantly to what is essential for a successful research and development unit and that is a creative corporate culture.

Those companies that are the fastest to translate the enormous scientific and technical progress of recent years into new products, these are the ones that will win out in the international competition among the powerful pharmaceutical companies for tomorrow's better and safer drugs. The decisive factor is excellent and highly motivated employees who feel at home in the multidisciplinary world of pharmaceutical research and are ready and able to operate creatively and in a goal-oriented manner. More than ever, successful research will depend on the quality of the corporate culture.

REFERENCES

Doll, R., Prospects for prevention, British Medical Journal, 1, 445-53 (1983)

Drews, J., Wie wird Pharmaforschung wieder rentabel? Neue Züricher Zeitung (15.10.1985)

Dunlop, D. (1980) J.Roy.Soc.Med., 71, 693-96 (1980)

Foster, R.N., "Innovation - Die technologische Offensive", Gabler, Wiesbaden (1982)

Humphrey, W.S., "Managing for innovation", Prentice-Hall, Englewood Cliffs (1987)

Kieser, A., Unternehmenskultur und Innovation, Blick durch die Wirtschaft (30.05.1985)

Miller, D.B., "Managing professionals in research and development", Jossey-Bass, San Francisco/London (1986)

Peters, T.J. and Austin, N. "A passion for excellence", Random House, New York, p. 437 (1985)

Peters, T.J. and R.H. Waterman, "In search of excellence", Harper & Row, Cambridge, p. 360 (1982)

Servatius H.G., "Methodik des strategischen Technologie-Managements", 2. Aufl., Erich Schmidt Verlag, Berlin (1986)

Wolff, M.F., "Attracting First-Class Scientists", Research Management, Vol. XXX No.6, p.9-10, (1987)

Abbreviations

2-AAF	2-acetyl-amino-fluorene
ACE	Angiotensin converting enzyme
AFGF	Acidic fibroblast growth factor
AIDS	Acquired immune deficiency syndrom
c-AMP	3',5'-cyclic adenosine monophosphate
ATP	Adenosine-triphosphate
BGHF	Basic fibroblast growth factor
BLP	Bombesine-like peptides
BRM	Biological reponse modifiers
CMP	Cytosine-monophosphate
DNA	Deoxyribonucleic acid
ECGF	Endothelial cell growth factor
EEC	European Economic Community
EGF	Epidermal growth factor
EPA	Environmental Protection Agency
FAD	Flavin adenine dinucleotide
FDA	Food and Drug Administration, USA
FRG	Federal Republic of Germany
GABA	Gamma-aminobutyric acid
GI	Gastrointestinal tract
GMP	Good Manufacturing Practice
GTP	Guanosine-triphosphate
HIV-1	Human immune deficiency virus
HMG-CoA	Hydroxymethylglutaryl-coenzyme-A
HPB	Health Protection Branch, Canada
HPLC	High Performance Liquid Chromatography
HTLV-3	Human T-cell lymphotropic virus
IGF	Insulin-like growth factor
IL	Interleukin
KDO	2-keto-3-desoxy-octonate
LAV-1	Lymphadenopathy associated virus
MAO	Monoamine oxidase
MOAB	Monoclonal antibodies
8-MOP	8-methoxsalen
MPTP	1-Methyl-4-phenyl-tetrahydro-pyridine
MR	Multiple resistance
NAD^+	Nicotinamide adenine dinucleotide (oxidized form)

NADPH	Nicotinamid-adenine-dinucleotide phosphate (reduced form)
NGF	Nerve growth factor
NMR	Nuclear magnetic resonance
P-450	Cytochrome P-450
PAF	Platelet activating factor
PALA	N-Phosphonoacetyl-L-aspartate
PAPS	3'-phosphoadenosine-5'-phosphosulfate
PDGF	Platelet-derived growth factor
PI	Phosphatidylinositol
PKC	Protein kinase C
PLC	Phospholipase C
PLS	Partial least squares
3-PPP	3-(3-Hydroxy-phenyl)-1-propyl-piperidine
QSAR	Quantitative Structure-Activity Relationships
RNA	Ribonucleic acid
SAR	Structure-Activity Relationships
SDAT	Senile dementia of the Alzheimer type
tPA	Tissue plasminogen activator
TGF	Transforming growth factor
TNF	Tumor necrosis factor
UDPGA	Uridine-5'-diphospho-@-D-glucuronic acid
UTP	Uridine-5'-triphosphate

Index

Absorption of drugs
from gastrointestinal tract, 94
and bioavailability, 92, 93
definition of, 78-80
dependence on lipophilicity, 267
and enterohepatic circulation, 139-140
fundamentals of, 94-97
improvement of, 97, 408, 432-434 (see also Pro-drugs)
rate constant, 430-432
rate profile, experimental determination of, 430-432
role of drug binding to plasma proteins, 102
selection of drug candidates, 148
Acetanilide, 247
Acetazolamide, 318-319
2-Acetylamino-fluorene, 135, 141
Acetylation:
of xenobiotics, 135, 142
Acetylcholine:
as a neurotransmitter, 6 29
Acetylcholinesterase:
as a target for drug therapy, 45
Acetylsalicylic acid, 129 154
Acid glycoprotein-α_1, 101
Active metabolites:
of xenobiotics, 140, 146, 149, 154
Activity-destroying features of a molecule, 290
Activity profile of drugs:
determination of, 313, 323-325, 330-348
drugs for the central nervous system, 348-349
strategy of drug development, 316
Acyclovir, 50-52

Adenosine, 262
S-adenosylmethionine, 136
Adenylate cyclase, 20-22
Adhesion molecules, 58
ADP-ribosylation of N-proteins, 22
Aerosol, 407
Agonists (see Receptors, see Receptor agonists)
AIDS, 47-53
  adsorption of virus to T-cells, 49
  frequency of, 47
  retrovirus, 48-53
  stages of infection by retroviruses, 48-53
Alcohol dehydrogenase, 122
Aldehyde oxidase, 122
Alinidine, 338
Alkylating agents, 59, 243
Allergy, 54
Alpha-1 antitrypsin, 234
Alpha-2 adrenergic agonists:
  theoretical map of macromolecular binding site, 201
Alzheimer's disease, 43-47
  abnormal proteins of, 46-47
  and acetylcholinesterase, 45
  cholinergic system, 44-45
  dependence on age, 44
  ethiology of, 44-46
  frequency of, 43
  histopathology of, 44
Amantadine, 49
Ambroxol, 124
Ames-test, 376, 392
p-Aminobenzoic acid, 129
c-AMP in signal transduction, 20-24
Amphetamine, 127
Ampicillin, 129
Amrinone, 137
Amyloid, 46
Analytical instrumentation:
  impact on biotransformation studies, 138, 152, 157
Anesthetics, 4, 27, 28-29
Angiogenesis, 69-70
Angiogenic factors, 69-70
Angiotensin I, 256
Angiotensin II, 256
Angiotensin converting enzyme, 256-257
Animal:
  experiments with, ethical considerations on, 309, 313-314, 328-329, 347, 443, 450
  models
    and pathological states in humans, 349
    in pharmacokinetics, predictive power of, 117

Animals:
  use in drug design, 443
  welfare of, 356-358, 362
  veterinary care of, 361-362
Antagonists, 311, 321-325 (see also Receptor antagonists)
Antianginal drugs, 309
Antiarrhythmics, 328, 330, 334
Antibiotics, 189, 309, 312, 329-330, 333 (see also specific chemical class)
Antibodies (see also Monoclonal antibodies):
  antiidiotypic, 67, 238-239
  chimeric, 67, 235
Antidepressants, 310
Antidiabetics, 244, 253-254, 316
Antihistamines, 54
Antihypertensives, 256-257, 309, 316, 339, 349
Antiinflammatory drugs, 31, 310
Antimycotics, 155
Antiviral compounds:
  profiling of, 333
Apomorphine:
  in D-2 binding site map, 201
  conformations of, 197-198
  molecular electrostatic potential of, 167
  sites of favourable interactions with O^-, 199
Aspartate transcarbamylase, 263
Asthma, 54, 55
Atherosclerosis, 37-39, 309
  description of, 37
  lipoproteins in, 37, 38
  macrophages in, 38
  pathology, theories of, 37
  PDGF in, 38
  risk factors of, 37
  smooth muscle cells in, 38
  therapy of, 37-39
Atomic motions:
  molecular dynamics calculation of, 195
ATP, 20, 137
Atropine, 6
AUC, 93
Avarol, 52
Azidothymidine, 50
Bendroflumethiazide, 331
Benzimidazoles, 282-288, 295, 299
Benzodiazepines, 29, 263-264
Benzothiadiazine, 319

B-HT 920, 40, 323
binding site map for D-2 receptor, 200
molecular electrostatic potential of, 167
sites of favourable interactions with O^-, 199
Bias in drug design, 247, 293
Bile, 111-116, 131, 156 (see also Clearance of drugs, see also Hepatic elimination)
Binding models in lead finding, 259-260
Binding site, macromolecular, 86-88, 101 (see also Plasma protein binding)
in lead finding, 190-203, 255-259
mapping shape of, 190-203
Bioadhesive materials in drug delivery, 436
Bioavailability:
definition of, 92-93
and dosage form, 408
estimates from dose-response relationships, 347
hepatic clearance, influence of, 115, 129
Biochemical mechanisms in lead finding, 260-263
Bioisosteric replacement, 172, 267-269
examples of, 172, 268, 269
groups for, 172, 268-269
Biological concepts in lead finding, 250-255
Biological membrane, 82-84 (see also Membranes)
Biological processes, regulation of, 4-5
Biomolecular processes in lead finding, 255-263
Biomolecular structures in lead finding, 255-263
Biopharmaceutics, 94
Biotechnology, 217-240
processes, patentability of, 228
products,
patentability of, 228
in therapeutic use, 218, 239
toxicology of, 389-392
Biotransformation of xenobiotics, 78, 106, 118-146
extrahepatic, 139-140
genetic factors in, 143
hepatic elimination, 111, 112
kinetics of, 143-146

phase I enzymes of, 120-123
phase I reactions of, 119, 124-131
phase II reactions of, 119, 131-139
role in the development of better and safer drugs, 146-158, 321
role in excretion of drugs, 108
species differences of, 141-142, 349
Bisacodyl, 140
Blood-brain barrier, 99-100
Blood level (see Bioavailability, see Plasma levels of drugs)
Blood pressure (see Antihypertensives)
Blood proteins, 236
Body fluids, 97, 99, 104 (see also specific fluids)
Brain:
hemoperfusion of, 80
Bromhexine, 124
Buccal absorption:
correlation of rate with octanol-water log P, 182
for drug delivery, 436
B-values, descriptors of substituent size, 182
Ca^{++}-antagonists, 322, 338
Calcium in signal transduction, 24, 28, 29, 30, 64
Calcium ionophore A23187, 29-30
Cancer, 59-71
and angiogenesis, 69-70
and biological response modifiers, 65-66
combination treatment of, 60, 66
and dihydrofolate reductase, 256
and electrophilic compounds, 245, 261
and growth factors, 63-64
metastatic spread of, 68-69
monoclonal antibodies, 63, 64, 65, 66-68, 69
multiple resistance to cytostatics, 70-71
oncogenes, 22-24, 61-63
and PALA, 263
selectivity of chemotherapeutics, 246
signal transduction, 64-65 (see also Oncogenes)
Capillary wall, 97, 104
Capsules, 402, 407
Captopril, 257

Carbamazepine, 126
Carboxypeptidase A, 256, 257
Carbutamide, 253, 254
Carcinogenicity, 362, 377-381, 387-388
  dosage of carcinogens, 378-380
  short term assays for, 381, 388
Cardiac glycosides, 255, 265
  biological test for, 326-327
  mechanism of action of, 242-243
Carrier-mediated transport, 85, 96-97, 110, 113
CD4 receptor, 49
Cell proliferation:
  and calcium ions, 64
  and inositol phosphates, 64
  and phosphatidylinositol, 64
  and protein kinase C, 64
Cells
  antibody producing, 67, 221-222,
  immortalization of, 67, 221-222, 223
Central nervous system:
  clonidine, 339
  diseases of, 39-47, 309
Cephalosporins, 110
Chance in drug discovery, 246-247, 348
Charge neutralization in binding a small molecule to a protein, 167
Chelate effect
  definition of, 164
  example of, 164
Chemical concepts:
  in lead finding, 263-264
  in lead optimization, 276-277
Chemoreceptors, 4
Chimeric proteins, 234-235, 236
Chiral compounds, 201-202, 291 (see also Enantiomeres)
Chloral hydrate, 128
Chloride ion, 28, 29
Chlorothiazide, 331
m-Chlorophenyl-piperazine, 154
Chlorpromazine, 127
Cholecystokinin, 43
Cholesterol, 37, 38, 82
Cholinergic agonists, 45 (see also Muscarine, Nicotine)
Chromosomal aberration, 377
Cimetidine, 122, 269
Clearance of drugs (see also Excretion of drugs):

biliary, 111-116
contribution to by biotransformation, 118
definition of, 82
excretion as a factor of, 106
relevance for AUC, 93
total, 116-117
Clenbuterol, 105, 268
Clonidine:
controlled release of, 339, 417
transdermal system for, 423-427
Cloning of mammalian genes, 219, 220-221
Cluster analysis in QSAR, 187-188
Coating, of dosage forms, 407, 420
Coeruloplasmin, 69
Colony stimulating factors, 54
Computer graphics (see Molecular graphics)
Computer models of molecules (see Molecular graphics)
Computers, application in pharmacokinetics and metabolism, 80, 152, 153, 157
Computer-aided drug design (see Mapping macromolecular binding sites, see QSAR)
Conformation of ligands bound to macromolecules (see Mapping macromolecular binding sites):
in analogue design, 300-303
based on conformationally constrained analogues, 197-198
example, 197
NMR studies, 197
X-ray crystallography as a source of, 164, 196-197
Conformation-activity relationships, 190-203, 300
Conformationally restricted analogues (see also Three-dimensional properties of molecules):
use in mapping macromolecular binding sites, 190, 197-198
Conformations of a molecule
definition of, 162
example of,
apomorphine, 197-198
dopamine, 162
KDO, 197
Conjugation reactions in biotransformations, 119, 131-139

Controlled drug release, 97, 148, 403-406
  benefits of, 406,
  chemical control of, 413-416
  diffusional system, 416-418
    single phase system, 416
    two phase system, 417
  membrane systems, 418-421
  minipumps for, 421
  polymers for, 412-413
Convolution technique for determination of absorption rate profiles, 431, 432-433
Cooperation of scientific disciplines, 151-152, 249, 436-437
Corporate culture, 442, 459-463, 467
Creativity, 174, 441, 458
Cromoglycate, 54
Crystal structures (<u>see</u> X-ray crystallography)
Curare, 6
Cyclooxygenase, 54
Cyclosporin A, 58
Cysteine, 274
Cytochrome P-450, 119, 120-122, 144, 155
Cytostatics, 406, 438
D-2 Agonists:
  theoretical map of macromolecular binding site, 200-202
DNA, 217, 219, 246, 312
DNA plasmids, 219
DNA recombination, 217
DNA-drug complex, 203
Dealkylation:
  biological, 126
Deamination:
  biological, 126
Debrisoquin, 143
Descriptors (<u>see</u> Variables)
Design of ligands of a macromolecule, 209-212
  examples, 209
Desimipramine, 126, 154
Diffusion, 80, 89, 90-91
  coefficient, 90-91, 416, 417, 418, 419
  through biological membranes, 84-86, 97, 99-100, 108, 112, 140, 407-408, 423, 434, 436
Dihydrofolate reductase, 256
  charge interactions with methotrexate, 167
  X-ray crystallography of complexes, design of compounds from, 164
Diltiazem, 322
Diphenylhydantoin, 125

Discriminant analysis in QSAR, 186-187
Dispersion interactions, 167-168
  definition of, 167
  example, methotrexate in dihydrofolate reductase, 168
  MR as a descriptor of, 180-181
Disposition of drugs, 78-79
Distribution of drugs in the body, 78-80, 97-106
Diuretic:
  design from QSAR, 188
Diuretics, 316, 318, 319, 331-332, 349
Dopamine:
  agonists, 40, 45, 326, 311, 325 (see also D-2 Agonists)
  antagonists, 40, 41, 326
  autoreceptor antagonists, 40-42
Dopaminergic system:
  and Parkinson's disease, 41, 45
  and schizophrenia, 40
Dosage forms (see also Drug delivery):
  excipients for, 404, 408
  general requirements, 402-403
  manufacturing of, 410-412
  manufacturing factors of, 411
  membranes for, 418-421
  optimization, Nelder-Mead method, 411
  poorly soluble drugs, 408
  for proteins, 227-228
  role in pharmacokinetic models, 403-404
  stability of, 409-410
Dose ratio between different drug effects, 335, 340-341 (see also Selectivity of drug action)
Dose regimen in pharmacological and toxicological studies, 153, 367-368
Dose-dependency:
  in pharmacokinetics, 91-92, 147
  in biotransformation, 145, 148
Dose-response curves, 8-10, 321, 331-333, 341, 347
Down's syndrome, 46
Draize test for irritation, 370-371
Drug candidate:
  selection of, 147-151
Drug combinations:
  toxicity studies with, 371-372
Drug delivery (see also

Bioadhesive materials, see also Buccal absorption, see also Intranasal drug administration, see also Minipumps, see also Oral systems, see also Perlongets[R], see also Transdermal systems):
cell specific, 438 (see also Liposomes, see also Monoclonal antibodies)
feedback design, 437
organ specific, 438
Drug design:
based on structure of the macromolecules, 209-212
and biological testing, 315-316
direct approach, 202-203, 245-246
systematic procedure, 248
QSAR, 188-190
Drug development, 151
program, 151-155
timing, 152
Drug interaction, 122
Drug metabolism (see Biotransformation)
Drug poisoning, 108, 116 (see also toxicity)
Drug safety and efficacy (see Safety and efficacy assessment, see Toxicological studies)
Drug transport, 80-86, 102, 104
carrier-mediated, 85, 96
endocytosis, 85
passive diffusion, 84-85
Duration of action, 89, 105, 328, 346 (see also Time dependence of effects)
E_s values:
QSAR descriptors of substituent size, 182
Eadie-Hofstee plot, 12-13
Effect kinetics, 154
Efficacy of an agonist, 8
Efficacy and safety assessment (see Safety and efficacy assessment)
Electrostatic potential, molecular:
definition of, 165
example, apomorphine, 167
Electrostatic interaction, 165-167
influence of water on, 167
Eliminating, organ:
efficiency of, 80-81
and first pass effect, 93
Elimination of a drug, 78-80, 93,

and plasma protein binding, 102
Enantiomers, 88, 102, 154, 201-202, 291 (see also chiral compounds)
Endocytosis, 85, 109, 115,
Endoplasmic reticulum, 112, 120-121, 133
Energy refinement of three-dimensional structures of molecules, 193-195
Enterohepatic recycling, 115, 139
Enzyme assays of inhibitors, 312, 318-320 (see also Radioligand binding)
Enzyme inhibition, 88, 145
  ACE, 256-258
  acetylcholinesterase, 45
  analysis of, 10-14
  in antibacterial therapy, 312
  example for series design, 293
  irreversible, 13, 18, 260-261
    k_{cat}, 260-261
    monoamine oxidase, 172, 261
    suicide substrates, 260-261
  non-competitive, 13
  P-450, 122, 155
  papain, 259
Enzymes:
  allosteric, 4
  effects of drugs on, 316, 318-320, 326, 330, 343
  interaction of drugs with, 8, 10
Epidermal growth factor (EGF), 27, 63
Erythromycin analogues QSAR, 189
Ethacrynic acid, 332
Ethics, 157, 313-314, 327
Ethoxycoumarin, 144
Eukaryotic cell cultures,
  expression of proteins in, 222-223
Excipients (see dosage form)
Excretion of drugs (see also Clearance of drugs):
  via feces, 139
  hepatic elimination, 106, 111-116
    kinetic aspects of, 113-116
    physiological basis of, 111-113
  via intestine, 116
  via lung, 116
  phase II metabolites, 131-139
  prediction of pharmacokinetics, 146
  renal, 106-110

rate of, 110
routes of, 106
of unchanged drug, 78
Extraction ratio of xenobiotics from blood, 82, 115, 148
Extrahepatic biotransformation, 139-140
Factorial design of analogues, 293-300, 301-303
Factorial schemes for analogue design, 294, 302
Feedback drug delivery, 437
Fermentation, 223-225
fermentor design, 224
host strain optimization, 224-225
Fertility,
effect of drugs on, 373
Fibrinolytic enzymes, 230-232
Fibronectin, 68
Film coating in dosage forms, 411
Financial constraints to drug research, 448-457
First-pass effect of a drug, 93, 115, 116, 148, 422, 425, 436
Fit between molecules, 169
Fluid mosaic model of biological membranes, 82-84
Forces between molecules (see Intermolecular forces)
Forskolin, 265
Fractional factorial schemes for analogue design, 298-299
Free-Wilson QSAR analysis, 184, 278
Furosemide, 331, 332
GABA, 29
Gallopamil, 322
Gastrointestinal tract
absorption of drugs from, 94-97
and drug clearing, 116
metabolism of drugs in, 131, 139
Gastrointestinal transit of oral dosage forms, 428-430
Gene mutation by drugs, 376
Gene expression in microbial cells, 220-221
Generic competition, 450-452, 456
Genetic engineering, 218-220, 233-234 (see also protein engineering)
products of, as screening models, 319
Genotoxicity, 151, 375-377, 387
models for, 376-377
Gingkolides, 57

Glass temperature of polymers, 418
Glibenclamide, 254
Glomerular filtration, 106-107, 147
Glutathione conjugation, 137-138
Glucagon, 20
Glucocorticoids
  mode of action, 30-31
Glucose-1-phosphate, 133
Glucuronidation, 133, 147
Glycogen:
  mechanism of breakdown, 20-21
Good laboratory practice in toxicological studies, 382
Good manufacturing practice of medicaments, 412
Gramicidin A, 29
Growth factors, 24, 27, 63-64, 69, 237 (see also individual growth factors)
Growth hormone, 229
GTP in signal transduction, 22-24
Half-life,
  biological of drugs, 92, 105, 140, 143
Hammett early work on LFER, 176
Hammett constant (see Sigma constant)
Hammett equation, 177 (see also QSAR)
Hansch early work on QSAR, 176
Hansch pi value, 181-182, 278
Hansch analysis (see QSAR)
Hansch equation, statistics of, 185-186 (see also QSAR)
Hemoglobin:
  design of effectors from crystal structure of, 209
Hemoperfusion, 80-82, 89, 90 (see also Perfusion of tissues by blood)
Hemophilic factor VIII, 223, 230
Hepatic elimination, 111-116
  kinetic aspects of, 113-116
Hepato-biliary transport, 113
Hepatocyte, 111-113, 145
Herpes encephalitis, 52
Herpes simplex virus, 52
Hill coefficient, 17
Hill plot, 17
HIV virus (see Retrovirus)
HMG-CoA reductase, 38
Homeostasis, 5
Homology of proteins:
  three-dimensional

examples, 207-208
Human cells for biological tests, 326
Human enzymes for biological tests, 319
Human receptors for biological tests, 321
Human tissues for biological tests, 327
Hyaluronidase, 68
Hybridoma cells, 222
Hydrochlorothiazide, 319
Hydroflumethiazide, 331
Hydrogen bonds, 169-170
  definition, 170
  effects of water on, 170
  role in molecular recognition, 170
Hydrolases, 123
Hydrolysis:
  biological, 128-131, 140 147
Hydrophobic interactions, 85, 170-171 (see also lipophilicity)
  definition of, 171
N-Hydroxy phenacetin, 127, 135
N-Hydroxy-2-acetyl-amino-fluorene, 135
Imidazo [4,5-b] pyridines, 123, 284-288, 295-297, 299-300
Imipramine, 126, 154
Imipramine N-oxide, 127
Immune modifiers, 230
Immunological diseases, 53-59
  and biogenic amines, 54
  causes of, 53-54
  and colony stimulating factors, 54
  phospholipid derived mediators of, 54-58
In vitro models:
  of drug metabolism, 117-118, 144-145, 157
  in pharmacological evaluation of drugs, 318-327
    enzymes, 318-320
    intact cell membranes, 325-326
    intact cells, 326-327
    isolated organs, 327
    receptors, 320-325
  in toxicological studies, 276-277, 381
Indicator variables in series design, 184
  expression of structure-activity hypotheses, example, 278, 282,284
  in factorial design, 294, 295-296
    example, 299-300
Inhalational toxicology, 359, 263-364
Inhibitors of enzymes, 12-13, 122, 155, 172-174

(see also Enzyme inhibition)
design of, 197
Inositol-1-phosphatase, 27
Insulin, 27, 229
Insulin-like growth factor, 27
Interferons, 52, 65, 229, 234, 390, 392
Interleukins, 59, 65, 66, 235, 390
Intermolecular forces, 162-171
complementarity between drug and receptor, 191, 198-203
dispersion interactions, 167-168
MR as a descriptor of, 180-181
and electrostatic interactions, 165-167
example of, dihydrofolate reductase and NADPH, 164, 168
example of, methotrexate bound to dihydrofolate reductase, 168, 170
importance of electrons in, 163, 165-167
quantitative description of, for QSAR, 176-185
shape complementarity, example, methotrexate bound to dihydrofolate reductase, 168
steric repulsion, 168-169
Intermolecular interactions:
favorable sites on proteins, 196, 211
molecular graphics displays, 164, 166, 168, 169, 196, 199
Intestinal flora, 311
and biotransformation, 139
Intransal drug administration, 436
Inulin, 98, 108
Inventor protection:
impact on drug research, 450-452, 453, 456 (see also Patent life time)
Ion channel blockers, 28, 30
Ion channels in signal transduction, 5, 27-30
allosteric effects of drugs on, 28
direct blocking by drugs, 28
local anesthetics, 28
receptor control of, 29
Ionic membrane fluxes, 334-335, 343
Ionophores, 29-30
Iproniazide, 348

Irritation potential of drugs:
  test models for, 370-371
Ischemic heart disease:
  drugs for, 335-338
Isolated organs as biological, test models, 117-118, 309, 327 (see also in-vitro models)
Isoniazid, 136, 143
Isoprenaline, 136, 250, 268
Isoproterenol (see Isoprenaline)
Isoxazol-oxazoline derivatives, 49
k_{cat} enzyme inhibitors, 260-261
KDO:
  bound conformation from NMR, 197
KDO-CMP-synthetase:
  bound conformation of KDO, 197
Kidney:
  hemoperfusion of, 80
Lassa fever virus, 52
Lead, 171 (see also Lead compound)
  bioisosteric replacement, 172, 267-269
  synthesis to explore substructures required, 171-172
Lead compound:
  definition of, 248
  sources of,
    binding models, 259-260
    binding sites, 255-259
    biochemical mechanisms, 260-263
    biological concepts, 250-255
    chemical concepts, 263-264
    natural sources, 264-265
    targeted screening, 265-266
  value of, 254-255
Lead finding, 250-266
Lead optimization (see also Series design):
  definition of, 248
  empirical, 152, 275-291
  peptidic leads, 273-275
  procedures for, 266-303
  targeted, 267-273
Lead structure (see Lead compound)
Lergotrile:
  molecular electrostatic potential of, 167
Leukotrienes, 31, 54, 55,58
  antagonists, 55
Linear free energy relationships (see QSAR)
Lipid bilayer (see Membranes, biological)

Lipocortin, 31
Lipophilicity (see also Hydrophobicity):
  binding of drugs to enzymes, 144
  drug absorption distribution and excretion, 267
  hepatic elimination of drugs, 113
  influence of biotransformation on, 128
  membrane transfer of drugs, 84-86
  pi as a descriptor of, 181-182
  protein binding of drugs, 102
  renal elimination of drugs, 110
  rules for selection of drug candidates, 146, 147, 148
  in targeted lead optimization, 267
  tissue binding of drugs, 105
Lipoproteins, 37, 38
Liposomes in drug delivery, 438
5-Lipoxygenase, 58
Lithium ion:
  significance for phosphatidylinositol pathway, 27
Liver
  drug penetration through capillary walls, 98
  elimination of drugs, 111-116
  hemoperfusion of, 80
  phase I enzymes, 120, 122
  phase II reactions, 131, 133, 137
Local anesthetics, 28, 309
Local tolerance studies, 369-371
Log P (octanol-water) (see also Lipophilicity):
  calculation of, 181
  CLOGP program, 182
  correlation with biological properties of molecules, 180, 182
  correlation with plasma protein binding, 101, 102
  values, 181
Lung:
  and drug absorption, 94
  and drug elimination, 116
  and drug metabolism, 139
  hemoperfusion of, 80
Lymphotoxins, 65
Macrophages, 38
Magnesium ion in signal transduction, 28
Mapping macromolecular binding sites, 190-203
  and design of selective

drugs, 192
general strategy for, 190-192
shapes of, 198-203
Master seed culture for biotechnology, 226
Maternotoxicity of drugs, 374, 386-387
Mean time of persistence of drug molecules in the organism, 92
Mechanism of drug action, 316
in biological profiling, 335, 338-339, 342-343
pargyline analogues as monoamine oxidase inhibitors, 172-174
QSAR prediction of, 118
and responses in isolated organs, 327
Medicinal chemistry:
structure-activity relationships analysis, 171-203
theoretical basis of, 161-212
Melanin, 105
Membrane proteins, 3
function of, 3
modes of association with the membrane, 3
Membrane systems for controlled drug release, 418-421
Membranes, biological:
associated proteins, 82-83
glycogen breakdown, 20-22
and ion channels, 28
separation from soluble cellular fraction, 14
structure of, 82-84
transport of drugs across, 84-86, 99, 108, 112
Mephenytoin, 143
Mercurial diuretics, 349
Metabolites of drugs (see also Pro-drugs):
biotransformation of xenobiotics, 118-146
clearance into bile, 115
excretion of, 106, 107, 110, 111, 115
recognition of caveats concerning efficacy and safety of a drug, 155-156
and selection of drug candidates, 152-153
significance for the biological effects of a drug, 146-147, 153, 344
significance for drug clearance, 117
Metastatic spread, 68-69
mechanism of, 68

Methiamin, 269
Methotrexate:
  charge interactions with dihydrofolate reductase, 167
Methoxamine, 323
Methylation:
  biological, 136-137
5-Methyl-tetrahydrofolic acid, 136
8-Methoxsalen, 145
Mexiletine, 334
  sustained release form, 433
Michaelis-Menten model, 12, 90, 109
Microbiology in the search for better and safer drugs, 310-312 (*see also* Antibiotics, *see also* Biotechnology)
Minimal toxic dose, 380
Minipumps for drug delivery, 421
Mitochondria, 120, 122, 135
Molar refractivity (*see* MR)
Molecular dynamics:
  calculations of atomic motions, 195
  calculations of proteins, 211-212
Molecular graphics, 192-193
  design of ligands from three-dimensional structure of macromolecule, 209-212
  displays of intermolecular interactions, 164, 166, 168, 169, 196, 199
  examples, 164, 166, 168, 169, 199, 200, 201, 202, 208
  with NMR of proteins, 207
  and protein crystallography, 205
  protein engineering, 235-236
Molecular mechanics:
  calculations of proteins, 211-212
  refinement of three-dimensional molecular structures, 193-195
Molecular recognition (*see* Intermolecular forces)
Molecular structure, basic skeleton of, 243, 289, 294
Monoamine oxidase, 122
Monoamine oxidase inhibitors, 261, 348
  pharmacophore of, pargyline analogues, 174
  structure-activity relationships of, 172-174
Monoclonal antibodies, 63, 64, 65, 67-68, 69, 217, 236
  cells for production of, 221-222

in drug delivery, 438
Motivation, 461, 463, 467
MPTP, 41
MR (molar refractivity)
as a descriptor of dispersion interactions, 180
correlation with molecular size, 180
in series design, 280, 287, 289, 296-297
table of, 181
Multi substrate inhibitors, 263
Multiple regression (see Regression analysis)
Muscarine, 6
Muscle relaxants, 309
Mutagenicity, 376, 387, 388
predictive value for carcinogenicity, 387
NADPH:
X-ray crystallography of binding to dihydrofolate reductase, 164
Natural products in lead finding, 264-265
Nelder-Mead method for optimization of manufacturing of dosage forms, 409
Nerve growth factor, 27
Neuritic (senile) plaques, 44
Neurofibrillary tangles, 44
Neurotoxin, 28
Nicotine, 6
Nifedipine, 30
N_i-protein, 22
Nitrendipine, 322
Nitroglycerin:
transdermal system, 422, 424-426
Nitrosamines, 151
NMR:
in establishing bound conformation of a ligand, 197
example, 197
of proteins, 206-207
Noradrenaline, 251, 339
Norethisterone:
controlled release form, 415-416
N-oxidation:
biological, 126-127
N_s-protein, 22
Oncogenes, 61-63
abl, 62
function of encoded proteins, 62
mechanism of tumor promotion, 61-62
myc, 62
ras, 24, 62
relation to growth factor receptors, 237
in signal transduction, 22-24
and tyrosine kinases, 27

Oral activity of drugs, 148, 328, 347
Oral systems for drug delivery, 427-434
  factors to be considered, 427-432
  sustained release forms, 432-434
Osmosis in drug delivery systems, 420-421
Oxidation:
  biological, 120-123, 124-126
P-170, 70
P-450 (see cytochrome P-450)
PAF, 54, 55-57
  antagonists, 57
  biosynthesis of, 56
  effects on cell types, 56
  pathophysiological role, 56-57
Paracetamol, 126, 137, 154, 246
Pargyline analogues:
  mechanism of action, 172
  pharmacophore of, 174
  structure-activity relationships of, 172-174
Parkinson's disease, 41, 45, 311-312
Partial agonists, 8
Partial atomic charges:
  role in hydrogen bonds, 165-167, 169
Particle size in inhalational application of drugs, 407
  significance of in toxicology, 359
pH partition in drug absorption and excretion, 96, 108, 428
Patent life time, 448, 450, 452, 456
Pathology underlying diseases:
  as a guide to effective drugs, 320
  imitation by animal models, 310, 349
Penetration (see Diffusion)
Penetration enhancers, 434
Penicillin, 100, 110, 144
Penicillamine, 274
Peptide mimetics, 275
Peptide synthesizer, 235
Peptide T, 49
Peptides
  backbone variations, 274-275
  bombesine-like, 64
  clearing mechanisms, 97, 109, 117
  as drugs, usefulness of, 246, 252
  peptidic modifications, 275

side chain variations, 274

Peptidic leads
  optimization of, 273-275

Peptidoleukotrienes, 55

Performance of scientists:
  evaluation of, 466-467

Perfusion of tissues by blood, 80 (see also Hemoperfusion)
  drug distribution to tissues, 99, 105
  effect on local drug concentration, 90
  gastrointestinal tract, effect on absorption from, 96
  rate of renal drug elimination, 110

Perinatal toxicity, 375

Perlongets[R], 432-433

Pethidine, 129

Pharmaceutical preparations:
  importance of in toxicological studies, 359
  toxicological studies with special preparations, 369

Pharmaceutical industry:
  role in research, development and production of dosage forms, 401, 410

Pharmaceutics:
  future contributions of to drug development, 434-438

Pharmacodynamic characterization of test compounds, 331-343

Pharmacogenetic differences in drug metabolism, 143, 154

Pharmacokinetics, 77-118
  absorption, 94-97
  biological systems for study of, 117-118
  concepts of, 78-80
  correlation of in vitro with in vivo biological data, 318
  data for toxicological studies, 360-361
  distribution, 97-106
  and dosage form, 403
  dose-independent, 92
  excretion, 106-117,
  in lead optimization, 267-273, 327
  models and oral pharmaceutical systems, 430-432
  parameters of, 92-93
  physiological considerations, 93-118
  rate-determining processes, 89-92
  role in the evaluation of drugs, 246, 315, 343-348

role in interdisciplinary drug development, 151-155
rules for selection of drug candidates, 147-151
Pharmacology:
role of in the search for better and safer drugs, 310-312
Pharmacophore:
defined by properties, 195-196
definition of, 174
example of, pargyline, 172-174
example, dopamine, 196
use of in mapping macromolecular binding sites, 196
Phase I reactions of drug metabolism, 124-131
Phase II reactions of drug metabolism (see conjugation reactions)
Phenacetin:
active metabolite of, 154
dealkylation of, 126
discovery of, 246
and genetic differences, 143
hydrolysis of amide group, 131
N-oxidation of, 126
sulfation of N-hydroxy metabolite, 135
Phenethanolamine derivatives, N-methyltransferase, 293
Phenetidine, 131
Phenobarbital, 341
Phosphatidylinositol pathway of signal transduction, 24-27, 64
Phosphodiesterases, 21
Phospholipase A_2, 58
Phospholipase C, 24, 26
Phospholipids, 24
Phosphonoacetyl-L-aspartate, 263
Phosphorylase, 20
Phosphorylase kinase, 20
Phosphorylation:
of phosphorylase, 20
of receptors, 18, 30
reversal of, 20
Phototoxicity, 371
Physical properties of molecules, 165-171, 176-185
Physico-chemical properties in series design, 176-186, 209-212, 278, 279-280, 286-287, 289, 293, 296-297
Pi value of hydrophobicity:
definition of, 181-182
in series design, 278, 279, 286-287, 289, 293, 296
table of, 181

Pilocarpine, 6
Pinocytosis, 85, 97
Pirenzepine, 323
Pivaloylampicillin, 129
pK_a's
  effect on pharmacokinetic properties of a drug, 96, 99, 108, 116, 118, 146
  effect of substituents on, 176, 177-180
  Hammett equation for, 177
  predicition of, 177-180
Plasma level of drugs:
  time course of, 79-80
  and dosage form, 404-406
Plasma protein binding, 86-88
  and biological activity, 344
  and distribution volumes, 92
  relevant plasma proteins, 100-103
  and selection of development compounds, 148
Plasminogen activator (see Tissue plasminogen activator)
Plasminogen activator inhibitor, 31
Platelet derived growth factor, 27, 38, 63
Plots
  QSAR, 185
  examples, 180, 183
Polio, 53
Polylactides, 414-415
Polymers for controlled drug release, 412-421
Population kinetics, 154
Postnatal toxicity, 375
Potassium ion, 28
Potency of agonists and antagonists, 8
  determination of, 8-10, 318-327, 331-333
  and undesired drug effects, 19
  values for QSAR, 184-185
3-PPP, 40
Prazosin, 321, 323
Probenecid, 110
Procainamide, 129
Procaine, 129
Processing enzymes of precursor proteins, 232-233
Pro-drugs, 119, 146, 269-273, 329, 359
Protonsil, 329
Propranolol, 252
Prostaglandines, 54, 392
Protein binding (see Plasma protein binding)
Protein kinase A, 20
Protein kinase C, 24, 64
  inhibition via surrounding phospholipids, 27
Proteins:
  abnormal, in Alzheimer's disease, 46-47

chemical synthesis of, 235
chimeric (see Antibodies, chimeric)
engineering, 235-236
expression of, 220-225
guanosine nucleotide binding, 22-24, 26
human via recombinant DNA, 220-221, 222-223
immune reaction to, 229
isolation of from fermentation broths, 225
large scale production via genetic engineering, 223-225
natural, as leads for modified proteins, 233-236
pharmaceutical formulation of, 227-228
quality control of biotechnological products, 224-227
as second messengers, 30-31
secretion of, 221
site-specific mutagenesis of, 233-234, 237
as targets for drug action, 203
three-dimensional structure of, 203-212
examples, 164, 169, 208
X-ray crystallography of, 204-206, 235
antiviral drug, 49
Protocol for toxicological tests, 381
Psychotropic agents:
examples for serendipidous drug discovery, 316, 347
Public opion:
and drug therapy, 442-448, 458
Purines, 123
Pyridines, 151
Pyridoxal phosphate dependent enzymes, 260-261
Pyrimidines, 151
QSAR, 176-190
biological potency, descriptors of, 184-185
cluster analysis, 187-188
correct predictions from, 188
correlation of buccal absorption with log P, 182
decision to stop synthesis in a series, 188
derivation of binding modes, 259-260
description of, 176
discriminant analysis, 186-187

early history, 176
electrostatic properties of molecules, descriptors of, 177-180
erythromycins, 189
factor analysis, 187
flawed, 189-190
Free-Wilson analysis, 184
requirements for, in series design, 278, 284
indicator variables, 184
in the specification of corresponding structural fields, 278, 279
methods of deriving relationships, 185-188
plots, 185
examples, 180, 183,
PLS, 187
principal component analysis, 187
regression analysis, 185-186
role in drug design, 188-190
significance of, 278-279
SIMCA, 187
substructural vairables, 184
synergism with molecular graphics and protein crystallography, 210-211
tables, 185
QSAR descriptors:
biological variables, 92-93, 184-185
substituent size, 182-184
limitations of, 183-184
table of, 181
Quality control of biotechnological products, 225-227
Quantum chemical calculations
refinement of three-dimensional molecular structures, 193-195
role in deriving parameters for molecular mechanics, 194
Radioligand binding:
assays, 14-15, 311, 320-325
role in mapping macromolecular binding sites, 191
Range finding of toxic dose, 364
Rate of success in drug discovery, 290
Rate-determining processes:
in drug metabolism and pharmacokinetics, 89-92, 146
oxidation by cytochrome P-450, 121
Reabsorption (*see also*

Enterohepatic recycling):
tubular, 107-109
Reactive intermediates in drug metabolism, 124, 135, 146, 148-151, 156, 260-261
Receptor mapping (see Mapping macromolecular binding sites)
Receptor:
agonists, definition of, 6, 7
antagonists, definition of, 6, 7
binding, cooperativity, 17
binding methods, 14-17 (see also Radioligand binding)
analysis, 15-17
choice of labeled ligand, 15
in drug discovery and evaluation, 265-266, 310-312, 320-325, 342
separation of bound from free ligand, 14-15
mimics, 239
theory, 5-7
Receptors:
adrenergic, 7, 321
α-adrenergic, 237, 245-246, 250, 321-323, 339
ß-adrenergic, 237, 245-246, 250, 251, 330, 338-340, 347
antiidiotypic antibodies, 239
CD_4, 49
cellular, and oncogenes, 237
cholinergic, 45
muscarinic, 6, 7, 237, 250, 311
nicotinic, 6, 237
dopaminergic, 40-42, 45, 195, 250, 312, 321, 325-326
effector systems of (see signal transduction)
EGF, 63
GABA, 29
for growth factors, in signal transduction, 27
histamin, 311
interaction of drugs with, 7-19
laminine, 68
opioid sigma, 43
peptidergic, 7
production by recombinant techniques, 236-237
serotonin, 43, 321
steroid, 30-31
three-dimensional structure of in lead finding, 258-259
and tissue distribution, 38

Reduction:
  biological, 127-128
Registration authorities, 355, 363, 364, 368, 372, 389, 394, 453-455,
Regression analysis (see also QSAR):
  QSAR, 185-186
  examples in QSAR, 179, 182
Renal excretion of drugs (see Excretion, renal)
Renin as a target of drug design, 256-257
Repeated dose toxicity (see toxicological studies)
Repressors, definition of, 5
Reproductive toxicology (see Toxicological studies)
Research and development:
  budget, 465
  capacity, 315
  climate of, 442, 458-468
  management of, 458-468
  organization of, 151-158, 442, 458, 466
  planning of, 151-158, 458, 462, 464
  strategy of, 151-158, 248-250, 315-316, 464-465
Respiratory syncitial virus, 52
Response-to-injury hypothesis, 38
Restriction endonucleases, 219
Retrovir, 50
Retrovirus, lymphotropic, 48-49
Reverse transcriptase, 49-50
Rheumatism, 54, 55
Rho of Hammett equation:
  values of, examples, 177-178
Ribavirin, 52
Rimantadine, 49
Risks:
  of dying due to drug therapy, 356
  of dying due to various factors, comparison of, 356
  of modern technology, 445-446
Safe structural features in drug design, 149-151
Safety and efficacy assessment, 330-348 (see also Species, differences)
  caveats, 155-157
  development time and costs, 448-457

and hepatic elimination, 115-116
and models of pharmacokinetics, 118
role of pharmacokinetics and biotransformation in, 146-158
Salicylic acid, 138, 154
Saliva:
excretion of drugs via, 106, 116
Scatchard analysis, 15-17
Scavenger receptor pathway, 38
Schild analysis, 9-10
Schizophrenia, 39-43
dopaminergic system, 40-42
pathophysiology, theories of, 39-40
serotoninergic system, 43
Screening, biological, 31-33, 155, 265-266, 315, 316
Second messengers (see also Signal transduction):
cyclic AMP, 20
diacyl glycerol, 24
definition of, 20
inorganic ions, 28-30
inositol triphosphate, 24
proteins, 30-31
Secretion of drugs (see Excretion of drugs)
Selective toxicity, 312,320
Selectivity of drug action, 335, 340-341 (see also Dose ratio)
Senile dementia (see Alzheimer's disease)
Series design, 277-288
cluster analysis, 187-188
conformational properties, 300-303
decision to stop synthesis, 188
importance of uncorrelated physical properties, 188
example of, 282-288
techniques for, 188, 209-212, 291-303
Serine proteases three-dimensional structures of, 207, 208
Serum albumin (see plasma protein binding)
Sets of compounds or structures (see structural fields)
Shape complementarity (see also Intermolecular forces):
in designing ligands to fit experimentally determined binding sites, 211
in mapping macromolecular binding sites, 191, 201-203

Shapes of molecules (see Three-dimensional properties of molecules)
Shelf-life of dosage forms, 409
Sickle-cell disease design of antagonists, 209
Side effects of drugs, 19, 33, 40, 45, 50, 154, 155, 340-342
Sigma constant for electronic effects of substituents:
  definition of, 177
  in series design, 278, 280, 287-288, 289, 296-297
  table of, 179
  types of, 178-179
Signal peptide, 221
Signal transduction, 19-31, 64-65, 238
  c-AMP, 20-24
  definition of, 20
  growth factor receptors, 27
  guanine nucleotide binding proteins, 22-24, 26
  ion channals, 27-30
  phosphatidylinositol pathway, 24-27
  in steroid action, 30-31
Sinusoids, 111-112
Site-directed mutagenesis, 233-234
Site-specific mutagenesis, 218
Size of substituents:
  described by molar refractivity, 180, 279-280, 287-288, 289, 296-297
  QSAR descriptors for, 182-184
Skin sensitization, 371
Small pox, 53
Sodium as second messenger, 28, 29
Sodium-proton pump, 64
Soft-drug concept, 147, 149
Solid solution of a drug:
  in delivery systems, 417-418
Solubility of drugs:
  lipid (see Hydrophobicity)
  membrane systems of controlled drug release, 419
  polymers, design of diffusional systems for controlled drug release, 417
  water
    consideration in drug design, 148

membrane transfer, 85
phase II reaction products, 119
pH-dependence in oral absorption, 430
poorly soluble drugs, enhancement of absorption rate, 408
significance for absorption from gastrointestinal tract, 94, 96
significance for enhancement of excretion, 108
Somatostatin, 252, 275
Somatostatin analogues, 252, 275
Sotalol, 334
Spasmolytics, 309
Species:
differences, 153, 361
due to competing biotransformation, 141, 142
due to defective metabolic reactions, 142-143
in drug metabolism, 133, 272, 273, 349
examples of, 272-273, 349
in pharmacokinetics, 117, 349
in pro-drug concept, 272-273
selection for toxicology, 360-361, 362-363, 370-371, 373, 374-375, 391-392
similarity to man, 141-143, 153, 349
Specific bradycardic actions and agents, 328, 338
ST 587, 323
Stability of drugs and dosage forms, 409-410
Statistics:
cluster analysis, 187-188
discriminant analysis, 186-187
factor analysis, 187
partial least squares, 187
principle components, 187
regression analysis, 185-186
Stereoisomers (see Chiral compounds, see Enantiomers)
Steric repulsion:
between molecules, 168-169
complexity of describing for QSAR, 183-184
Strategy:
of binding site mapping, 190-192

of the SAR approach in lead optimization, 288-291
in pharmacological studies, 315-318
in toxicological studies, 393-394
Stress test of dosage forms, 409
Structural fields:
characterization of, 243, 277
examples of, 287, 289, 291, 293, 294, 296
and specific therapeutic effects, 243-245
specification of, 297-280
Structure-activity hypotheses:
derivation of, 280-282
description of, 277
examples of, in lead optimization, 283-287
in factorial design, 295, 298
from images of enzymes and receptors, 165-167, 255-259, 303
specification of a corresponding structural field, 279-280
as starting point in lead optimization, 277
strategic aspects of formulation, 288-291
Structure-activity-relationships, 171-190
accomplishments of, 175
description of, 171-175
empirical nature of, 174-175
limitations of, 175
in lead optimization (see also Structure-activity hypotheses)
pitfalls, 290
and metabolism, 120, 143, 144
necessity for synthesis of analogues, 171
pargyline analogues, 172-175
pharmacokinetics, 77, 89, 91, 97, 110, 115, 148
relevance of, 249, 278-279
role of creativity in success, 174
"safe" structural features, 149-151
in terms of sub-structural features, 171-175
Substituents, standard sets of, in series design, 292
example of, 292-293
Sub-structural descriptors

(see Indicator variables)
Substrates, choice of concentration in the analysis of enzyme inhibition, 11-14
Suicide substrates, 260-261
Sulfanilamide:
metabolism of, 136
from pro-drug prontosil, 329
Sulfation in drug biotransformation, 133-135
Sulfinpyrazone:
biological reduction of, 128
Sulfonamides:
antibacterial potency as a function of pK_a, 179
binding to albumin of, 102
discovery of, 329
diuretics, relevance of rats for man, 349
diverse biological activities, 243-244
metabolism of, 136
Sulfonyl ureas, 253-254
Sulmazole, 121, 287
Suramine, 49
Tablets as drug dosage forms, 402, 407, 432
Tachyphylaxis, 18
Targeted lead optimization, 267-273
Targeted screening, 265-266, 316
Teratogenicity:
testing for, 373-375, 386-387
Test protocol (see Protocol for toxicological tests)
T-cells in AIDS, 49
Tetrachlorodibenzodioxins, 121
Tetrodotoxin, 28
TGF-alpha, 69
TGF-beta, 39, 69
Theophylline, 404
Theoretical chemistry (see Medicinal chemistry, theoretical basis of)
Therapeutic goal:
and duration of drug effect, 346
requirements for lead optimization, 266-267
and test strategy, 315
Therapeutic monitoring:
of drug levels, 88, 116,
of metabolism, 154
Therapeutic range:
definition of, 340-342
Thiazides, 331-332
Thiazoles, 151
Thiopental, 105
Three-dimensional proper-

ties of molecules, 162-163, 188, 190-203
drug desing based on, 192, 255-259
example, methotrexate-dihydrofolate reductase, 168, 169
role in dispersion interactions, 168
use of in mapping macromolecular binding sites, 190-203
Three-dimensional structures of molecules:
building with molecular graphics, 192-193
displaying with molecular grpahics, 192-193
refining with energetics, 193-195
in series design, 300-303
Three-dimensional structures of proteins, 203-208
from comparative modelling, 207-208
examples, 164, 169, 208
from NMR, 206-207
from X-ray crystallography, 204-206
Thromboxane, 55
antagonists, 55
Thymosine, 65
Thyroxine analogues:
design from molecular graphics, display of X-ray results, 209
Time dependence of effects, 18-19, 154
Time effect curves of drug action, 347
Tissue binding of drugs, 86-87, 104-106
Tissue cultures:
for expression of proteins, 222-223
optimization of, in genetic engineering, 224-225
as pharmacological tools, 309, 326
for prediction of metabolic elimination, 117, 145
for tPA production, 232
Tissue distribution of drugs, 104-106
Tissue plasminogen activator, 223, 229, 230-232, 234
Tolbutamide, 254
Toxic dose, minimal, 380
Toxic moieties, 245
Toxicity:
acute, 362-365, 384
chronic, 360-361
recovery from, 368

repeated dose, 362, 365-368, 385
selective, of chemotherapeutics, 312, 320
therapeutic range, 340
Toxicological studies
in animals, 362-382
model selection, 153, 370-371, 373, 378
drug combinations, 371-372
evaluation of, 384-389
extrapolation from animals to humans, 385-389
with inhalational dosage forms, 363-364
local tolerance, 369-371
parameters, 381-384
prerequisits for,
analytical chemistry, 358-359
dosage form, 359
pharmacokinetics and metabolism, 360-361
pharmacology, 358
regulations for, 355, 363, 364, 365, 368, 372, 373, 374, 376, 378, 380, 382, 387-388
repeated dose, 365-368, 385-389, 391
dosage, 367-368, 391,
duration in relation to length of human trials, 365-367
evaluation of, 385-389
reproduction, 372-375
single dose, 362-365, 384-385
evaluation of, 384-385
for industrial chemicals, 364-365
special pharmaceutical preparations, 369
test protocol, 381
Toxicology (see also Toxicological studies):
of biotechnological products, 389-392
expenditure in, 392-394
and pharmacological and microbiological profile in drug evaluation, 315
Transcription of genes, 219
Transdermal systems for drug delivery, 421-427
advantage of, 422
examples for, 423-427
improvement of, 434
requirements for, 423
Transformation of host cells, 219

Transgenic animals, 238
Transition state analogues, 260, 261-263
Tranylcypromine, 261
Trazodone, 154
Triazolobenzodiazepines, 57
Trimethoprim analogues
QSAR, molecular graphics, and protein crystallography, 209
Tubular reabsorption of drugs in the kidney, 107-109
Tubular secretion of drugs in the kidney, 109-110
Tumorigenesis, 362
Tyrosine kinases, 27

Vaccines, 53, 67, 222, 229, 232
Valinomycin, 29
Valium[R], 29
Van der Waals radius
definition of, 168
of hydrogen-bonded pairs, 170
Variability in drug response, 88, 96, 122, 147, 148
Variables:
continuous,
as descriptors of physical properties, 176-185
characterization of structural fields, 279-280
expression of structure-activity hypotheses, 278, 282, 289, 293
in factorial design, 294, 296-297
for the properties of a dosage form, 410-411
Varicella, 52
Vasocortin, 31
Venture capital, 462, 463
Verapamil, 30, 71, 322
Verloop L and B values:
QSAR descriptors of substituent size, 182
table of, 181
Vesicular processes in drug transport, 85
Viral infections:
stages of, 48-52
Volume of distribution, 92-93
definition of, 92
WEB 2086-BS, 57
Working climate in drug research, 458-468
X-ray crystallography
and design of hemoglobin effectors, 209
and design of thyroxine analogues, 209

dihydrofolate reductase, NADPH, and methotrexate, 164, 169
DNA-drug complex, 203
example of drug-protein complex, 164, 169
of proteins, 204-206,
  synergism with QSAR, 210-211
of protein-ligand complexes, 206
  in drug desing, 190, 259
  examples, 164, 169
  as source of bound conformation of a ligand, 169, 196
  as a source of starting structures for energy refinement, 193
Xanthine oxidase, 122-123
Xenobiotics:
  biotransformation of, 118-146
  definition of, 118
  studies in the Ames test, 151
Zidovudine, 50